INNOVATION IN ASTRONOMY EDUCATION

Astronomy leads to an understanding of the history and nature of science, and attracts many young people to education in science and technology. But while in many countries astronomy is not part of the standard curriculum, many scientific and educational societies and government agencies have produced materials and educational resources in astronomy for all educational levels. This volume highlights the general strategies for effective teaching and introduces innovative points of view regarding methods of teaching and learning, particularly those using new technologies. Technology is used in astronomy, both for obtaining observations and for teaching. The book also presents ideas for how astronomy can be connected to environmental issues and other topics of public interest. This valuable overview is based on papers and posters presented by many of the world's leading astronomy educators at a Special Session of the International Astronomical Union General Assembly in Prague in 2006.

Jay M. Pasachoff is Field Memorial Professor of Astronomy at Williams College, and was President of the Commission on Education and Development of the International Astronomical Union.

Rosa M. Ros is Professor of Mathematics at the Technical University of Catalonia in Barcelona and Vice-President of the International Astronomical Union's Commission on Education and Development.

Naomi Pasachoff is a Research Associate at Williams College and an author of science textbooks and biographies of scientists.

Cover: The Astronomical Clock of Prague, one of the main tourist sites in this city that hosted the 2006 International Astronomical General Assembly, in which the Special Session on which this book is based was included. The clock shows the Sun's position in the sky, the lunar phase, the zodiac, the positions of the Sun and Moon on the ecliptic, and other items of interest to astronomers. The oldest part of the clock dates back to 1410, though the clock's current appearance comes from major repairs after World War II devastation. Moving statues, for which tourists gather on the hour, were added in the seventeenth century. (Richard Nebesky/Lonely Planet Images/Getty Images)

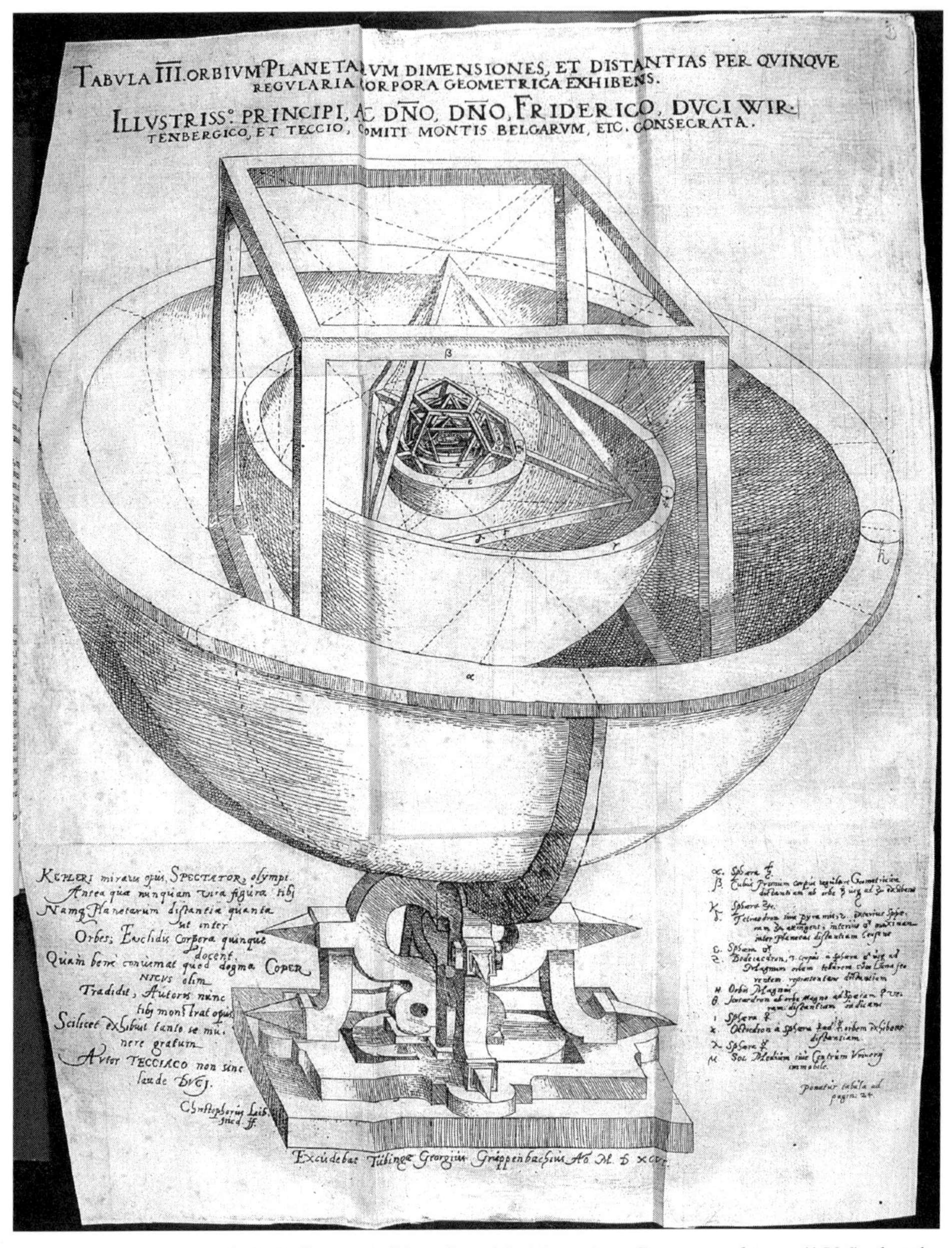

Johannes Kepler's heliocentric idea, from his *Mysterium Cosmographicum* (1596), that the planets' spacing was determined by the Platonic solids. Kepler moved to Prague, the site of the International Astronomical Union's 2006 General Assembly at which this Special Session on Innovation in Teaching and Learning Astronomy was held, to work with Tycho Brahe, leading to Kepler's three laws of planetary motion. (Photo courtesy of Jay M. Pasachoff with the assistance of Wayne Hammond, Williams College's Chapin Library.)

INNOVATION IN ASTRONOMY EDUCATION

JAY M. PASACHOFF
Williams College, Massachusetts, USA

ROSA M. ROS
University of Catalonia, Barcelona

NAOMI PASACHOFF
Williams College, Massachusetts, USA

CAMBRIDGE UNIVERSITY PRESS
Cambridge, New York, Melbourne, Madrid, Cape Town, Singapore, São Paulo, Delhi

Cambridge University Press
The Edinburgh Building, Cambridge CB2 8RU, UK

Published in the United States of America by Cambridge University Press, New York

www.cambridge.org
Information on this title: www.cambridge.org/9780521880152

First published 2008

Printed in the United Kingdom at the University Press, Cambridge

A catalog record for this publication is available from the British Library

ISBN 978-0-521-88015-2 hardback

Contents

Preface

This book is based on the proceedings of a conference on education in astronomy. On August 17 and 18, 2006, the International Astronomical Union's Commission on Astronomy Education and Development held a Special Session at the IAU General Assembly in Prague. The session, on Innovation in Teaching/Learning Astronomy, was organized around four themes: (1) general strategies for effective teaching, (2) connecting astronomy with the public, (3) effective use of instruction and information technology, and (4) practical issues connected with the implementation of the 2003 IAU Resolution that recommended including astronomy in school curricula, assisting schoolteachers in their training and backup, and informing them about available resources. Approximately 40 papers were presented orally; in addition, 60 poster papers were displayed.

Some of these topics had been considered in the Special Session at the 25th General Assembly in 2003 in Sydney, the subject of a book published in 2005, Jay M. Pasachoff and John R. Percy, *Teaching and Learning Astronomy: Effective Strategies for Educators Worldwide* (Cambridge University Press, 2005). But it is necessary to continue and extend the work started then in order to increase the quality and quantity of astronomy in schools.

The Organizing Committee for the conference consisted of:

Rosa M. Ros (Spain, co-chair), *Spanish National Liaison to IAU Commission 46; Vice-President of the Commission 2006–2009*

Jay M. Pasachoff (USA, co-chair), *President, IAU Commission 46*

Michael Bennett (USA), *Executive Director, Astronomical Society of the Pacific*

Julieta Fierro (Mexico), *Former President of IAU Commission 46*

Michele Gerbaldi (France), *Chair, IAU International Schools for Young Astronomers Program Group*

Petr Heinzel (Czech Republic), *Astronomical Institute of the Czech Academy of Sciences*

Bambang Hidayat (Indonesia), *Bosscha Observatory, Institute of Technology Bandung, Past Vice-President of the IAU*

Syuzo Isobe (Japan), *Former President of IAU Commission 46*

Edward Kononovich (Russia), *Russian National Liaison to IAU Commission 46*

Margarita Metaxa (Greece), *Greek National Liaison to IAU Commission 46*

John R. Percy (Canada), *Former President of IAU Commission 46*

Magda Stavinschi (Romania), *Astronomical Institute of the Romanian Academy of Sciences; President of the Commission 2006–2009*

Richard West (Germany), *Former Chair, Department of Outreach and Education, European Southern Observatory*

Lars Lindberg Christensen (Germany, webmaster), *PIO/Head of Communication, ESA Hubble/James Webb Space Telescope, IAU Press Officer*

The meeting was supported not only by Commission 46 on Education and Development but also Commission 41 on the History of Astronomy and Division XII on Union-wide Activities.

Over 400 astronomers and educators from 63 countries registered for this conference. The conference was part of the triennial General Assembly of the International Astronomical Union, which this year gained notoriety from the resolution defining the word "planet" and putting Pluto and some other objects in a new category of "dwarf planet." One of the papers in this symposium, by Lars Lindblad Christensen, dealt with public-information aspects of that situation – which may wind up continuing until the next IAU General Assembly to be held in Rio de Janeiro in 2009.

We thank all the authors and other contributors.

Prague was a particularly apt site for a conference on astronomy, since Tycho Brahe and Johannes Kepler did so much important work there. It is particularly suitable that this book is available in time for the International Year of Astronomy (www.astronomy2009.org), which commemorates the 400th anniversary of the 1609 work of Galileo and Kepler.

We acknowledge the generous support of the International Astronomical Union and its Executive Committee, both in the form of travel grants for some participants and in the form of moral support for the importance of education. Many other participants received support from their institutions or countries, and we are grateful to those who made sure that these individuals could attend and participate.

We dedicate this book to Syuzo Isobe of Japan, past President of Commission 46 (2000 to 2003), who died on 31 December 2006. Accounts of his life and work appeared in the March 2007 edition of the Commission 46 Newsletter, which is accessible through the Commission's Website at www.astronomyeducation.org.

We thank Javier Moldón of the University of Barcelona for helping to organize the Special Session. We thank Madeline Kennedy for her assistance at Williams College with the preparation of this book and for compiling the index. We are grateful for some financial support from Williams College for work carried out on the educational activities of our International Astronomical Union Commission on Research and Development. The participation of one of us (J.M.P.) in the Prague General Assembly was supported in part by a research grant from the Planetary Sciences Division of NASA and by a travel grant from the US National Science Foundation through the American Astronomical Society.

At Cambridge University Press, we thank our acquiring editor, Vince Higgs, for his support of this project. We are pleased with the excellent assistance there of Lindsay C. Barnes, Eleanor Collins, and Bethan Jones. Frances Nex has proven to be a most capable copy editor.

Jay M. Pasachoff
Rosa M. Ros
Naomi Pasachoff

Attendees photographed during the meeting in Prague. (Photo by Robert L. Hurt, Spitzer Science Center, Caltech.)

Part I

General strategies for effective teaching

Introduction

The 20 papers constituting the first day's proceedings were devoted to the first theme, "General strategies for effective teaching." After welcoming remarks, Jay M. Pasachoff, commission president, and Rosa M. Ros, who later became the commission vice-president, briefly summarized the main objectives of the Special Session, mentioning a variety of new methods of information dissemination (e.g., World Wide Web, Astronomy Picture of the Day, podcasts), the role astronomy can play in attracting young people to careers in science and technology, and the usefulness of technology both to observers and to teachers.

Former commission president John Percy of Canada next spoke about "Learning Astronomy by Doing Astronomy." Percy contrasted learning astronomy facts from lectures and textbooks and "doing astronomy" in a more intellectually engaging fashion, emulating the actual scientific process, than called for in standard activities culminating in a predetermined result. While recognizing the value of "hands-on" activities, such as making scale models of the Solar System, Percy argued that "minds-on" activities – such as involving students in meaningful ways in their teachers' professional research (even if the result is not publishable) – are more valuable. He pointed out that in labs, which should mirror actual research, students can manipulate actual data, using the same computer languages and software used by real researchers. These activities help them grasp that astronomy facts do not emerge full-blown from textbooks but are figured out by astronomers based on ongoing research. Percy spoke also of the value of having students themselves assume the role of astronomy communicators by tutoring peers or younger students or making classroom or public presentations. Percy referred to conference participant Richard Gelderman's assertion that students should be exposed to recreational science, such as science fairs, "for curiosity, interest, and . . . for fun, fellowship, and . . . mental well-being," just as they are encouraged from their earliest years to participate in recreational sports for physical and mental health. He noted, too, that even urban students can learn to understand and make the astronomical observations that underlie the religious observances of major world faiths. He concluded with the thought that "the ultimate goal of astronomy education" is to reach every student. While only a fraction will become professional astronomers, every student may become an amateur astronomer, with a lifetime passion for astronomy.

Like Percy, Roger Ferlet of France underscored the importance of engaging pupils in "observing, arguing, sharing, discussing and interpreting real astronomical data, in order to enhance autonomy and reasoning; in brief, learning science by doing science." Ferlet discussed the European Union's Hands-On Universe project as a tool to reverse the "clear disaffection for scientific studies at universities" by convincing middle and high school students that scientific understanding "can be a source of pleasure." The project, a partnership of eight European countries under the auspices of the French University Pierre and Marie Curie in Paris, invites these younger students "to manipulate and measure images" in class, using "real observations acquired through an Internet-based network of robotic optical and radio telescopes or with didactical tools such as webcam," assisted by scientific experts and by a select group of teachers who are trained in special workshops. The newly trained teachers go on to become "resource agents" for their own countries, with a mandate to train other educators. This European undertaking is important not only to the international scientific community but also to society as a whole, since not only does a "sustainable economy" depend on innovations by "a critical number of scientists and engineers," but also societies that undervalue science "regress to more primitive and much less attractive" conditions. An additional social benefit of the project is that it encourages communication among students from different countries. The positive reaction to Hands-on Universe has aroused hopes that similar European initiatives will result in "'Hands-on life' for biology, 'Hands-on Earth' for geology/ecology, etc."

Former commission president E. V. Kononovich of Russia described a manual about the Sun, the Earth, and their interactions assembled by the Sternberg Astronomical Institute of Russia for older high school students and for "students of natural faculties of universities and teachers' colleges" with an interest in solar problems. The manual is also a useful teaching aid for courses covering solar physics and solar–terrestrial relations. The manual, which makes use of both ground-based observations and results from SOHO, Yohkoh, Ulysses, TRACE, CORONAS, and other space telescopes, is divided into three sections. The first considers the Sun as a star, solar activity, and helioseismology. The second describes the Earth in space, its structure, its atmosphere, its magnetic field, its weather and climate, and active phenomena on the terrestrial surface. The third section considers such solar–terrestrial relationships as ionospheric disturbances, solar cosmic rays, and aurorae.

Bill MacIntyre of New Zealand presented "A Model of Teaching Astronomy to Pre-Service Teachers that Allows for Creativity in Communicating Students' Understanding of Seasons." (In MacIntyre's paper, "students" are teacher trainees, not the young people they will go on to teach.) MacIntyre began by distinguishing among mental models ("cognitive notions held by individuals"), expressed models (mental models that have been communicated to others), consensus models (expressed models valued by a social group and widely used by it), scientific models (consensus models that are used by scientists for further scientific developments), and teaching models (used to provide opportunities for teachers-in-training to develop their understanding of basic astronomy along with pedagogical skills). A goal of the "investigating with models" approach is to have students understand "the limitations and strengths" not only of their own models but also of scientists' models. Since expression of a scientifically inaccurate mental model has the potential to embarrass the student, teachers must make sure to have the revelations take place "in a small group" and "in a non-threatening way." The exercise requires students to collect evidence that will either support or lead them to change their mental models, a requirement that guarantees they will practice "aspects of the nature of science (observations, inferences, creativity and empirically based knowledge) that relate to the systematic nature of investigating." MacIntyre described in detail the process by which two teachers-in-training creatively developed expressed models after identifying specific aspects of their original expressed model that other members of their group had difficulty comprehending. Why is such an exercise for teacher trainees useful? "If we expect classroom teachers to cater for the creative–productive gifted students during astronomy teaching in primary and secondary schools, then pre-service teacher training must model the appropriate classroom environment that allows it to occur."

In "How to Teach, Learn About, and Enjoy Astronomy," Rosa M. Ros of Spain described "what I learnt after 10 years of the [European Association for Astronomy Education] Summer School," drawing on the questionnaire responses of approximately 600 opinions of teachers of secondary school students from over 20 countries. To make students – in this case, the teachers – "feel like actors in the teaching–learning process," she, like other Special Session 2 participants, advocates "learning by doing." During the summer sessions the teachers are exposed to a variety of approaches to the teaching of astronomy, including model-making, drawing, and playground activity. Just as a classroom teacher must be prepared to answer spontaneous questions from pupils about an astronomy topic that interests them, Ros as summer school director had to modify the activities to accommodate "the topics, matters, and methods" about which the teacher participants wanted to know more, bearing in mind always that astronomy concepts must be presented in some context, not in isolation. Instead of presenting a body of facts for students to memorize, "it is important to connect the concepts with the personalities related to the topics, with the scientific situation in the past or maybe the social implications of the subjects." One goal of the summer school is to encourage more inspired and more passionate teaching of math and physics. "If the teachers enjoy teaching, the students will also enjoy their classes." Ros lamented the fact that science museums tend to mount exhibits about science that stress the "spectacular and funny," leaving to schools "the boring science area." Nonetheless, she noted that "not everything can be fun at school," arguing that teachers must also "introduce the culture of making an effort to students." Even under-funded schools with limited resources can include creative astronomy activities in the curriculum. "All schools have a sky over their buildings. It must be used to observe and take measurements. If the school does not have tools and devices for making observations, we can encourage the students to produce their own instruments." Whatever is lost in precision by doing so is more than compensated for in student commitment.

Based on his experience as director of astronomical laboratories at the University of Colorado in the United States, Douglas Duncan advocated the use of "clickers" – wireless student response systems – as "the easiest interactive engagement tool" in teaching large lecture classes. Studies show that students enjoy using clickers, which transform students from "passive listeners" to "active participants" in the learning process. The use of clickers also enables teachers to determine their level of comprehension without waiting for test-time. Studies show that the average student in a large physics lecture course, regardless of the effectiveness of the lecturer, truly comprehends at most 30% of newly introduced concepts. To master a new concept, students "must think about the idea and its implications, fit it into what they already know, and use it." They must dislodge the misconceptions with which they enter the lecture hall. While professional scientists bounce ideas back and forth among themselves and within their own minds, students often believe "that taking notes, memorizing, and repeating material on an exam is all there is to learning." Duncan also advocated using clickers to facilitate peer instruction, since studies show that "when comparable numbers of students start with right and wrong conceptions, peer instruction usually results in students agreeing on the correct answer, not the wrong one." He cautioned, however, that "like any technology, clickers can be misused, and it is important to practice and to explain their use to students before starting."

In "Educational Opportunities in Pro-Am Collaboration," Richard Fienberg, editor of *Sky and Telescope* magazine, echoed other symposium participants in asserting that "the best way to learn science is to do science," and called, as Michael Bennett would do in a later paper, for collaborations among amateurs, professionals, and educators. "Amateurs will benefit from mentoring by expert professionals, pros will benefit from observations and data processing by increasingly knowledgeable amateurs, and educators will benefit from a larger pool of skilled talent to help them carry out astronomy-education initiatives." Noting the important contributions amateur astronomers have historically made to the field, the loss of access of professional astronomers to mid-sized meter-class telescopes, and the need to follow up on "countless interesting objects" being discovered by automated all-sky surveys, Fienberg recommended that serious amateurs – many of whom have access to digital imagers on computer-controlled mounts – be given the opportunity to do some of this monitoring. He identified the American Astronomical Society Working Group for Professional–Amateur Collaboration as a "forum for collaboration between amateur and professional astronomers." Fienberg noted that amateurs continue to make important contributions to astronomy in areas including occultations; variable stars; meteor showers; CCD photometry and astrometry; and the search for and discovery of novae, supernovae, GRB counterparts, comets, and asteroids. In addition to this important work, amateurs can also help with astronomy outreach within their local communities and over the Internet with other researchers and educators around the world.

José Maza reported on his two decades of experience "Teaching 'History of Astronomy' to Second-Year Engineering Students at the University of Chile." The course partly fills the two-course "Humanistic Studies" requirement that each engineering graduate must complete. As a result of the course, men and an increasing number of women who will go on to work at, and often become senior executives at, major Chilean companies are exposed to the basics of astronomy and to its development over history. The first part of the course, "a tour to the scientific revolution," begins with the ancient civilizations; leads up to Newton's construction of modern science; and ends with the contributions of Euler, Clairaut, Lagrange, D'Alembert, and Laplace to celestial mechanics. The second part begins with William Herschel and the discovery of the Milky Way and proceeds over several weeks to a discussion of the big bang, the cosmic background radiation, and dark energy, before culminating with a lecture on the history of astronomy in Chile. Maza would be happy to exchange ideas with other astronomy educators.

Gilles Theureau reported on his experience with coauthor L. Klein teaching a two-semester course for students with varied backgrounds and interests at the University of Orléans (France) on the "history and epistemology of the concepts of stars and galaxies" from antiquity to the early twentieth century. The cross-disciplinary approach is unusual in France, "where pupils start to be specialized" from the age of 15. The course opens with a study of world systems from the pre-Socratics to the philosophers of the Middle Ages, moves on to concepts of mechanics and planetary motion from Aristotle to Newton (and a little beyond), proceeding then to a discussion of spectroscopy and the nature of stars, and concluding with descriptions of the Milky Way and the nature of nebulae. The course emphasizes "mechanisms of knowledge," including observation, experiment, and theory, as well as mythology, theology, philosophy, metaphysics, physics, mathematics, and instrumentation, in order to demonstrate that human ideas of the Universe evolved "as a part of human history and culture," that "science belongs to the patrimony of humanity and that it has no frontiers," and that astronomy draws on both the humanities and the sciences. The course makes use of original documents that are considered in their historical context, including Aristotle on meteorology (350 BCE), Nicolas Oresme's challenge to Ptolemy's view of celestial motion (c. 1380), William Herschel's *On the Construction of the Heavens* (1785), and Agnes M. Clerke's *Problems in Astrophysics* (1903). The class also paid a visit to the Nançay Radio Observatory Astronomy Museum and Visitor Center; for the majority of students, who have "never looked at the sky," this visit provided "an impression of the questions of interest to past generations of philosophers and astronomers." Although in such a course it would have been easier to evaluate students through exams exclusively, the professors opted for the more difficult choice of assigning individual written projects in the first semester and a comprehensive exam in the second semester. The professors broadened their own cultural outlook by teaching the course, but found the presentation of material to and evaluation of such a heterogeneous group of students very challenging.

Jay M. Pasachoff spoke about "Educational Efforts of the International Astronomical Union." He described how the work of the commission, which resulted from a merger of the commissions on education and on astronomy in developing countries, is carried out in ten program groups. These groups include the world-wide development of astronomy, which sends some of its members to visit countries interested in learning about advances in carrying out astronomy and perhaps even becoming members of the International Astronomical Union; teaching for astronomy development, which provides visiting experts or lecturers to help advance a country's astronomical education; exchange of astronomers, which arranges international visits of several months or longer for people from developing countries to visit major research institutions; the International School for Young Astronomers that is held every non-General-Assembly year for some dozens of new astronomers or graduate students; a semiannual newsletter; a group charged with coordinating with international institutions such as UNESCO, and that will now work with the International Year of Astronomy scheduled for 2009; a group involved in international exchanges of journals that could aid developing countries; and a group related to taking advantage of public interest at the times of solar eclipses to spread astronomical knowledge, including but not limited to the eclipse itself. All these groups, the newsletters, and other related activities are accessible through the commission's website at www.astronomyeducation.org.

Magda Stavinschi of Romania, who became commission president at the end of the Prague general assembly, argued that astronomy is an integral part of human culture, in the evolution of which it often played a significant role. The discoveries of archaeoastronomers have proven that prehistorical civilizations pondered cosmological questions and wondered about the place of humankind in the Universe. From the beginnings of history people across cultures have recorded significant events through markers that include not only human events, such as wars and the births and deaths of leaders, but also cosmic events, notably comets and eclipses. Stavinschi called attention to the often overlooked relationship between politics and astronomy, noting the post-World War I confirmation of Einstein's general theory of relativity by the Englishman Sir Arthur Eddington. During the war, Germany, Einstein's native land and the country that employed him, was a bitter enemy of England. This scientific collaboration not only served as "a perfect proof of scientific internationalism" but also helped reincorporate German scientists into the scientific community after the war. After briefly mentioning the links between astronomy and geography, mathematics, physics, chemistry, meteorology, technology, medicine, and pharmacology, Stavinschi pointed out some famous examples of the incorporation of astronomy in art, music, heraldry, folklore, and literature. She spoke of the importance of astronomy education: those with an appreciation of the Universe understand the need to protect Earth from manmade devastation "much before its natural end."

While the mass media have succeeded in educating the public through coverage of space missions, and television programs featuring scientists like Carl Sagan have popularized astronomy, the media also are responsible for disseminating misinformation about astronomy. After briefly reviewing the history of astrology, she pointed to the astronomers' "moral duty" to "prove the quackery of astrology." She concluded by arguing that "there is no conflict between science and religion," since they are "two different ways of considering the world," and dismissed the usefulness of arguing the relationship between astronomy and philosophy, since "all that defines philosophy intimately contains the Universe and especially man in the Universe." Astronomy can continue to play a role in the development of culture in pointing humanity in the direction of "what it has to do from now on."

Margarita Metaxa of Greece, where she teaches at the Arsakeio High School in Athens, spoke about "Light Pollution: a Tool for Astronomy Education," which can help motivate not only students but also "the public, government officials and staff, and lighting professionals." Like Duncan, Metaxa noted that students "hold misconceptions about the physical world that actually inhibit the learning of scientific concepts" and that "students can remember less" than their teachers often assume. She called attention to a two-year program on light pollution sponsored by the Greek Ministry of Education and Religion, with backing from the EU; to the Internet Forum on Light Pollution, sponsored by the netd@ys Europe project, a European Commission initiative in the area of education, culture and youth for the promotion of new media; and to a UNESCO-backed conference on "Youth and Light Pollution" held in Athens in autumn 2003. Outside of Europe, Chile has played a significant role in educating students about "the effects of light pollution on the visibility of stars in the night sky." She concluded by emphasizing that bringing both astronomy and light pollution to the world's attention in order "to protect the prime astronomical places and the 'dark skies' as a world heritage" represents a significant challenge. It is one, however, that astronomers can meet by working "together with interested organizations."

On behalf of a group of collaborators, S. P. S. Eyres of the UK described worldwide online distance learning from astronomy courses prepared by the University of Central Lancashire. The student subscribers to these online courses range in age from 16, though most are over 21. They include a retired industrial professional with a doctorate in chemical engineering, an English teacher with a deep interest in astronomy, an employee of an examinations board responsible for school astronomy curricula, a high school student preparing for university entrance exams, a primary school classroom assistant working toward a degree, among others. While all students share an interest in astronomy, they differ in what they hope to achieve through participating in the online distance learning program. Although distance learning in the UK is often associated primarily with the Open University, the University of Central Lancashire (UCLan) has been offering adult education courses in astronomy for about a decade. It is now possible to earn an honors bachelor of science degree in astronomy through UCLan. Eyres explained why the traditional "teacher-focused" astronomy education, in which an "expert in the subject decides what the student needs to know at the end of the course, and works backwards from there to determine where they must start," is inappropriate to distance learning. The framework of modules in the UCLan program enables students to determine if they are capable of higher-level work before they sign up for a lengthy course of study. Students are able to study modules provided by other institutions and use them for UCLan credit. They may also receive credit for skills they may already have in such fields, for example, as IT or math. Students correspond with their tutors both through e-mail and through "discussion and chat tools on UCLan's virtual learning environment." With the introduction of competitive tuition fees at UK brick-and-mortar universities, the honors bachelor of science degree available through UCLan has the potential "to attract students from the traditional 18- to-21-year-old UK degree market." Even students who can pay the tuition fees at traditional universities may choose to take distance courses as a way of drawing attention to their qualifications and making themselves stand out from other applicants for admission. UCLan is also thinking of entering the teacher training market.

Donald Lubowich of Hofstra University, on New York's Long Island, USA, demonstrated how he has successfully used "Edible Astronomy Demonstrations" to motivate students of all ages and to enhance their understanding of such varied concepts as differentiation, plate tectonics, convection, mud flows on Mars, formation of the galactic disk, formation of spiral arms, curvature of space, expansion of the Universe, and radioactivity and radioactive dating. His materials have included chocolate, marshmallows, candy pieces, nuts, popcorn, cookies, and brownies. Echoing other participants' comments that passionate and joyful teachers are effective teachers, he urged symposium participants to be "creative, create your own edible demonstrations, and have fun teaching astronomy."

In "Amateur Astronomers as Public Outreach Partners," Michael A. Bennett, executive director of the Astronomical Society of the Pacific (ASP), which is based in San Francisco, USA, identified "a huge, largely untapped source of energy and enthusiasm to help astronomers reach the general public" as volunteer science educators and urged astronomers and astronomy educators around the world "to consider more formal cooperation with amateurs." The ASP has estimated that, if one defines "amateur astronomer" as one who has joined a club of like-minded people, there are over 50 000 "affiliated amateur astronomers" in the US alone. It has also estimated that US amateur astronomers "reach some 500 000 members of the general public every year" through public star parties, classroom visits, community fairs, and museum/science center events. In March 2004 the ASP, with funding from the Navigator Public Engagement Program at NASA/JPL, launched its NASA Night Sky Network (NSN) to provide amateurs "with tested Outreach ToolKits on specific topics that can be used in a wide variety of ways with many different types of audiences," as well as training in how to use these resources. Amateur clubs must meet certain criteria in order to become members of NSN, and approximately 200 clubs had joined by summer 2006, representing approximately 20 000 amateur astronomers who have participated in over 4500 public outreach events. NASA funding is expected to continue for this effective program. Bennett urged astronomers around the world to identify ways to engage "outreach amateur astronomers" in their own countries.

Underlying the paper of former commission president Syuzo Isobe of Japan about "Does the Sun Rotate Around Earth or Does Earth Rotate Around the Sun?: an Important Aspect of Science Education" is the conviction that effective astronomy teaching must begin with the consideration of four variables: the pupil's class year, ability, level of interest, and future career. While of course it is more accurate to teach that the Sun does not rotate around Earth, it is not quite correct to teach that Earth rotates around the Sun, since "Solar System bodies rotate around a gravity center different from the center of the Sun." While for most students the assertion that Earth is a sphere is adequate, students who have a higher interest level and students who may go on in scientific professions should know that Earth "is an ellipsoid or a geoid."

Fernando J. Ballesteros and his colleague Bartolome Luque, both of Spain, made a case for "Using Sounds and Sonification" – and not merely impressive astronomy photographs – "for

Astronomy Outreach." The authors have a successful weekend radio program, "The Sounds of Science," broadcast on the national radio station of Spain. Sometimes, in fact, as with pulsars, "the images are not very spectacular but the sounds are strangely attractive." Astronomical sounds are also available to blind people in a way that images are not. [Editors' note: At least four books of astronomy images are available to the blind: Noreen Lawson Grice's *Touch the Stars, Touch the Universe: a NASA Braille Book of Astronomy, Touch the Sun: a NASA Braille Book and Touch the Invisible Sky*. Pasachoff reviewed the first three in the US college honor society Phi Beta Kappa's *The Key Reporter*, spring 2006, pp. 15–16, downloadable at www.pbk.org.] Ballesteros identified a number of Internet resources for astronomy sounds, the computer software "Sounds of Space," available for both PCs and Macs, and the possibility for professional astronomers to "sonificate" their own data, by passing them "to an audible format." Addressing the issue that there is no sound in the vacuum of space, Ballesteros notes that this fact represents a teachable moment in itself, since "in many cases the sounds will be radio signals passed to sound," as is the case with both pulsars and aurorae. Similarly, black holes, lightning storms on Saturn, and ionization tracks from shooting stars also emit radio signals. Ballesteros noted that in some cases there are real sounds, such as "when a shooting star crosses the sound barrier"; in other cases the sound may be inaudible but can be indirectly reconstructed, as in the case of "sound waves crossing the solar surface," which the vacuum of space prevents from reaching Earth, but which SOHO instruments can record indirectly and reconstitute after the fact.

Basing their argument on successful activities offered for school students at Sydney (Australia) Observatory, Nick Lomb and Toner Stevenson contend in "Teaching Astronomy and the Crisis in Science Education" that the trend in some countries to shun careers in math and science can be overcome by using astronomy "as a tool to stimulate students' scientific interest." The matter is of some importance, since if the trend is not offset, "there may not be enough people with Science, Engineering and Technology (SET) skills to satisfy the demand from research and industry." Not only will it be necessary to replace retirees from the "baby boomer" generation but also burgeoning industries including nanotechnology, biotechnology, and information technology will require workers with SET skills. In Australia, however, studies show a decline in the number of high school students studying advanced mathematics and both the life and physical sciences. Data from other countries are similarly dispiriting. Unless their parents have a positive attitude toward science, students often shun the physical sciences because they think of them as boring and irrelevant. Many students perceive science to be so difficult that only highly gifted students can succeed at them. Even those with high aptitude sometimes enroll in science courses only to improve their chances at excelling in university entrance exams. Lomb and Stevenson assert that planetaria and public observatories can improve student attitudes to science by engaging students' interest in a personal way.

In "Astronomy for All as Part of a General Education," J. E. F. Baruch *et al.* discussed the pluses and minuses of using www.telescope.org, a Web-based education program in basic astronomy available free of charge to anyone with Internet access. The authors explained the advantages of truly autonomous robotic telescopes, such as the Bradford Robotic telescope, which can "deliver the initial levels of astronomy education to all school students in the UK" over remotely driven telescopes that reach only "a tiny percentage . . . of students." They also discussed "practical solutions" for assisting teachers lacking not only a deep knowledge of basic astronomy but also confidence in working with information technology.

See also short descriptions of posters, pages 132–143.

1

Main objectives for the meeting on innovation in teaching/learning astronomy

Jay M. Pasachoff and Rosa M. Ros

Williams College, Williamstown, MA 01267, USA, and Technical University of Catalonia, Barcelona, Spain; President and Vice-President, respectively, of the Commission on Education and Development of the International Astronomical Union at the time of the 2006 General Assembly

Abstract: The aim of Special Session 2 on "Innovation in Teaching/Learning Astronomy" was to contribute to the implementation of the recommendations included in the 2003 International Astronomical Union resolution on the Value of Astronomy Education, passed at the General Assembly held in Sydney, Australia. These recommendations introduced innovative points of view regarding methods of teaching and learning. Astronomers from all countries – developed or developing – will be interested. Astronomy attracts many young people to education in important fields in science and technology. But in many countries, astronomy is not part of the standard curriculum, and teachers do not receive adequate education and support. Still, many scientific and educational societies and government agencies have produced materials and educational resources in astronomy for all educational levels. Technology is used in astronomy both for obtaining observations and for teaching. In any case, it is useful to take this special opportunity to learn about the situation in different countries, to exchange opinions, and to collect information in order to continue, over at least the next triennium, the activities related to promoting astronomy throughout the world.

In 2003, after much effort by members and officers of the International Astronomical Union's commission on education and development, and spurred on by Magda Stavinschi of Romania, the IAU was placed on record as endorsing worldwide astronomy education in a variety ways that we will specify in Part IV of this book. At the Prague General Assembly of the International Astronomical Union, held three years after the passage of the resolution, it seemed to be time to assess what progress had been made in the intervening years. We were also able to bring together a wide variety of astronomers and astronomy educators from around the world, including many who had not been able to go to Sydney. We were also glad to have the participation of some astronomers from developing countries who were new to our educational projects.

(1) In the IAU resolution on the Value of Astronomy Education, passed by the IAU's General Assembly in 2003, it was recommended:
   - to include astronomy in school curricula,
   - to assist schoolteachers in their training and backup,
   - to inform teachers about available resources.

(2) New methods of dissemination of information are making big changes in the opportunity of spreading astronomical knowledge.
   - The World Wide Web continues to expand its reach.
   - The Astronomy Picture of the Day (antwrp.gsfc.nasa.gov/apod/) reaches the home-page of millions.
   - The new phenomenon of podcasts is spreading rapidly.

Innovation in Astronomy Education, eds. Jay M. Pasachoff, Rosa M. Ros, and Naomi Pasachoff. Published by Cambridge University Press.

(3) Astronomy attracts many young people to education in important fields in science and technology. But in many countries, astronomy is not part of the standard curriculum, and teachers do not receive adequate education and support. Still, many scientific and educational societies and government agencies have produced materials and educational resources in astronomy for all educational levels.

(4) Technology is used in astronomy both for obtaining observations and for teaching. In any case, it is useful to take advantage of this special opportunity to learn about the situation in different countries, to exchange opinions, and to collect information in order to continue, over at least the next triennium, the activities related to promoting astronomy throughout the world.

In particular, all participants were invited to explain their experiences so they can be adapted for other regions of the world. Everyone was invited to exchange his or her initiatives and to try to involve other countries in common projects.

2

Learning astronomy by doing astronomy

John R. Percy

Department of Astronomy and Astrophysics, University of Toronto, Toronto ON, Canada M5S 3H4

Abstract: In the modern science curriculum, students should learn science knowledge or "facts"; they should develop science skills, strategies, and habits of mind; they should understand the applications of science to technology, society, and the environment; and they should cultivate appropriate attitudes – positive ones, we hope – toward science. While science knowledge may be taught through traditional lecture-and-textbook methods, theories of learning (and extensive experience) show that other aspects of the curriculum are best taught by *doing* science – not just hands-on activities, but "minds-on" engagement. That means more than the usual "cookbook" practical activities in which students use a predetermined procedure to achieve a predetermined result. The activities should be "authentic"; that is, they should mirror the actual scientific process.

In this paper, I describe several ways to include science processes within astronomy courses at the school and university level. Among other things, I discuss: the general philosophy of curriculum, and effective teaching and learning; strategies for developing science process skills through projects and other practical work; topics that expose students to astronomy research within lecture courses; topics that reflect cultural diversity and "the nature of science"; activities based on those developed and carried out by amateur astronomers; and topics and activities suitable for technical-level or "applied" courses.

2.1 Introduction

This book concerns *innovation* in teaching. Very little of what I shall say is innovative. Most of it is well known and accepted by education professionals, including schoolteachers. But it may not be well known, or implemented, by astronomers, and by university instructors – most of whom are amateurs in the sense that they have received little or no pre-service or in-service training in teaching and learning.

Many people associate the teaching of astronomy (or any other subject) with the transmission of knowledge, through lectures and textbooks. In this paper, I examine the role of activities and projects in which students learn astronomy by doing astronomy, or by being exposed to the practice of astronomy. The focus will be on the high school and undergraduate levels. And I shall interpret "doing astronomy" very broadly, since there is much more breadth to astronomy, and astronomy careers, than forefront research. This has been discussed eloquently by Tobias *et al.* (1995); there is a perception that science education prepares students to be research scientists; in fact, their knowledge and skills potentially prepare them for a very wide variety of careers. And that is a good thing in many ways. It provides science graduates with a wider range of job opportunities. It infuses society with well-trained science students. It connects science with society in many more ways.

Innovation in Astronomy Education, eds. Jay M. Pasachoff, Rosa M. Ros, and Naomi Pasachoff. Published by Cambridge University Press.

Therefore, "doing astronomy" can include a wide variety of activities, including teaching or communicating astronomy, developing astronomical instruments and software, etc.

General principles apply: astronomy teaching should begin with a definition of the *objectives* of the teaching, followed by a choice of the *curriculum*, followed by a choice of effective *teaching strategies and methods*. Every part of this process should be subjected to *assessment* – evaluation and feedback in the form of research or reflection – the results of which are used to continually improve the teaching (e.g., Hodson, 2001).

It goes without saying that a great deal of useful information is available in the proceedings of previous conferences on astronomy education: Pasachoff and Percy (1990), Percy (1996), Gouguenheim, McNally, and Percy (1998), Percy and Wilson (2000), and Pasachoff and Percy (2005). In this paper, I have therefore concentrated on some general issues which are not often raised at conferences such as this one.

To some extent, I have concentrated on the situation in North America, but I am fully aware of the excellent student activities and projects which have been developed by, for example, the European Southern Observatory (Madsen, 2005).

2.2 The curriculum, broadly defined

There is more to astronomy than factual knowledge. In my province (Ontario, Canada), the curriculum (Ministry of Education and Training, 1999) includes (i) knowledge or content, (ii) skills, (iii) applications of the content to the real world, and (iv) attitudes that the students should cultivate towards science. Skills would include making and recording observations; applications would include the contributions of astronomy to modern technologies such as space exploration; attitudes should include an appreciation of the wonder of the Universe, and the beauty of the night sky. Another education concept that expresses the breadth of the curriculum is "STSE": science, technology, society, and environment. "Society" would include the many ways in which cultures, past and present, have used astronomy for practical purposes such as navigation and timekeeping; "environment" would include an understanding of topics such as light pollution.

And, of course, there is a difference between the curriculum that is mandated by the government, the curriculum that is taught by the teacher, and the curriculum that is learned by the students. The curriculum that is taught depends on how well the teacher has been educated and supported. The curriculum that is learned depends on the use of appropriate teaching methods.

2.3 Depth of understanding: Bloom's and SOLO taxonomies

The learning of factual knowledge does not necessarily lead to deep understanding of the subject. Two common measures of depth or quality of understanding are Bloom's taxonomy (Bloom, 1984) and the Structure of Learning Outcomes or SOLO taxonomy (Biggs and Collis, 1982). The application of Bloom's taxonomy to astronomy is discussed in Slater and Adams (2002), along with many other useful and important topics; I have used Slater and Adams's examples in the list below.

Bloom's taxonomy encompasses six levels:

- knowledge: observation and recall of information; knowledge of dates, events, and places; awareness (but not necessarily understanding) of major ideas; and mastery of content (e.g., student states when the first day of spring is);

- comprehension: understanding of information; grasp of meaning; translation of knowledge into new contexts; ability to interpret, compare, and contrast facts; order, group, and infer causes; predict simple consequences (e.g., student describes the significance of the summer solstice);
- application: use of information; use of methods, concepts, theories in new situations; solution of problems using required skills or knowledge (e.g., student explains why the seasons are reversed in the southern hemisphere);
- analysis: seeing patterns; organization of parts; recognition of hidden meanings; identification of components (e.g., student can analyze what Earth's seasons would be like if Earth's orbit were perfectly circular);
- synthesis: using old ideas to create new ones; generalization from given facts; relation of knowledge from different areas; predicting and drawing conclusions (e.g., given a description of a planet's seasons, student can propose plausible orbital and tilt characteristics);
- evaluation: comparing and discriminating between ideas; assessing the value of theories and presentations; making choices based on reasoned arguments; verifying the value of evidence; recognizing subjectivity (e.g., student can distinguish what would be the important and non-important variables for predicting seasons on a newly discovered planet).

The SOLO taxonomy is based heavily on understanding the connections between different pieces of knowledge, and is divided into five levels:

- pre-structural: acquisition of bits of unconnected information, which have no organization and make no sense;
- unistructural: simple and obvious connections are made, without understanding of their significance;
- multistructural: multiple connections are made, still without understanding of their overall significance;
- relational: the significance of the parts, in relation to the whole, is grasped;
- extended abstract: the student can generalize, both within and beyond the subject area, and can generalize and transfer his or her understanding.

There is somehow a belief that, with good standard teaching, students will reach the highest level of understanding, but research shows that the average university graduate reaches only the middle levels. In my university, for instance, only the best students are accepted into medical school, but these students function only at middle levels in these taxonomies (C. Boyd, personal communication).

The reason for introducing these taxonomies, and their levels, is the following: while students can master the lower levels through rote memorization of lecture or textbook material, they can master the higher levels much better through minds-on engagement with real astronomical problems. Or, to paraphrase Tim Slater: in teaching, *what the students do is much more important than what the instructor does.*

2.4 Effective learning: hands-on/minds-on

For many years, the belief was that "hands-on" activities and experiences – making scale models of the Solar System, for instance – were the most effective tool for learning, and this is certainly so – especially for younger students. It is even more important, however, for the activities to be "minds-on." And there are many topics which do not lend themselves to hands-on activities, and there are settings such as large lecture courses where hands-on activities are simply not possible.

There are many ways to ensure that learning is minds-on. For instance: if you do a demonstration, ask the students to predict what will happen. If possible, choose a demonstration in which the result will conflict with their prediction, such as asking the class what color a blue object will appear, when illuminated by pure yellow light (a sodium vapor lamp, for instance); most will incorrectly say "green." Green (2003) presents a comprehensive list of "conceptests": questions that can be used to engage the students' minds, even in large lecture courses.

2.5 Effective learning: linking teaching and research

In universities today, *linking teaching and research* is a high priority. It refers to providing undergraduates with opportunities to carry out real research projects, and to be exposed to research-related concepts and skills in their lecture courses. In the US, these have routinely been done in many undergraduate colleges – editor Jay Pasachoff 's Williams College, for instance – for decades, but not necessarily in research-intensive universities. But the 1998 report of the Boyer Commission (Boyer Commission, 1998) has led to a re-thinking of how research universities should educate undergraduates.

The advantages of linking teaching and research are that students learn about the *process* of science; they learn research skills; they have higher levels of interest, excitement, and motivation; they are exposed to up-to-date knowledge; they can learn about their instructors' areas of research expertise and interest; and they may receive experiences that will be useful when they apply to graduate or professional school. Linking teaching and research connects the two main missions of a modern research university.

One method of linking teaching and research is to enable students to carry out research projects, within or outside of courses. My university has a Research Opportunities Program (ROP) in which second-year students may carry out a supervised research project for course credit. Students may also carry out research as summer research assistants, or in work-study programs. Most of my research students analyze archival measurements of variable stars, using various forms of time-series analysis (e.g., Percy and Ursprung, 2006 describes an ROP project done by Ursprung). They are motivated by the excitement of doing real science, with real data, leading to a publishable result.

2.6 Effective learning: linking teaching and teaching

Students can also learn astronomy effectively by having to teach it or communicate it. I therefore recommend that your students have opportunities to teach or tutor their peers, or teach younger students, or give presentations for their class or for the public. During the last decade, I have successfully built an undergraduate Science Education program at the University of Toronto, in which science and math students can develop knowledge, skills, experiences, and satisfaction through teaching and outreach. They carry out useful projects and placements that interest and motivate them. In developing this program, I was motivated by the advantages of introducing students to *education* – not just as consumers, but as reflective practitioners.

2.7 School science skills and "habits of mind"

As mentioned above, the science curriculum includes science skills and "habits of mind." In the school science curriculum in my province (Ministry of Education and Training, 1999), those skills include:

- application of the skills and strategies of scientific inquiry: initiating and planning; performing; recording; analyzing; interpreting; problem-solving – planning an observing session, for instance;
- application of technical skills and procedures: using tools, equipment, and materials safely and correctly – setting up a telescope for solar viewing, for instance;
- communication of information and ideas, in different ways, to different audiences, for different purposes, using different technologies – always using scientific language, symbols, conventions, and units correctly – teaching their peers about a challenging topic in astronomy, or writing a research report, for instance.

2.8 Science fair projects

The most obvious form of "research" at the school level is science fair projects. In North America, many elementary schools organize science fairs within the school. By the high school level, there are both district and national science fairs. The highest-level science fairs in the US have top prizes with values of several tens of thousands of dollars.

There are several potential problems with science fairs. If they are unduly competitive, they may be a negative experience for the student. And if they are unduly competitive, there may be temptation to cheat. For instance, the projects may end up being mainly the work of parents! Students who have access to a university scientist or laboratory may also have an unfair advantage. There may be difficulty finding qualified judges, and this can be a special problem with an astronomy project, since most teachers have little or no content background in the subject.

There is another potential problem that is also specific to astronomy. The evaluation, in some science fairs, is inflexible. Students are expected to have a hypothesis, and to test it in a way that controls variables. This may be appropriate in an experimental science, but it is not appropriate in an observational science like astronomy. Astute judges can allow for differences of this kind – if they are available.

There is a related problem in a country (which I shall not name) in which one requirement of senior high school courses is to do a research project. The requirements for this project are also phrased in terms of making a hypothesis and testing it by controlling variables. Clearly it is important for students, teachers, and science fair judges to be aware of the nature of astronomy.

Gelderman, at this conference, made an interesting analogy between participating in science and participating in athletics. From an early age, children should be encouraged to participate in recreational athletics for health and fun. Likewise, they should be encouraged to participate in science, for curiosity, interest, and fun. Older children can participate in more competitive athletics, as a way of developing their individual and team skills, but the emphasis should continue to be on health and fun. So too in science. Some children will go on to participate in athletics at the highest, most competitive level, just as some students will compete in high-level science fairs. But we hope that *every* student will continue to participate in recreational athletics – and science – for fun, fellowship, and physical or mental well-being.

2.9 Is it possible for students to "do astronomy"?

My research on variable stars lends itself to undergraduate projects. I even supervise outstanding senior high school students in my university's Mentorship Program, which allows such students to work on research projects at the university (e.g., Percy *et al.*, 2006 describes a project done by high school student Molak; three of the other authors are amateur astronomers who supplied the data that were used). The goal, for each student, is to complete a self-contained project which can be presented or published, with the student as co-author. In carrying out these projects, each student goes through the steps of the research process, developing and using the skills mentioned below.

There is a great deal of useful information and resources available for undergraduate research students, especially in the US where the National Science Foundation funds Research Experiences for Undergraduates. The Boyer Commission has undoubtedly had an influence also. There is a National Council on Undergraduate Research (www.cur.org), National Conferences on Undergraduate Research (www.ncur.org) and – especially useful – WebGURU (www.webguru.neu.edu) which is specifically for students in science, technology, engineering, and mathematics. This site provides a wealth of useful advice. There is even a *Journal of Undergraduate Research* (www.scied.doe.gov/scied/JUR.html).

At a somewhat higher level: my colleague Professor Chris Matzner has a website full of useful information for beginning undergraduate or graduate researchers in astronomy at www.cita.utoronto.ca/~matzner/svc/resources.html.

Most of my colleagues argue that it is not possible, in their "big-science" field of research, for students to be meaningfully involved in a real project. This may be partly a shortage of imagination on their part, but it is still possible that the student could carry out a meaningful project, *but not necessarily a publishable one*. After all, most university students have carried out science fair projects while in school (see above). So it should be possible for students to develop research skills by completing a project if they carry out steps such as the following:

- development and understanding of the objectives and strategies for carrying out the project;
- writing a project proposal and progress reports;
- critical background reading of print and electronic sources;
- description of the origin, accuracy, and peculiarities of the data being used;
- description of the method(s) of analysis, including any software being used;
- testing the method(s) on trial data, to make sure a reasonable result is the outcome;
- analysis and interpretation of new data;
- writing, editing, and defending a project report;
- preparing a poster, display, or talk on the project.

I give my research students this list as a framework for developing their research project and their research skills.

A second objection to student research projects is that research supervision is a time-intensive process. This is true, but if it results in the student's gaining high-level skills (as well as motivation), then perhaps it is worth it – especially if it enables the students to become lifelong self-learners. An alternative is to develop ways in which *groups* of students can develop research skills, under supervision, together. An example, at my university, is First-Year Seminars. These courses are restricted to 20 students. They involve extensive research, presentation, and discussion, under the supervision of a research-active instructor. Elsewhere in this book Mary Kay Hemenway describes a similar program of first-year seminars at the University of Texas. The papers produced in such courses are not necessarily publishable in research journals, but the best are increasingly published on-line. At my university, the library has a system called *TSpace* in which I can archive and disseminate the best student papers. There is no reason why a high school astronomy class could not take a similar approach.

It is highly desirable for students to present their research work to their peers, either orally, or through publication, or ideally both. In addition to developing communication skills, and giving students the opportunity to critically evaluate their peers' work, it introduces an important social aspect to research. In the US, many Research Experiences for Undergraduates programs include symposia at the end of the year, at which student projects are presented as talks or posters. In my university's Research Opportunity Program, projects are presented as posters at an end-of-term Research Fair. It is exciting to see the range of projects that are done in the many science, social science, and humanities departments across my university.

2.10 Linking teaching and research in lecture courses

It's rarely possible to "do astronomy" in a large lecture course, though one example I have used successfully is based on my research interest in variable stars, and the fact that amateur astronomers and students can measure variable stars visually. I show a sky image of a variable star such as δ Cephei, along with non-variable comparison stars. I ask each member of the audience to estimate the magnitude of the variable star, relative to the comparison stars, to a precision of 0.1 magnitude. I then tally and show the results, and determine the mean and standard error. The students see the distribution of the measurements and the typical measurement error. They see the power of multiple or repeated measurements. A frequent question is "what's the answer supposed to be?" The questioners are obviously missing the point!

But there are many ways of developing research skills, or making research connections, even in large lecture courses, and many advantages of doing so. At some point, students should read simple examples of research papers, perhaps historical ones – there are lots on-line – and be asked to summarize or critique them. They should, of course, learn to use print and electronic sources critically and effectively. In my undergraduate Science Education course, when I ask the students to think of their best experience in being exposed to research, they almost always refer to their Introductory Psychology course and laboratory, in which they observe experiments, design experiments, participate in experiments, access and analyze archival data, read and summarize research papers, etc. The advantage in psychology, of course, is that the students have readily available research subjects – themselves or their classmates. In astronomy, the problem is that, with the notable exception of the Sun, "the stars come out at night, the students don't."

In laboratory work: students can use real-life data; they can use the same computer languages and software packages that researchers use. The labs should be "authentic" – mirroring real research, rather than aiming for a pre-determined "cook-book" result. This, by the way, is strong motivation for including a well-designed practical or lab component for any Introductory Astronomy course in high school or university.

In lectures, students can certainly be exposed to current issues and research, discuss their instructor's research, and evaluate the astronomical facilities that they are funding as taxpayers. They can learn that astronomical knowledge is not produced by textbooks but by astronomers – perhaps even by themselves! If possible, they can tour the astronomical research facilities at their university; if not, there are "virtual" webcam tours of major astronomical facilities available to any student. And there are remote, robotic telescopes – many of which can be publicly accessed; these are described elsewhere in this book.

Remote, robotic telescopes can be used by students for astronomical research projects but, as McKinnon (2005) has emphasized, such telescopes only lead to significant *learning* when the technology is integrated with curriculum development, teacher education, and ongoing evaluation. Students may, however, learn skills by using these telescopes, and may experience pleasure by obtaining or ordering images of astronomical objects. They can do science with these telescopes through curriculum-linked programs such as *Hands-On Astrophysics* (hoa.aavso.org), mentioned below.

2.11 Doing astronomy: connections with multicultural astronomy

The first people to "do astronomy" were the pre-technological cultures who were able to determine time, date, and direction from the sky. They made and recorded observations, archived and used them, and passed their knowledge down through the generations. Astronomy is part of almost every culture, and *multicultural science* is emphasized in many curricula, including the one in my province. Astronavigation is closely connected with sky motions and observations that any student can make; these observations can be extended and enhanced by using sky-simulation software. Likewise, the Islamic calendar, the lunar new year, and Easter can be linked to observations that even city-bound students can make. And timekeeping can be done with a simple stick or pole that casts a shadow in the schoolyard through the day. See Chandra and Percy (2001) for a teachers' guide to multicultural astronomy.

One could argue that these simple observations are the most relevant kinds for most students to make (as compared, for instance, with the types of observations that professional astronomers make). In fact, at the first IAU Colloquium on Astronomy Education, almost two decades ago, Wayne Osborn (1990) argued that they were much more relevant for most students than plotting Hertzsprung–Russell or period–luminosity or Hubble diagrams!

2.12 Doing astronomy through amateur astronomy activities

But we would also like students to carry out astronomical observations and activities that are more representative of modern astronomy.

Professional astronomers are expert at the academic or "why" side of astronomy. Amateur astronomers come in many "flavors," but most are interested in the descriptive, practical aspects. And many of them are interested in the hands-on aspects of astronomy: telescopes, instruments, and computers. Indeed, many of them seek the same kind of excellence in their use of tools as a professional astronomer would. See Percy and Wilson (2000) for a comprehensive discussion of the role of amateur astronomers in research and education.

What are some of the activities, based on amateur astronomy, that might be part of an astronomy course?

- Making, demonstrating, and using telescopes and other tools.
- Making careful observations or measurements of astronomical sky phenomena.
- Making and processing digital images of astronomical objects.
- Using sky-simulation computer programs to predict or explain views of the sky.
- Making classroom models or demonstrations of astronomical phenomena.
- Making attractive posters or displays of astronomical concepts.
- Giving non-technical presentations to other students or to the public.
- Writing non-technical articles on astronomical topics.

Note that models, demonstrations, posters, presentations, and articles all provide material that is useful, above and beyond its immediate role in the course. It can be retained in the classroom, for future use, proudly labeled with the name of the student who created it.

One obvious practical activity is observing the Sun over many weeks or months. This can be done using images on the Internet, but students should observe the real Sun (safely, by projection, of course) at least once.

A less-obvious activity is observing variable stars, and analyzing variable star data. Lots of archival data is available, on the website of the American Association of Variable Star Observers (AAVSO), for instance (www.aavso.org). The AAVSO's *Hands-On Astrophysics* project (hoa.aavso.org) provides all the materials and curriculum links that are needed. But it would be best if each student made at least one measurement on his or her own, for comparison with the archival data. Betelgeuse is one bright variable star that is visible, even from cities.

2.13 Relevance to "applied" courses

In my province's curriculum, courses are divided into *academic* and *applied*. The academic courses tend to be taken by academically-strong students who are bound for university. They are heavy with content and theory. The applied courses are intended for students with more practical interests, who are bound for technical school, trades, or the workplace.

There are many problems connected with the applied courses. They were not as carefully thought-out as the academic courses. They are mistakenly considered to be low-level, diluted courses, rather than courses for students who could demonstrate their excellence in practical work, as opposed to academic work. Disadvantaged students are often relegated to these courses, even though they could probably succeed in academic courses if they had the necessary scaffolding.

Astronomy would seem to be ideally suited to an applied course because astronomy naturally divides into professional (or academic), and amateur (or practical or applied). So one strategy for teaching an applied course or unit would be to emphasize the kinds of topics and activities that amateur astronomers would enjoy – the ones that I have previously listed. In these courses, practically-oriented students can demonstrate excellence in different ways than simply achieving high marks in academic courses. After all, academic students might possibly become professional astronomers, but *all* students might become amateur astronomers – individuals who love astronomy and adopt it as a hobby. They can become lifelong learners and lovers of astronomy. That is the ultimate goal of astronomy education.

References

Biggs, J. and Collis, K., 1982, *Evaluating the Quality of Learning: the SOLO Taxonomy* (New York: Academic Press).

Bloom, B. S., 1984, *Taxonomy of Educational Objectives* (Boston: Allyn and Bacon).

Boyer Commission, 1998, *Reinventing Undergraduate Education: a Blueprint for America's Research Universities*, naples.cc.sunysb.edu/Pres/boyer.nsp/.

Chandra, N. and Percy, J., 2001, *Using Multicultural Dimensions to Teach Astronomy* (San Francisco: Astronomical Society of the Pacific).

Gouguenheim, L., McNally, D. and Percy, J. R. (eds.), 1998, *New Trends in Astronomy Teaching* (Cambridge: Cambridge University Press).

Green, P. J., 2003, *Peer Instruction in Astronomy* (Upper Saddle River NJ: Pearson Education Inc.).

Hodson, D., 2001, *OISE Papers in STSE Education*, Ontario Institute for Studies in Education, at the University of Toronto, **2**, 7.

Madsen, C., 2005, in J. M. Pasachoff and J. R. Percy (eds.) *Teaching and Learning Astronomy* (Cambridge: Cambridge University Press).

McKinnon, D. H., 2005, in J. M. Pasachoff and J. R. Percy (eds.) *Teaching and Learning Astronomy* (Cambridge: Cambridge University Press), 104.

Ministry of Education and Training, Ontario, 1999, *Science: The Ontario Curriculum Grades 9 and 10* (Toronto: Queen's Printer for Ontario).

Osborn, W., 1990, in J. M. Pasachoff and J. R. Percy (eds.) *The Teaching of Astronomy* (Cambridge: Cambridge University Press), 152.

Pasachoff, J. M. and Percy, J. R. (eds.), 1990, *The Teaching of Astronomy* (Cambridge: Cambridge University Press).

Pasachoff, J. M. and Percy, J. R. (eds.), 2005, *Teaching and Learning Astronomy* (Cambridge: Cambridge University Press).

Percy, J. R. (ed.) 1996, *Astronomy Education: Current Developments, Future Coordination* (San Francisco: Astronomical Society of the Pacific, Conference Series), **89**.

Percy, J. R. and Wilson, J. B. (eds.), 2000, *Amateur–Professional Partnerships in Astronomy* (San Francisco: Astronomical Society of the Pacific Conference Series), **220**.

Percy, J. R. and Ursprung, C., 2006, *J. Amer. Assoc. Variable Star Observers*, in press, www.aavso.org/publications/ejaavso/ej36.shtml.

Percy, J. R., Molak, A., Lund, H., Overbeek, D., Wehlau, A. F., and Williams, P. F., 2006, *Publ. Astron. Soc. Pacific*, **118**, 805.

Slater, T. F. and Adams, J. P., 2002, *Learner-Centered Astronomy Teaching: Strategies for Astro 101* (Englewood Cliffs, NJ: Prentice Hall).

Tobias, S., Aylesworth, K., and Chubin, D. F., 1995, *Rethinking Science as a Career* (Tucson AZ: The Research Corporation).

3

Hands-On Universe – Europe (EU-HOU)

Roger Ferlet

Institut d'astrophysique de Paris, UMR7095 CNRS, Université Pierre et Marie Curie, 98bis Bd. Arago, 75014 Paris, France

Abstract: The EU-HOU[1] project aims at re-awakening the interest for science through astronomy and new technologies, by challenging middle and high school pupils. It relies on real observations acquired through an internet-based network of robotic optical and radio telescopes or with didactical tools such as webcams. Pupils manipulate and measure images in the classroom environment, using the specifically designed software SalsaJ, within pedagogical trans-disciplinary resources constructed in close collaboration between researchers and teachers. Gathering eight European countries coordinated in France, EU-HOU is partly funded by the European Union. All its outputs are freely available on the Web, in English and the other languages involved. A European network of teachers is being developed through training sessions.

3.1 Introduction

A clear disaffection for scientific studies at universities, in particular mathematics and physics, is currently observed in the occidental world, including most of the European countries. Very worrying for the future, such a complex situation cannot be reversed through any single undertaking. Renewing the teaching of science, however, might significantly improve the situation by making science courses more attractive. By promoting the learning of the scientific method, teachers can not only re-awaken student interest in science but also help pupils feel that understanding can be a source of pleasure. Bringing the excitement of scientific discovery to the classroom is of utmost importance for a sustainable economy, which depends on a critical number of scientists and engineers. Moreover, unless our society keeps science in its heart, it risks regressing to a more primitive and much less attractive state. These assumptions summarize the overall philosophy of the eight European partners – universities, research institutions or science centers from France (F), Greece (Gr), Italy (It), Poland (Pl), Portugal (Pt), Spain (Sp), Sweden (S) and the UK – who are affiliated under the leadership of the French University Pierre and Marie Curie (UPMC) in Paris. Their very ambitious but also very timely goal may be fulfilled through astronomy, a well-established source of motivation for science and technology learning, and by using new technologies for high and middle schools, including Internet-controlled optical and radio telescopes, webcam systems, didactic software, multimedia, etc.

These eight partners were already connected with the Global-HOU (G-HOU) project (Ferlet and Pennypacker, 2006), which enabled them to benefit from sharing world-wide

[1] On behalf of the EU-HOU partners: R. Ferlet and A.-L. Melchior (F), M. Metaxa (Gr), A. Zanazzi (It), L. Mankiewicz (Pl), R. Doran (Pt), C. Horellou (S), A. I. Gomez de Castro (Sp), R. Hill (UK).

Innovation in Astronomy Education, eds. Jay M. Pasachoff, Rosa M. Ros, and Naomi Pasachoff. Published by Cambridge University Press.

experiences. A major goal of G-HOU is to give all the children of the world the ability to look at the stars meaningfully, deepening their scientific education and inspiration at the same time. As Oscar Wilde put it in *Lady Windermere's Fan* (1892), "We are all in the gutter, but some of us are looking at the stars." Children of the world want to feel connected to each other, and love and need humanity's common relationship to the heavens. The revolutionary new ease of working together with telescopes and of sharing astronomical data make astronomy a viable means of world development and growth. The need to address the crucial problem of disaffection for scientific studies within the European educational system and to strengthen European collaborations between middle/high schools and university systems, however, led to the launch of the EU-HOU project.

3.2 Objectives

The gap between major scientific research and the general public is obviously large and increasing. The educational consequences are twofold. First, it is our duty to fill this gap for the sake of democracy. Second, the gap is increasingly difficult to overcome because important scientific experiments and observations are not easily demonstrated or even simulated in a classroom. Thus, to maximize the chance of success, any project should fulfil at least the following few basic conditions.

- To be based on hands-on activities: in order to enhance their autonomy and reasoning skills, pupils should directly participate as much as possible through observing, arguing, sharing, discussing, and interpreting real astronomical data; in brief, pupils should learn science by doing science.
- It should be possible to begin at a low level, as far as pre-required knowledge and financial expenses are concerned; complexity can be increased subsequently if kids prove interested.
- Cross-curricular approaches that address various interests are advisable.
- Easy access to knowledge and databases as well as a self-sustained system for teacher training should be available.

In order to move forward toward our overall goal, we aimed at identifying, gathering, organizing and producing pedagogical resources ready to be used in European classrooms. This initiative has been formatted into several specific objectives.

- Continuous production of new innovative pedagogical resources: astronomical data, exercises, multimedia supports. Constructed in close collaboration between researchers and teachers, these resources are trans-disciplinary in essence (astronomy, physics, mathematics, history, language, ...), and available in English. Most of them are translated into the relevant European national languages, adapted to national curricula, and tested by teachers/educators.
- Production of a dedicated users-friendly didactical software in Java – called SalsaJ (*Such A Lovely Software for Astronomy*) – for image and spectra manipulation and analysis in classrooms; it is multiplatform, with a multilingual interface, and freely downloadable.
- Pedagogical use of the optical 2-m Faulkes Telescopes (in Hawaii and Australia) operated remotely through the Internet, which enables direct observations in classrooms in Europe, thanks to the longitude effect.
- Development of a 2.3-m radio antenna (at Onsala observatory) operated remotely through the Internet, for 21-cm neutral hydrogen observations of the Milky Way, even

with poor meteorological conditions. Pedagogical use of the 7-m radio telescope of the Jodrell Bank observatory.

- Production of a webcam observing tool to be used directly by pupils – through a plug – within SalsaJ – with either a photo lens or a small telescope, for astronomy, physics, etc., initially distributed to 20 schools in each country free of charge.
- Creation of a European network of researchers with scientific expertise and middle and high school teachers with pedagogical expertise, to promote science and technology education. This network should gear the education system to research and development at both European and national levels.
- A specific website at UPMC offering a free multilingual portal (www.euhou.net) to all EU-HOU resources, consisting of a central European site in English and eight national sites for translated resources and additional national material, together with communication forums.
- Dissemination through workshops and teacher training sessions. Three such sessions have been held (in Poland, France, and Italy) during the second year of the project, gathering almost 100 European teachers, who are expected now to share their expertise within their own countries.

Relying on real astronomical data and the expertise of scientists, EU-HOU is providing classrooms tools to explore central concepts in science, math, and technology. By visualizing and analyzing their own data with SalsaJ, similar to the software professional astronomers use, students become more engaged and more excited about math and science.

3.3 Organization

The project was submitted to the European Community in 2004, accepted as a MINERVA action (within the SOCRATES framework), and funded at a level of about 380 000 Euros (44% of the total budget, the rest being provided by the different partners) for two years ending September 30, 2006. It is coordinated and managed by UPMC, with the help of a Pedagogical Coordinating Committee composed of representatives from each country, which meets twice a year. A Scientific Advisory Committee with personalities from each country, including ESO (European Southern Observatory), external to the project and known for their involvement in innovative education in science, has performed a very positive mid-term evaluation of the project.

Each partner has produced at least one resource, in his national language and in English. Here are examples of the first available resources (with responsible countries): from the Doppler effect to extrasolar planets (F); distances to Cepheids (F, Pl); the life of stars and stellar population (Gr); how to weigh a distant galaxy (It); webcam astronomy (Pl); variable stars (Pl); what is a star? (Pt); radio astronomy in the classroom and hydrogen in the Milky Way (S); the Solar System as a math laboratory (Sp); voyage through space (EduSpace/ESA), Faulkes telescopes and Jodrell Bank radio telescope (UK); SalsaJ and the website (F). The reader is invited to visit the EU-HOU website www.euhou.net for more details.

In each country, a pool of five to fifteen so-called "resource agents" has been set up; they are teachers eager to learn about the various EU-HOU outputs, adapt them to national curricula, use them in their schools and train other teachers/educators about these outputs. A pool of about 20 pilot schools per country has also been built, in which the different outputs are tested and validated.

The expected overall impact is to answer the demand from teachers willing to introduce in their classrooms a new way to teach science in order to stimulate pupils. In addition, the project encourages international communication among pupils. By the end of 2006, more than ten thousand pupils in eight European countries will have used EU-HOU (and many more for G-HOU)! Feedback from users has always been very positive, supporting reasonable hopes for the future expansion and diversification of Hands-on Universe into "Hands-on life" for biology, "Hands-on Earth" for geology/ecology, etc. This expansion of the Hands-on program should help mold open-minded pupils able to think on their own.

Reference

Ferlet, R. and Pennypacker, C., 2006, in A. Heck (ed.) *Organizations and Strategies in Astronomy* (New York: Springer), **6**, 275.

Comments

Nigel Douglas: The term "robotic" you have used is normally used to describe "queued" observing, which is *not* real-time. Real-time corresponds to "remote telescope."

Roger Ferlet: The robotic Faulkes telescope we presently use in Hawaii within the EU-HOU project is allotting 30 min. slots for *real-time* observations in schools. The Onsala radio-antenna of the EU-HOU project also provides *real-time* observations from Sweden, furthermore, without cloudy sky constraints.

John Mattox: What is the relationship between EU-HOU and Global-HOU?

Roger Ferlet: EU-HOU is part of Global-HOU, which meets every year. (The most recent G-HOU was held in France at OHP only 10 days ago!) The relationship is more in the spirit, as there is no formal agreement.

4

Life on Earth in the atmosphere of the Sun: a multimedia manual

E. V. Kononovich, T. V. Matvejchuk, O. B. Smirnova, G. V. Jakunina, and S. A. Krasotkin

Sternberg Astronomical Institute and Skobel'tsyn Institute of Nuclear Physics, Moscow State University, Moscow, 119992, Russia

Abstract: The purpose of this manual is to illustrate the major physical processes occurring in the Sun–Earth system and to determine the ecology of the planet's life. The material is divided into three separate parts: "The Sun," "The Earth," and "Sun–Earth Relationships." These parts do not require cross-references since each is self-contained. Within each part the material is organized in sequences according to the well-known methodical principle: from simple to complex. "The Sun" is subdivided into three sections: "The Sun as a Star," "Solar Activity," and "Helioseismology." "The Earth" is devoted to the description of the basic characteristics of the planet: internal structure, magnetic field, lithosphere, and atmosphere, together with the tectonic, hydro, and atmospheric processes occurring in them. The top layers of the Earth's atmosphere (ionosphere, the aurora zone, radiating belts, magnetosphere) are also considered. "Sun–Earth Relationships" utilizes the previous information to present the influence of the active solar processes on various aspects of terrestrial life: ecological, biological, economic, and so forth. Solar activity forecasting as the key parameter determining properties of so-called space weather is also considered. The manual is addressed to students of the science faculties of universities and teachers' colleges, as well as to pupils in the upper classes of high school who are interested in solar problems. Teachers of solar physics and Earth sciences courses can use it as a teaching aid to develop the corresponding sections of lecture courses and methods manuals and to prepare small discussion sessions, etc. All materials related to each of the three parts are found in the corresponding directories under the same titles presenting the available information in PowerPoint format.

4.1 Introduction

Significant recent changes in our conceptions concerning solar activity are due, first of all, to the new data about the Sun. If earlier understanding of the Sun was based mainly on ground observations of the visible and radio parts of the solar spectrum, nowadays the situation has changed dramatically. On board the spacecraft SOHO, Yohkoh, Ulysses, TRACE, CORONAS, etc., spectral tools with high spatial and time resolution are installed. These tools make possible observations of solar activity phenomena in spectral regions inaccessible to observers from the Earth's surface, including the ultraviolet, X-ray, and gamma-ray radiation of the Sun. Significant progress has been made in the equipment techniques of solar atmosphere ground observations, especially of the magnetic fields and plasma motions of the Sun.

Part One of the manual, **"The Sun,"** consists of three sections: **"The Sun as a daytime star," "Solar activity,"** and **"Helioseismology."** The first section describes solar instruments, space research, and the basic phenomena visible on the Sun. The main parameters of the Sun and its internal structure are considered. The most impressive demonstrations are presented.

Innovation in Astronomy Education, eds. Jay M. Pasachoff, Rosa M. Ros, and Naomi Pasachoff. Published by Cambridge University Press.

Ground-based spectral observations of the solar atmosphere, as well as those made by space telescopes, are also described.

The second section of Part One, **"Solar activity,"** describes the active events in the photosphere, chromosphere, and corona, including problems concerning solar flares and coronal mass ejections. Data on indexes of helio- and geophysical activity are also considered. The nature of solar cyclicity and the feasability of forecasting solar cycle parameters are discussed. Dynamo theories of the solar cycles, based on the interaction between convection and turbulent motions in the convective zone with the differential rotation of the Sun, are briefly outlined. The important, if not the main, role played by the magnetic field in the cyclicity of solar activity is described.

The last section of Part One, **"Helioseismology,"** briefly outlines the youngest area of solar physics.

Part Two of the manual, **"The Earth,"** includes a description of the Earth's structure, its atmosphere, and its magnetic field. Problems related to the Earth's weather, climate, water, and energy balances are also treated. Descriptions of the active phenomena on the Earth's surface, such as volcanoes, earthquakes, continental drift, etc., are also given.

Part Three of the manual, **"Solar–terrestrial relationships,"** introduces a topic whose rapid development is a consequence of its direct practical value. The discussion includes results obtained in the past, as well as modern conceptions concerning the terrestrial atmosphere and the magnetosphere. Solar activity displays itself in such geophysical phenomena as magnetic storms, polar lights, and ionosphere disturbances. The influence of solar activity on biological systems and on climate and the Earth's weather are also considered. The importance of solar influence on the biosphere and on ecology is increasingly apparent.

4.2 Contents of the manual

4.7 Indexes of magnetic activity
5 Weather and climate
5.1 Basic conceptions
The weather
Basic climate types
The solar constant
Thermal balance of the Earth
The thermal balance equation
Thermal balance of the Earth's atmosphere
Radiation balance of the Earth
Greenhouse effect
Water balance
Energy balance
5.2 Physical and chemical processes in the Earth's atmosphere
Passage of radiation through the Earth's atmosphere
Interaction of sunlight with the terrestrial surface
Mechanism of greenhouse effect formation
5.3 Climatic zones and local properties of climate
History of the problem
Temperature of the air
5.4 Seasonal climatic phenomena
5.5 Circulation forms
5.6 Meteorological phenomena
5.7 Weather forecasting
5.8 Optical phenomena in the atmosphere
6 Active phenomena on the Earth's surface
6.1 Continental drift
6.2 Seismology and tectonics
6.3 Volcanoes
6.4 Earthquakes
6.5 Origin of earthquakes
Part 3 SOLAR–TERRESTRIAL RELATIONSHIPS
1 Solar heliosphere and terrestrial magnetosphere
2 Solar activity displays in geophysical phenomena
3 Ionospheric disturbances
4 Solar cosmic rays
5 Magnetic storms
6 Night sky airglow and polar lights
7 The Sun, weather, and climate.
8 Biological effects caused by solar activity
9 Historical survey of solar–terrestrial effects

Futher reading

Brandt, J. C. and Hodge, P. W., 1964, *Solar System Astrophysics* (New York: McGraw-Hill, reprinted Moscow, 1967).

Brückner, E., 1890, *Klimaschwankungen* (Moscow).

Bruzek, A. and Durrant, C. J. (eds.), 1980, Illustrated Glossary for *Solar and Solar-Terrestrial Physics* (Dordrent, The Netherland Reidel).

Ellison, M. A., 1955, *The Sun and its Influence* (London: Routledge and Kegan Paul, reprinted Moscow, 1959).
Kononovich, E. V., Khramova, M. N., and Krasotkin, S. A., 2002, The sun as a variable star, *Astronomical and Astrophysical Transactions*, 2002, **21**, 293–303.
Lang, K. R., 1995, *Sun, Earth and Sky* (New York and Heidelberg: Springer).
Menzel, D. H., 1963, *Our Sun* (Cambridge, MA: Harvard University Press).

Comments

Kala Perkins: How is it possible for other teachers to access you and your materials?
Edward Kononovich: sai.msu.ru "Astronet."
Barrie W. Jones: How do you (do you?) promote *safe* observations of the Sun by students?
Edward Kononovich: I did not show a slide that summarized our safety advice.

5

A model of teaching astronomy to pre-service teachers

Bill MacIntyre

School of Curriculum & Pedagogy, Massey University College of Education, Palmerston North, New Zealand

Abstract: This paper details a model of teacher development that allows students to demonstrate their understanding of basic astronomy concepts as well as communicating that understanding in creative ways. Key features of the model include starting from students' initial understanding about astronomical concepts and providing time for students to assess first their mental models and then their astronomical understanding, in both cases using 3-D models. These key features appear collectively to provide an appropriate creative environment for students. Two students adapted their original 3-D models in order to communicate specific aspects of seasons – different solar inputs and varying lengths of day/night throughout the year. Interviews of the two students highlighted the thinking that led to the creation of the new components. The uptake by other students, during the modeling assessment task, demonstrated the effectiveness of the new components in communicating an astronomical understanding of seasons. The model of teacher development illustrates how teacher educators can teach for astronomy understanding as well as allow for creative ways to communicate that understanding to others – an essential aspect of being an effective astronomy educator in the classroom.

5.1 Introduction

Models and modeling research in science education have steadily developed over the last decade thanks in part to the collaborative work of members of the Centre for Models in Science and Technology: Research in Education (MISTRE) Group, Duit and Glynn (1996), Harrison and Treagust (2000), and others. Specific research on teachers' knowledge (Justi and Gilbert, 2002) and the professional development of science teachers' knowledge (Justi and van Driel, 2005) are providing the foundation blocks for incorporating a study of models and modeling in science education '*methods*' papers as well as for science '*content knowledge*' papers.

"A model is a representation of an object, event, process, or system" (Gilbert and Boulter, 1998). Gilbert *et al.* (2000) provide a practical way to classify models based on their ontological status. *Mental models* are the cognitive notions held by individuals. Once a mental model is communicated to others it is known as an *expressed model*. If the expressed model is deemed "of value" or "worth" by a social group and used extensively by the group then it is called a *consensus model*. Within the scientific community, consensus models can become the *scientific model* that is used for further developments in science. As consensus models are dependent on the context at the time of their genesis, many are changed, modified, or superseded by other, later consensus models. The teaching model – "investigating with models" approach, used in the Spaceship Earth and Beyond course – is an example of a *hybrid*. It was intended to provide opportunities for primary (elementary) teacher trainees to

Innovation in Astronomy Education, eds. Jay M. Pasachoff, Rosa M. Ros, and Naomi Pasachoff. Published by Cambridge University Press.

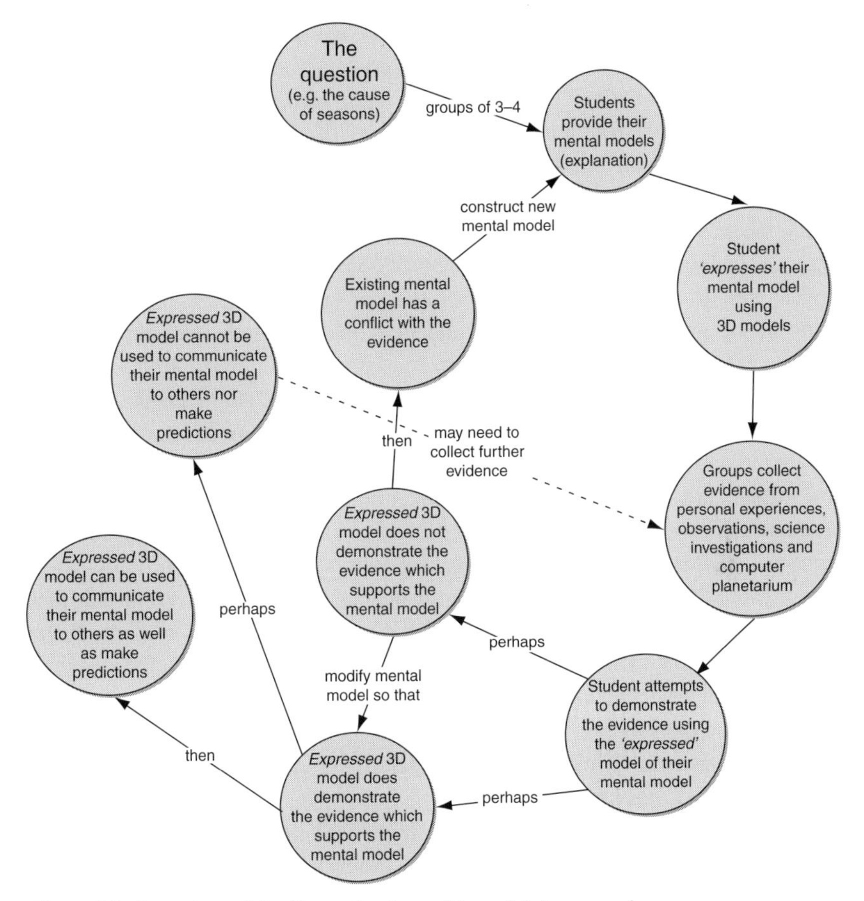

Figure 5.1 Overview of the "investigating with models" approach.

develop their basic astronomy understanding (content knowledge) and pedagogical content knowledge (PCK) of astronomy using 3-D models. It also attempted, in an indirect way, to improve their curricular knowledge, content knowledge, and PCK of models and modeling in a collaborative group. Hipkins *et al.* (2002) identified three aspects of effective science teaching. They have been contextualized for this paper. Effective teaching of science (astronomy) concepts occurs when (a) teachers emphasize the place of *mental* models and *expressed* models in their teaching; (b) students understand the role of *mental* models in the communication of astronomical explanations; and (c) students are able to critique their own models and scientists' models, understanding the limitations and strengths of each.

The "investigating with models" approach (Figure 5.1) ensured that students started their learning journey with the opportunity to *express* their existing *mental* model, using 3-D models in a group environment. In the collaborative group, students recognize that there are alternative *mental* models when modeling their 3-D models. The alternative *expressed* models are good starting points for teaching, yet are very seldom used by the classroom teacher as a platform upon which to build (Hubber and Tytler, 2004). In this approach, the student's original *expressed* (3-D) *mental* model becomes the "central focus" of the investigation and explanations. The fact that the process occurs in a small group recognizes the importance of alternative *mental* models in a non-threatening way, opens the way for "shared meaning," and allows for active involvement in the (re-) (de-) construction of the *mental* model.

As students collect evidence that relates to their *mental* model (e.g., explanation of an astronomical concept), they are practicing aspects of the nature of science (observations, inferences, creativity, and empirically based knowledge) that relate to the systematic nature of investigating. This approach leads the group away from a fact-based content approach to an approach where students are collecting evidence to assess their *mental* models – a key step for conceptual change and an important element of constructivist learning. Koschmann, Myers, Feltovich, and Barrows note that when students articulate their understanding to other students in the group, the articulation enhances the knowledge construction process and sets the stage for future learning (1994, cited in Hmelo and Ferrari, 1997).

Demonstrating the collected evidence in an *expressed* (3-D) model is the key to keeping the original *mental* model (explanation), rejecting it, or modifying it. Students become critical observers of whether the modeler is able or not able to model all of the evidence. Competing *mental* models are tested as *expressed* models using the evidence collected by the group. Throughout the investigation students are reminded of the limitations of using 3-D modeling to represent *mental* models. The *investigating with models* approach allows the teacher to learn alongside the students in a meaningful way. The following section will focus on two teacher trainees enrolled during 2005 in the course Spaceship Earth and Beyond. Their involvement with the *investigating with models* approach provides some insights into their *mental* models of modeling and at the same time opportunities for creative *expressed* models.

5.2 Insights into thinking about modeling *expressed* models: links to creativity background

Kerry and Mark were members of two different collaborative groups. The groups were provided with a semi-structured three-week (12-hour) block of class contact time to collaborate, develop, and demonstrate their astronomy *mental* models using *expressed* 3-D models. The author became a mentor, coach, and facilitator to the groups during this block of time. Each group was provided with a range of questions that focused their thinking and learning toward explanations (expressed models) of day/night, seasons, Moon phases, eclipses, Earth's place in the Solar System, and the zodiac backdrop. Students were assessed individually when they felt ready to model their astronomy understanding (*mental* models) with the *expressed* 3-D models. So it was important for students to communicate their *mental* models via the modeling process. The modeling of their *mental* model to other members in the group would assist with the development of appropriate pedagogical skills for the teaching of astronomy.

During the three-week block, Kerry and Mark created two new *expressed* models in addition to those used by the members of their respective groups. Their use by other students in the class during the modeling assessment task demonstrated their usefulness and authenticity in communicating evidence for two aspects of seasons – differing solar inputs and varying lengths of day/night throughout the year. Kerry and Mark were interviewed four months after completing the course in an attempt to gain some insight into the creative thinking behind the two new *expressed* models. How did the "investigating with models" approach allow Kerry and Mark to be creative?

Amabile and Hennessey's (1998, cited in Hennessey, 2004) research has focused on the background of creative performance rather than on individual skills or talent. They believe that for a creative product, idea, or solution to be generated, one must approach the

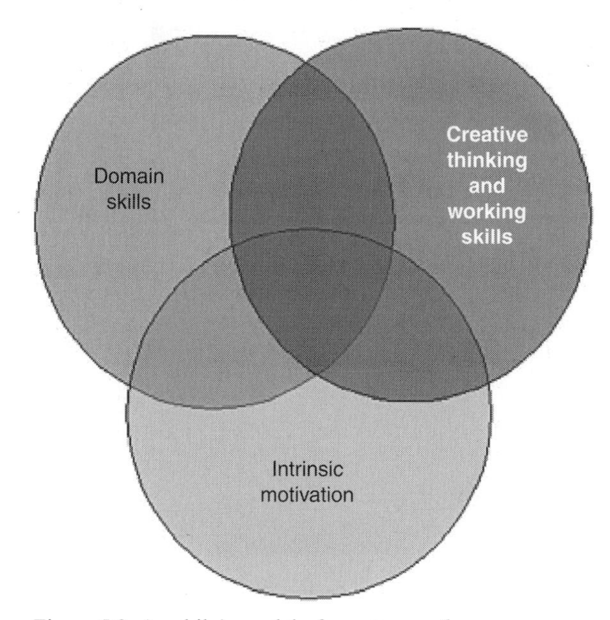

Figure 5.2 Amabile's model of *creative performance*.

problem/issue with the appropriate domain skills (background knowledge), creativity skills (willingness to take risks, experiment, etc.), and task motivation. They present their ideas as a three-ring model (Figure 5.2). The intersection of the three factors is what Amabile (1997, cited in Hennessey, 2004) terms the "creative intersection" – when creativity occurs. Their research suggests that it is possible to teach the creativity skills and the domain skills to students but that the motivational state (intrinsic or extrinsic) is more transient and largely dependent on the situation at hand. Amabile and colleagues were able to identify intrinsic motivation as conducive to creativity and extrinsic motivation as usually detrimental. They believe that the creative product/solution is dependent on the motivation state of the person at that particular time. As intrinsic motivation comes from within the individual and is temporary, there are many external factors that can dampen that motivation or speed up the temporary nature of it.

5.3 The creative process thinking: Kerry and Mark

Surprisingly, both students followed the same journey with their creative processes. The start for both was a sound knowledge or understanding of the specific astronomical concept associated with seasons. Kerry understood that solar input varies throughout the year, and Mark was also sure of his understanding of the variable daylight/nighttime hours during the year. The next step with their thinking that led to the new *expressed* model was that both recognized that other students in their respective groups were having difficulty modeling the consensus *expressed* model. After recognizing that other members in the group were having difficulty, both identified the specific aspect of the *expressed* model that was causing the problem. Once the specific aspects were identified, Kerry and Marks's intrinsic motivation helped them to create new *expressed* models (see Figures 5.3 and 5.4) that allowed the students to communicate their *mental* models more effectively.

Figure 5.3 Kerry's *expressed* model illustrating different areas of light being produced at one time on the surface of the globe.

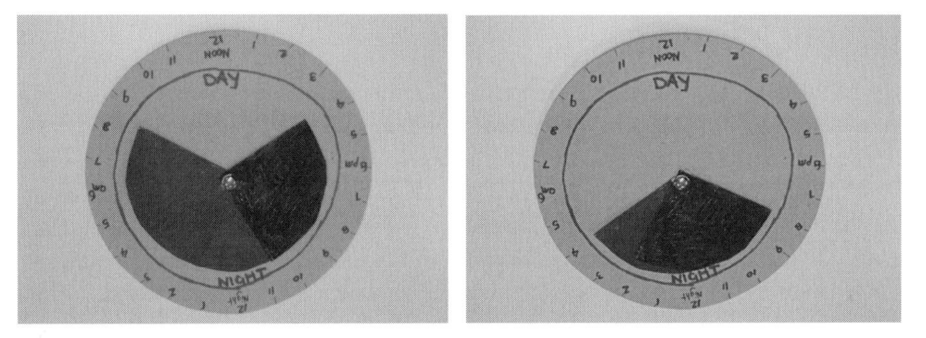

Figure 5.4 Mark's *expressed* model illustrating the movable parts that can demonstrate the increasing or decreasing daylight hours when modeling the Earth as it orbits the Sun. View from the southern hemisphere,

5.4 Discussion

The *investigating with models* approach provided opportunities for both students to demonstrate their domain knowledge in an environment that provided motivating opportunities to demonstrate their understanding and where both could feel safe taking risks and using their working skills. Kerry's and Mark's intrinsic motivational states during the *investigating with models* approach was evident in their interview statements about the development of their personal understanding of astronomy and their role as a teacher in the classroom.

In addition to their subject and professional development, Kerry's and Mark's interview statements indicated an excitement and interest about their own learning. I believe the creative model generated by Kerry and Mark demonstrated working within Amabile's *creative intersection*, i.e., the coming together of their domain skills, creative thinking, and working skills together with a high level of intrinsic motivation. Why a creative product with Kerry and

Mark and not with the other students? Perhaps some students were not risk-takers; perhaps others desired a "transmission" approach; for still others, there was definitely a lack of astronomical knowledge to manipulate and/or critique the 3-D models.

5.5 Conclusion

The *investigating with models* approach will allow learners to engage with new astronomical concepts by focusing on the existing mental model at the start. It will also provide opportunities for students to create *expressed* models, as well as communicate astronomy understanding. This approach is one way to support the creative–product giftedness in our pre-service students. If we expect classroom teachers to cater to the creative–productive gifted students during astronomy teaching in primary and secondary schools, then pre-service teacher training must model the appropriate classroom environment that allows it to occur. Tertiary educators can create that learning environment by fostering intrinsic motivation in their students, using appropriate pedagogy (investigating with models) to develop students' *mental* models, and by allowing students to (1) engage in collaborative and interdependent work, (2) safely express their *mental* models, (3) develop and communicate their understanding, (4) take control of their own learning with support from the tertiary educators. This approach may mean taking risks – favoring quality over quantity and a depth of understanding over surface learning – but I do believe creativity needs to be nurtured at the tertiary level as well as in our primary and secondary schools.

Acknowledgments

I would like to acknowledge the collaboration of Kerry and Mark in this paper. Their honesty, humor, and forthright comments made for wonderful reading. I know that they will become effective primary teachers, creating an environment where constructive and facilitated learning occurs.

References

Duit, R. and Glynn, S., 1996, Mental modelling, in G. Welford, J. Osborne and P. Scott (eds.) *Research in Science Education in Europe: Current Issues and Themes*, 166–176 (London: Falmer).

Gilbert, J. K. and Boulter, C., 1998, Learning science through models and modelling, in B. Fraser and K. Tobin (eds.), *International Handbook of Science Education*, 53–66. Dordrecht, The Netherlands: (Kluwer Academic Publishers).

Gilbert, J. K., Boulter, C. J., and Elmer, R., 2000, Positioning models in science education and in design and technology education, in J. K. Gilbert and C. J. Boulter (eds.), *Developing Models in Science Education*, 3–17. (Dordrecht, The Netherlands: Kluwer Academic Publishers).

Harrison, A. G. and Treagust, D. F., 2000, A typology of school science models, *International Journal of Science Education*, **22**, 1011–1026.

Hennessey, B. A., 2004, *Developing Creativity in Gifted Children: the Central Importance of Motivation and Classroom Climate* (RM04202). (Storrs, CT: The National Research Center on the Gifted and Talented, University of Connecticut).

Hipkins, R., Bolstead, R., Baker, R., *et al.*, 2002, *Curriculum, Learning and Effective Pedagogy: a Literature Review in Science Education*, (New Zealand: Ministry of Education).

Hmelo, C. E. and Ferrari, M., 1997, The problem-based learning tutorial: cultivating higher order thinking skills, *Journal for the Education of the Gifted*, **20** (4), 401–422.

Hubber, P. and Tytler, R., 2004, Conceptual change models of teaching and learning, in G. Venville and V. Dawson (eds.) *The Art of Teaching Science*, 38–51 (Crows Nest, NSW: Allen and Unwin).

Justi, R. and Gilbert, J., 2002, Modelling, teachers' views on the nature of modelling, implications for the education of modellers, *International Journal of Science Education*, **24**, 369–387.

Justi, R. and van Driel, J., 2005, The development of science teachers' knowledge on models and modelling: promoting, characterising, and understanding the process, *International Journal of Science Education*, **27**, 549–573.

Comments

George Greenstein: Did you find that the misconceptions of the pre-service teachers you are training significantly impede their learning? Is it possible that their (erroneous) feeling that "this is a simple problem that I understand" prevented their learning?
Bill MacIntyre: In fact pre-service teachers were much more aware of other students (and children) in the class and that they would be explaining possible alternative conceptions that are not scientific. The students' understanding of alternative conceptions (misconceptions) was transferred to other subject disciplines in interviews with the students in the course.

Jean-Claude Pecker: First a comment about models. I remember that I suggested a "model" to a Science Minister in Paris. He told me: "My friend, the Sun is rising in the East. Right?" "Right." – "It is setting in the West? Right?" "Right." – "But this is true in the Northern Hemisphere? In the South, it is the reverse?" !!! So I suggested that his "mental model" was: equator!
Bill MacIntyre: So as a teacher I would ask him to demonstrate his "expressed model" so that one could *see/observe* that it was not correct. Then ask him to model two people on earth (S. hemisphere) (N. hemisphere) to demonstrate the evidence of the direction of the sunrise.

Jay Pasachoff: Phases have been studied for thousands of years. Are you trying to inspire with more recent materials – pulsars and black holes?
Bill MacIntyre: The use of the *investigating with models* approach allows the teacher to assist the children with the learning or interest in a specific topic/area such as pulsars and black holes.

Richard Gelderman: Could you comment further on how you clarified the relationship between the student's creative solutions and the creative process as part of the scientific process? How does emphasis on creativity fit into the overall course?
Bill MacIntyre: Creativity is an integral part of communication, and this course has a focus on communicating students' understanding in astronomy and earth science via various means. So if students create "expressed models" that enhance the communication, it helps demonstrate the notion that teachers can be effective in classrooms when explaining to children.

M. Vlahos: Are teachers aware of children's preconceptions before they plan lessons?
Bill MacIntyre: Students are provided with engaging activities (concept cartoons, photos, etc.) that elicit their understanding before teachers plan. Sometimes teachers will change their planning in response to students' alternative understanding as they *model* astronomy concepts.

Rosa M. Ros: Did you organize similar courses for in-service teachers?
Bill MacIntyre: Yes. We organize courses for teachers who are working in schools and we use models in order to explain astronomical concepts.

6

How to teach, learn about, and enjoy astronomy

Rosa M. Ros

Applied Mathematics 4, Technical University of Catalonia, Jordi Girona 1-3, Modue C3, 08034, Barcelona, Spain

Abstract: This contribution deals with the author's decade of experience organizing summer schools for European teachers of secondary school students. Teachers' main interests are similar to those of students. The goal of the summer school is to provide them with:

- answers to their questions;
- practical activities, to facilitate learning by doing;
- different approaches to the study of astronomy: making models and drawings, playing in the playground, and feeling like the principal actors in the teaching/learning process;
- astronomical activities that can help teachers/students to teach/learn mathematics or physics in a more appropriate way to attract young people to science;
- information through simple and clear language, minimizing specialized language, which acts as a barrier to close interaction with students;
- some methods that promote rationality, curiosity, and creativity, making use for observations and measurements of the sky that lies above every school;
- a contextualized approach to astronomy, to avoid presenting concepts in an isolated way and to connect the school with the place where the students live.

In summary, students who are exposed to opportunities to connect astronomy with their lives are more likely to have a positive reaction to astronomy in general. This paper will present some concrete examples of all these ideas.

6.1 Introduction

The European Association for Astronomy Education (EAAE) organizes an annual summer school for European teachers. Every year, at the end of the course, participants answer a questionnaire. Over a 10-year period we have collected approximately 600 opinions from teachers from over 20 European countries. This essay reflects upon "what I have learned from my 10-year involvement with the summer school." I would like to invite the reader to take part in this exercise with the author.

6.2 Answering questions

The majority of teachers who participate in our course do so for professional improvement. They are also interested in exchanging information and opinions with other teachers from other countries.

Teachers are professionals who want to invest their time in useful courses. The most important aspect of the questionnaire is to discover how useful participants find our course. Responding to participants' opinions, we changed the structure of the course after the first few

Innovation in Astronomy Education, eds. Jay M. Pasachoff, Rosa M. Ros, and Naomi Pasachoff. Published by Cambridge University Press.

Figure 6.1 Teachers preparing their future classes during a practical workshop on spectroscopes.

efforts. We reduced the number of "General Lectures" and "Working Groups," which were too theoretical and linked to the general contents of astronomy, and increased the number of "Practical Workshops" that enable participants to learn by doing. (see Figure 6.1.)

To communicate appropriately with our teacher participants, we must select topics, matters, and methods that reflect *their* needs and interests. (Of course we also can and do expose the participants to new ideas and especially new methodologies that can be implemented in European schools.)

What is true for teachers is also true for students. It is important to answer their questions – the matters that spontaneously interest them. After some astronomy classes, the children ask questions such as "What would happen if Mars crashed into the Moon?" By making a model of the Solar System with the students, they normally succeed in discovering the answer for themselves. They observe that the Solar System is much more of an "empty" area than they had imagined.

6.3 Practical activities: learning by doing

In many cases, science museums focus attention overly on spectacular and "funny" science, leaving the schools to cover "boring" science. This approach is dangerous for science and bad for schools.

It is good to promote the study of astronomy by means of different approaches, using more than just blackboards and computer screens. We promote the active participation of students by having them make models and drawings, play in the playground, and in general by making them feel like actors in the teaching–learning process.

For instance, designing and building an analemmatic sundial in the school is a big undertaking in which to involve all students and as many teachers of different subjects as possible. It can be a gigantic project! (see Figure 6.2.)

Figure 6.2 Teachers making a simplified version of an analemmatic sundial.

Not everything can be fun at school. It is necessary to introduce students to the **culture of making an effort**. In some cases, however, our astronomical workshops can help teachers to teach mathematics or physics in a more appropriate way to attract pupils to science. If the teachers enjoy teaching, the students will also enjoy their classes. Teachers can explain mathematics, particularly trigonometry, by using real problems, such as calculating the Earth's radius using Eratosthenes's experiment or the Earth–Sun distance by using Aristarchos's method.

An important part of teaching is communication. All instructions for every activity should be presented in clear and simple language. In order to make the students (the teacher participants in the summer school) feel fully committed to the task, it is advisable to avoid specialized language, which distances the non-professional from the professional. The main objective is to engage students actively in doing something, beyond merely watching or listening to the teacher talking.

Verbal communication should be supported by visuals. **Figures, drawings, and photos are important to capture everybody's interest and to simplify the teacher's task**.

6.4 Promoting rationality, curiosity, and creativity

All schools have a sky over their buildings. It must be used for making observations and taking measurements. If the school lacks tools and devices for making observations, **students can be encouraged to produce their own instruments**. What we lose in precision is made up for in student interest and involvement, which are much more important. Students learn by doing, observing, and reasoning.

6.4.1 Example

Students acquire practice by measuring times and positions of celestial bodies using devices prepared in an ad hoc way. We show students how to produce a collection of devices and assemble them into a minimum kit for observations. The kits are assembled with simple and

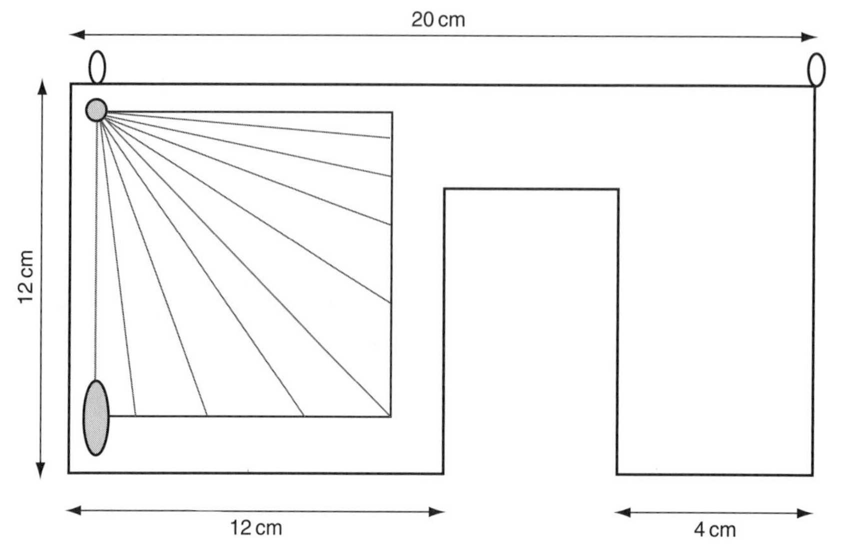

Figure 6.3 Simplified version of a "gun" quadrant.

readily available materials, including cardboard, glue, and scissors. This experience can offer the possibility of investigating both many old devices and some modern instruments. Students should be encouraged to use their imaginations and artistic ability to personalize their kits. This activity can be modified, depending on the age of students, to produce more sophisticated instruments. In particular, this kit contains:

- a ruler that measures angles,
- a simplified quadrant,
- a horizontal goniometer (an instrument for measuring solid angles, as of crystals),
- a star finder,
- a night timer,
- an equatorial sundial.

In particular, as an example, we introduce here a **simplified quadrant (a "gun" quadrant)**, which can be used to obtain the height of the stars. When students see an object through the visor, the rope indicates the angular position referring to the horizon. A very simplified version of a quadrant can be useful in measuring angles. We present here a "gun" version that is very easy to use. The students prefer to use this model rather than other more classical designs.

In order to make the device, you need a heavy rectangular piece of corrugated cardboard (about 12 cm × 20 cm). You cut out the rectangular area according to Figure 6.3, in order to hold it with one hand. You can put two metal eyelet-screws on the longer side in the middle corrugated layer.

A square of paper with the angles indicated (or a protractor) has to be glued onto the cardboard according to Figure 6.3 in such a way that one of the screws is on the 0 degrees position. A string is attached to the top and, at the other end, a little weight, such as a bolt, is fixed.

When you see an object through the two eyelet-screws, the string indicates the angular position referring to 0 degrees of your horizon (see Figure 6.4). Using this device it is easy to determine the latitude of the location (equal to the altitude of the polar star). It is also possible

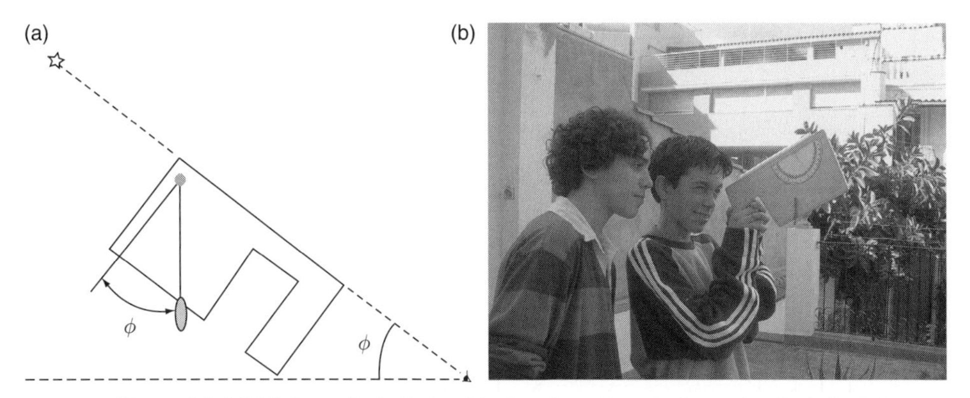

Figure 6.4 (a) Mathematical relationship that shows how to determine the latitude by means of a quadrant. (b) How to use the quadrant for observation.

to use this quadrant in a mathematics class to calculate the height of the school or another building well known to the students. A straw that passes through the eyelets is a viewfinder that enables us to measure the height of the Sun by projecting the image on a white piece of cardboard.

Be Careful: Never Look at the Sun Directly!!!

It is really important to promote rationality, curiosity, and creativity. In our society, where the pseudo-sciences are all too attractive, it is necessary to promote the scientific spirit.

Astronomy, of course, is not only about observation. Students' romantic preconceptions of astronomers peering through telescopes should be enlarged to include more innovative and technologically dynamic aspects of astronomy. We must demonstrate how new discoveries (e.g., exoplanets) and methods (e.g., the search for extraterrestrial intelligence) can impact the future of humanity.

It is also important for instructors to try to pass on to students of all ages the enthusiasm they feel for their subject. It is not always possible, but we have to present our lessons in an attractive format.

6.5 Catching the attention of the public

It is important to capture students' attention. If possible, we use surprises (or changes of activities) as a means of getting and maintaining attention. Observations can be a good resource, but it is necessary to alternate the kinds of activities.

In the classroom we can introduce some surprises. For instance a simple rocket made using an effervescent aspirin can be useful for the teacher in order to maintain the students' attention.

6.5.1 Example

To prepare the following model, the only equipment necessary is scissors, glue, adhesive tape, a film container with a lid, water, and one-quarter of an effervescent aspirin. The paper must be cut according to the model included in Figure 6.5 (cut for the straight continuous line and glue in the dotted line).

Take care that the rectangle that will be the rocket's central cylinder can contain the film capsule (verify that the small side of the rectangle is a little bigger than π times the diameter of

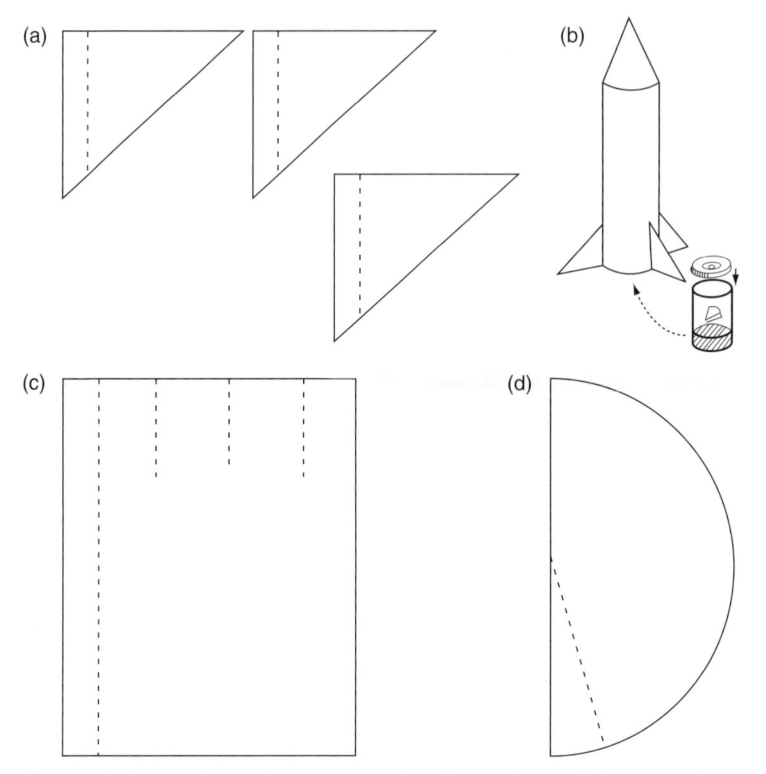

Figure 6.5 Model for cutting out the rocket. Cut on the solid lines and glue on the dotted lines.

the capsule). Also, be sure to glue the three triangles as a support for the body of the rocket. Do not forget to fix the conical nosecone on the top with tape.

When you have finished the rocket, you have to prepare for the launch. Put water inside the film capsule up to one-third of its height (approximately 1 cm). Add one-quarter of the effervescent aspirin (or another kind of effervescent tablet). Put on the lid and place the rocket over the top. After approximately 1 minute the rocket will take off. See Figure 6.6.

You can, of course, repeat the experiment. Don't forget that you have three-quarters of an aspirin left to play with – that's three rockets more! Enjoy launching your rocket again.

You will need:

1 scissors to cut out the model,
2 glue and tape to stick the model together,
3 film capsule where you put 1 cm water,
4 one–quarter of an effervescent aspirin,
5 put on the lid and place the rocket over the top,
6 wait 1 minute and the rocket will launch.

6.6 A contextualized approach to astronomy

Astronomy is not a part of the standard curriculum in practically all European countries. It is therefore a good idea to offer our material and our activities to science teachers so that they

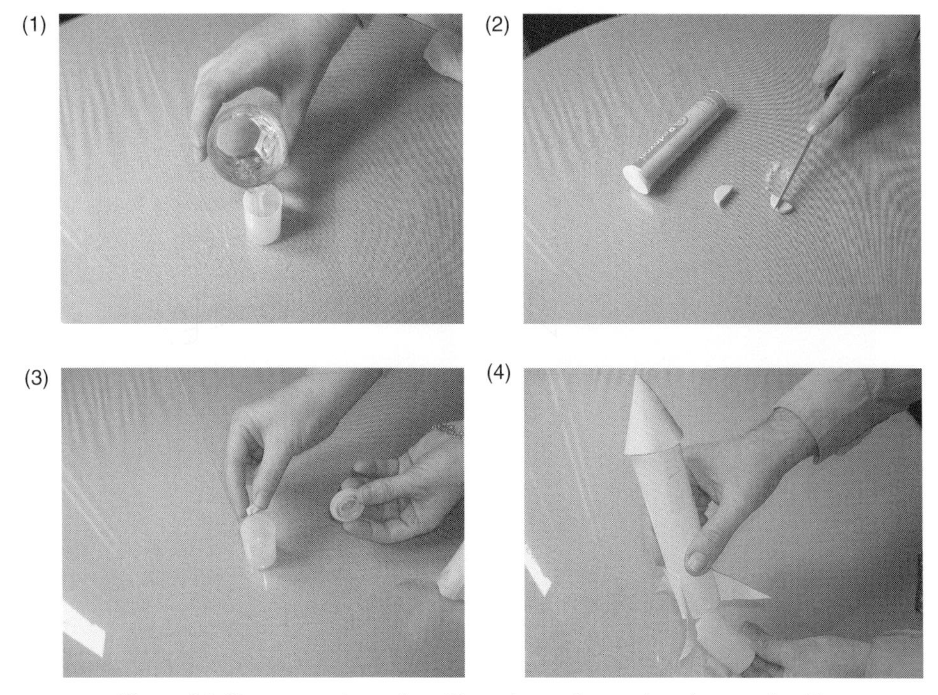

Figure 6.6 How to use the rocket: (1) put 1 cm of water into the capsule; (2) cut a quarter of an effervescent aspirin; (3) add the aspirin to the capsule; (4) put the rocket on the capsule.

can introduce and teach astronomy by means of other subjects. To avoid presenting concepts in isolation, it is important to contextualize the contents of our papers.

It is not good to introduce only a collection of facts. It is important to connect concepts with related personalities, with the history of science, and with current social implications. The idea is to humanize the material so that students do not feel like mere spectators, because in this case it is very difficult to maintain their interest. Teenagers, for instance, love to know the history of Neptune's independent discovery by the young English astronomer/mathematician John C. Adams and the young French mathematician Urbain J. J. Leverrier. Adams, who predicted that the planet would be about 1.6 billion kilometers farther from the Sun than Uranus, sent his calculations to the astronomer royal of England, Sir George B. Airy, who, lacking confidence in the young man, failed to verify the predictions with a telescope. Only after Leverrier sent his similar predictions to Johann G. Galle, director of the Berlin Observatory, did Galle and his assistant discover Neptune near the position both mathematicians had predicted. Students seem to identify with Adams: "My family routinely ignores my good ideas, too."

The school must be connected with the place where students live so that students feel more interested in the application of the topics they study.

It is good to organize visits to astronomical sites, observatories, telescopes, radio-telescopes, or historical buildings connected with astronomy. If it is possible it is good to organize some visits to places far away from school. Why not? (see Figure 6.7.)

Astronomy, which figures so prominently in the history of science, appears in the media more than many other sciences. All of us have heard of the ISS, ESA, NASA, GPS, etc. The future of humanity will probably be deeply affected by the future of this science. Astronomy is the most attractive science for young people, probably for everybody. It is easy to introduce

Figure 6.7 A boy observing the Venus Transit of 2004.

astronomy in the context of general audiences. It is much easier to contextualize astronomy than many other topics. We have to do it.

6.7 Be passionate!

We must take into account that if we want to communicate successfully it **is necessary to be passionate!** People who feel a passion related to some astronomical experience are likely to remember it, and astronomy will take on a positive connotation for them. This positive relationship is likely to stay with them for some time and to forge a tie that will not be broken easily.

Students can become passionate about astronomy by observing the transit of Venus or solar and lunar eclipses or by participating in student competitions. Students from different parts of the world can also use the Internet to cooperate in a common objective: astronomy. **We must get students to feel the thrill!** To reach this goal, every school should use strategies such as inspiring enthusiasm for the sky and for astronomy and encouraging unanimous participation in astronomical activities.

References

Palici di Suni, C., Ros, R. M., Viñuales, E., and Dahringer, F., 2006, Astronomy kit for young astronomers, *Proceedings of Tenth EAAE International Summer School: Workshops, Posters*, Barcelona, 39, 68.

Ros, R. M., 2004, A simple rocket model, *Proceedings of Eighth EAAE International Summer School*, Barcelona, 249, 250.

Ros, R. M., 2006, Balance sheet over ten years training to teach astronomy, *Proceedings of Tenth EAAE International Summer School: Lectures, Working Groups*, Barcelona, 17, 33.

Comments

Jan Vesely: You have mentioned a simple spectroscope. It looks to be very small and really very easy to construct. Is it possible to resolve spectral lines in the spectra?

Rosa M. Ros: Of course, this is a very simple apparatus. It is meant only for observing the phenomenon of making a spectrum, which, of course, is similar to a rainbow. The activity is very nice for the youngest pupils and they can do it by themselves.
Jay Pasachoff: I have used such simple spectroscopes with university classes and, most recently, with a workshop for elementary-school teachers. Seeing emission spectra from fluorescent lamps is easy and with careful adjustment of the slit one can even see absorption lines from a bright sky or from the Moon. (One shouldn't use it directly on the Sun.) Simple kits with spectroscopes out of cardboard, slides of diffraction grating, and a movie suitable for students ("Colors of the Sun") are available from solar-center.stanford.edu. I have developed some labs suitable for high-school or university students about the solar spectrum and about looking at the Sun through various filters, using simple telescopes, from supplied data, and from current data on the World Wide Web.

Jan Vesely: The analemmatic sundial is an enjoyable activity for students, but secondary school students are not able to clearly understand the principle of the analemma (equation of time [see, for example, www.analemma.com. Ed.]). They have fun walking through the dial casting their shadow, but when they had to make a presentation of their work to other students, they completely misinterpreted the function of the sundial and analemma. Did you test the understanding of the function of the analemmatic sundial? I mean not the function of the sundial generally, but the importance of the analemma in the center of the dial.
Rosa M. Ros: I agree with you. The analemma is not a simple concept for secondary school students. This model that I showed is a very simplified version and I think that the most important point is that students can understand that the "student-gnomon" change his/her position according with the altitude of the Sun. The altitude changes during the year and then the position of the gnomon has to change if we want to have the shadow final point over the same curve.

Julieta Fierro: I agree with all you said. Nevertheless I believe a *small* amount of fear and competition can increase meaningful learning: that is to say, overcoming challenges and comparing what you do with others.
Rosa M. Ros: I agree with you. Competition can be a possibility to take into account in any case. Really this is a good point to promote astronomy that can be carried out by the institutions in order to support the work of teachers at the school.

Robert Stencel: How has *light pollution* affected student awareness of and interest in the night sky over the years?
Rosa M. Ros: This is a serious problem in several countries, especially in Europe. At present, it is very difficult to observe from the schools in the center of our cities. We must work to promote measures by governments and municipalities to control light pollution. In any case, at present we can more or less observe in the cities and, when possible, promote an excursion to the countryside. The students greatly enjoy going outside their cities, spending one or two nights observing. Teachers can provide a set of activities and assist with different kinds of observations.

7

Clickers: a new teaching tool of exceptional promise

Douglas Duncan

Director of Astronomical Laboratories, University of Colorado, Boulder, CO 80309, USA

Abstract: The use of wireless student response systems – "clickers" – is growing remarkably fast in astronomy classes throughout the US. That is because clickers address two of the oldest and most fundamental challenges in teaching: how to engage students, and how to determine if they are learning what you are teaching. Clickers are relatively low cost and easy to use. RF clickers (some systems use infrared rather than radio) require no classroom wiring. Data show that when clickers are used well in large lecture classes they increase student engagement and learning. This is particularly true if students are encouraged to debate answers with their neighbors before answering. Students overwhelmingly like using clickers, and their use also increases class attendance. As is the case with any technology, it is possible to misuse clickers. Common mistakes made by new clicker-users and how to avoid them are described.

7.1 What are "clickers"?

Clickers are wireless student-response systems. Transmitters (Figure 7.1) look like a small TV remote control, but the buttons are labeled A, B, C, D, E (some have more buttons). At any time during class the instructor can ask a multiple–choice question; the students choose an answer and "click" a button on the transmitter. A receiving unit counts all the answers and displays them on the instructor's computer, usually as a histogram that may be projected for the class to see (Figure 7.2). Based on the results, the instructor can decide whether to proceed or to spend more time on a particular topic. Equally valuable, the student learns immediately whether he or she understands the concept the teacher is presenting, without waiting for a test.

Clickers from different companies do not work together.

7.2 Why you *should* use clickers

Lectures have been used in teaching for a very long time. If you are a good lecturer you probably watch your students' faces. "Do they understand?" "Are they enthusiastic about what I'm saying?" You stop and ask, "Does anybody have any questions?" Students nervously look at each other. No one raises a hand. So you continue . . ..

If you're experienced you know this isn't enough. Many students will not call attention to what they don't know, especially in a large class. Some students, particularly those at the back of the class, may not be engaged at all with your presentation.

There's an even more important reason to use clickers. Lectures, even when clear, well organized, and interesting, are not as effective as we think. Over the past two decades the physics community has developed a subspecialty of PER: Physics Education Research. These researchers study the effectiveness of different methods of teaching physics, in much more detail than the average teacher can do. Another way to put this: they apply the scientific

Innovation in Astronomy Education, eds. Jay M. Pasachoff, Rosa M. Ros, and Naomi Pasachoff. Published by Cambridge University Press.

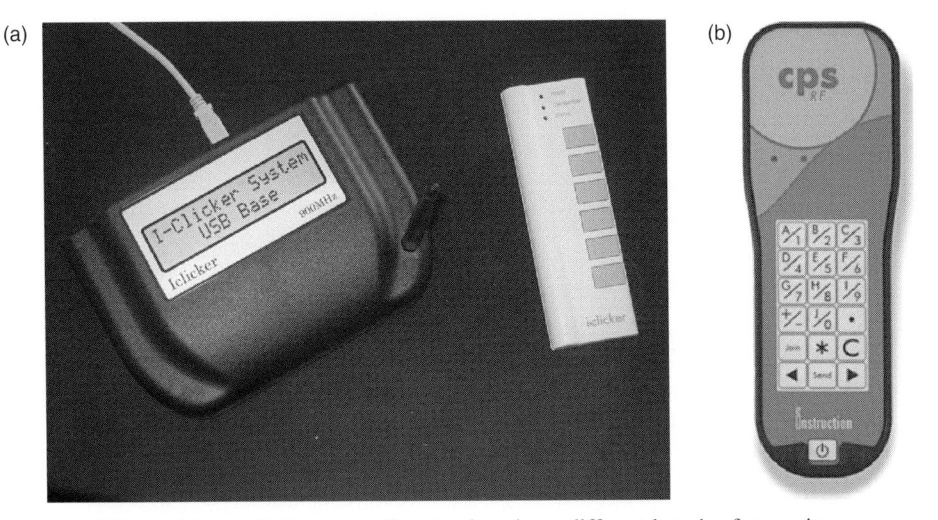

Figures 7.1a and b Receiving (base) unit and two different brands of transmitters.

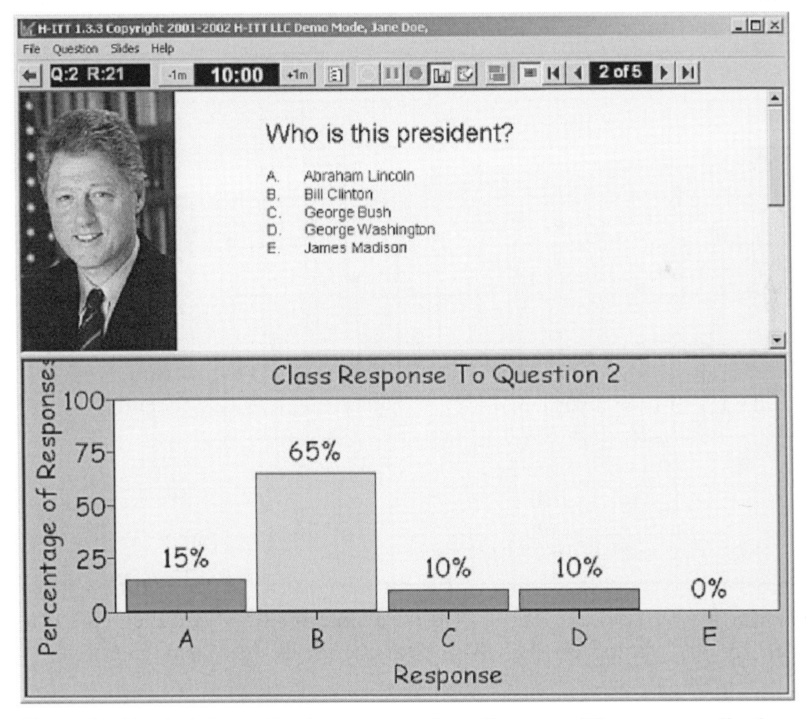

Figure 7.2 Typical data collection screen and results screen. These are usually shown to the class with an LCD projector.

method to the teaching of science. Careful oral interviews of large samples of students form the basis of studies of the depth of students' physics learning (e.g., Weiman and Perkins, 2005). The results are important and challenging. In a landmark survey of 6000 physics students in 52 different courses, Hake (1998) found that in *no lecture class* did the average student learning exceed 30% of the new concepts taught – not in any class.

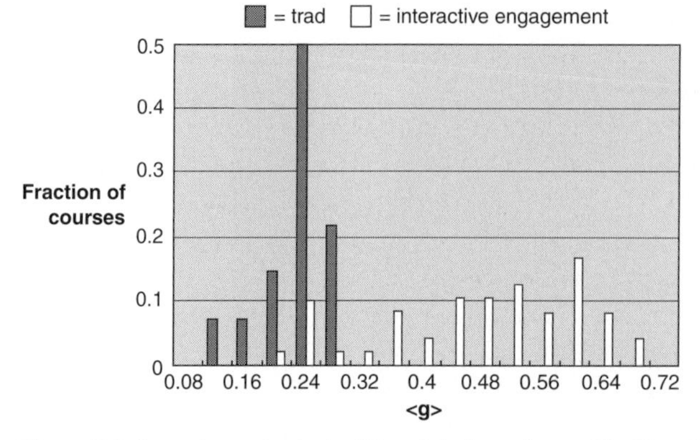

Figure 7.3 Learning gains in traditional lecture classes (dark) and in interactive courses (light). Since students in different classes start with different levels of knowledge, results are reported as normalized learning gains $< g >$, which are the fraction of possible improvement a student achieves; 1.00 means he or she learned everything that was taught.

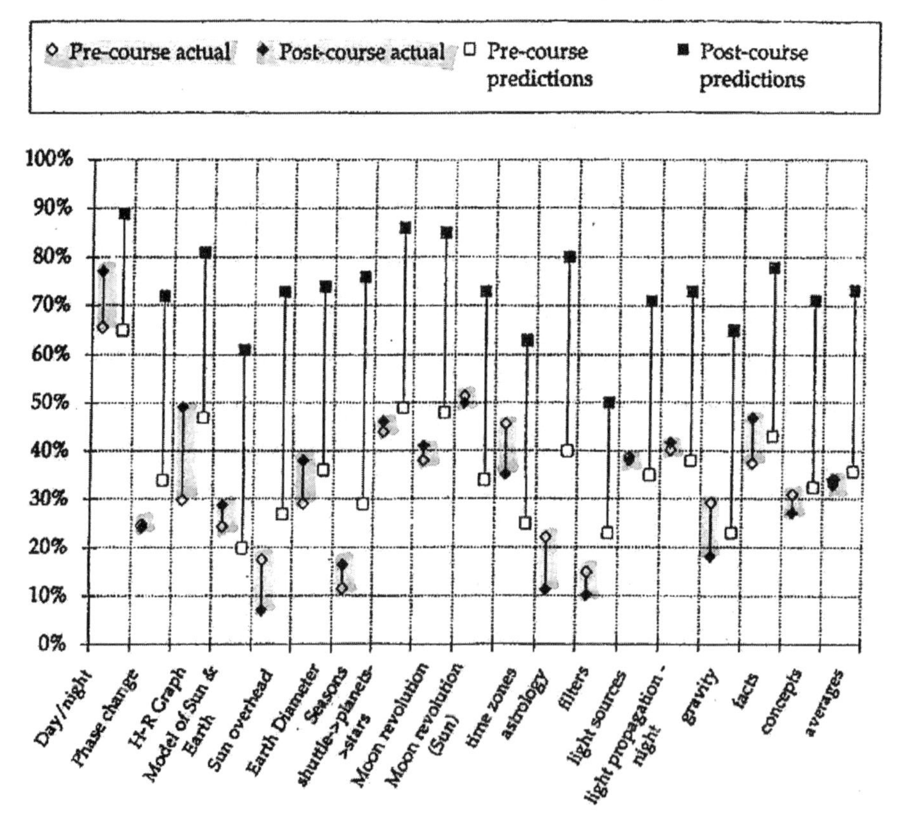

Figure 7.4 Lightman and Sadler, 1993. Students' actual gains (highlighted) were much less than instructors' predictions.

The reason for this limitation is the way people learn. Learning is an active, not a passive process. It is not enough for students to simply hear an idea and write it down. To learn it they must think about the idea and its implications, fit it into what they already know, and use it. This is why the present author could listen to Feynman lecture his freshman physics class, think that the lesson was wonderfully clear, yet find when confronting the homework problems that a lot more work was required before concepts truly became clear. Many listeners have probably had similar experiences.

Practicing scientists routinely debate ideas with each other and "play around" with ideas in their mind or in the lab. However, students often do not. Many actually believe that taking notes, memorizing, and repeating material on an exam is all there is to learning (Hammer, 1994).

Physics education researchers have developed numerous ways to make learning more interactive, such as tutorials. *Clickers are the easiest interactive engagement tool* if they are used in a mode where students have to convince their neighbors in the classroom whose answer is correct (see "peer instruction," below).

Surveys such as those of Figure 7.3 measure only learning, not inspiration or other possible goals of lectures. Nevertheless, the results are striking. The limitation on learning from lectures is *not a limitation of the lecturer*. It is a limit of the lecture approach itself. Astronomers are generally not aware of this limitation, and routinely overestimate how much their students learn. Figure 7.4 shows results from Lightman and Sadler (1993) at Harvard, who interviewed a number of astronomy instructors and then interviewed students. Their results show that instructors predicted students' initial (pre-class) astronomy knowledge well, but grossly over-predicted how much students learned in their courses. These results are also a warning about how thoroughly students learn concepts from our lectures.

7.3 A remarkable anecdote about learning in lectures

Carl Weiman is a very good teacher, as well as being a Nobel Prize winner in physics. In his "Physics of Everyday Life" course, he teaches how a violin works. Here is how he does it:

> "I show the class a violin, and tell them that strings alone cannot move enough air to make the sound loud enough to hear throughout a concert hall. I . . . show the bridge and soundpost and explain how the strings make the bridge move, and the soundpost transfers the vibrations of the top of the violin body to the back. It is the vibrating wood that makes most of the sound the students hear."

Fifteen minutes later in the same class, Weiman asked the students what causes the sound they hear from a violin. Is it (a) mostly the strings, (b) mostly the wood, (c) both equally, or (d) none of the above. *Only 10% of the students gave the correct answer*!

The working of a violin, like the cause of the seasons, is a topic where students *think* they know the answer and enter with a misconception in mind, not a topic for which they start with no idea. As Schneps and Sadler showed dramatically in the highly recommended video *A Private Universe* (1987), lectures are very ineffective at dispelling misconceptions. After Weiman began using clickers and asked students to debate their ideas of what caused the sound of a violin after he discussed the topic, he found that a majority of students retained the correct answer.

7.4 What difference does the use of clickers make?

With clickers you can easily ascertain whether students understand what you've just taught. You may choose to spend more time on a topic, or re-explain it. Even more powerful, however, is to address misunderstandings with *peer instruction*.

In peer instruction you ask conceptual questions, sample the student answers with clickers, and when you see a difference of opinion have your students discuss with their neighbors who is correct. Tell them that they should explain, debate, and agree before answering. This approach has been used by Mazur at Harvard since the early 1990s. His book, *Peer Instruction in Physics, a User's Manual* (1997), is highly recommended. Mazur attributes his students' impressive learning gains to this procedure in which they teach and debate with each other. Duncan (2005; 2006a) experimented with peer instruction in astronomy classes at the University of Chicago in the mid-1990s and surveyed student attitudes, documenting very positive results.

Conceptual questions usually require no calculation. They focus on one basic concept. Here is an example. Use a prism to make a spectrum at the front of your class. Take a red filter in hand, and ask the following question:

What happens to the spectrum in the front of the room if I put a **red filter** into the beam?

(a) Blue gets through, the other colors disappear.
(b) Red gets through, the other colors disappear.
(c) All the colors turn red.
(d) It depends on which side of the prism I put the red filter.

You will find that substantial numbers of students choose each answer. However, after discussion many more students choose the correct answer! Mazur finds that when comparable numbers of students start with right and wrong conceptions, peer instruction usually results in students agreeing on the correct answer, not the wrong one.

Many instructors now use clickers to facilitate peer learning. Although there are many things you can do with clickers, many consider this to be the most important use. *Clickers and peer instruction are the easiest interactive engagement technique known to the author.* Figure 7.3 shows how important students' engagement is in increasing learning.

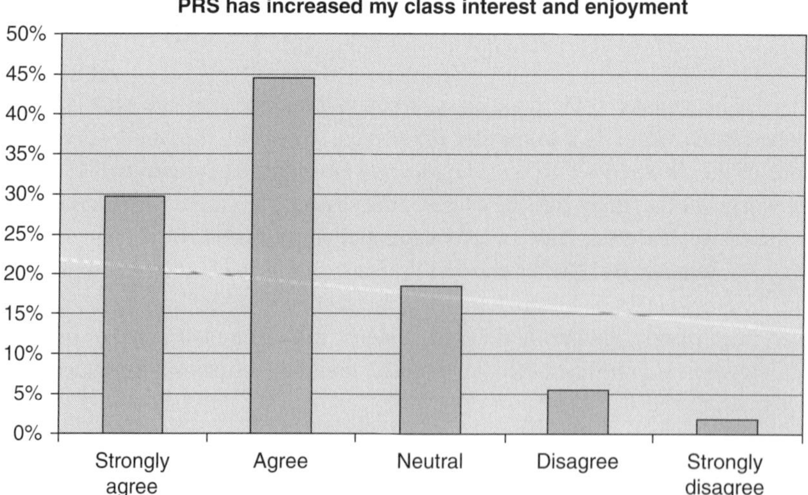

Figure 7.5 Results from a large survey of student attitudes about personal response systems, also known as clickers, show that students over whelmingly like them (Duncan, 2005). Attitudes also improve as instructors become more familiar with clicker use.

Practice with clickers before you use them. You may ask a TA or other student to run the "start" and "stop" counting software until you are comfortable with it. You may start with just a few questions. Once you make more regular use of clickers you will find they *transform* the classroom in a very positive way. Students become active participants, not merely passive listeners to a lecture. They ask more questions. They are lively! Most clicker users give a small amount of credit, say 10% of the class grade, for answers to clicker question. Many give 1 point for a wrong answer and 2 points for a right answer to encourage student participation.

You must explain why you are using clickers. Most students are used to years of passive learning in class. If you use clickers as suggested above you will make them work harder. You need to explain that their learning will increase, and what this will do to their grade. Students have made comments such as (this is an actual quotation), "I expected you to teach me – I didn't expect to have to learn." Failure to explain adequately the reasons for clicker use and failure to practice are two of the most commonly reported causes of trouble from clicker users.

7.5 Other things clickers can do

(a) Measure what students know before you start to teach them (pre-assessment),
(b) measure student attitudes,
(c) find out if students have done their assigned reading,
(d) get students to confront common misconceptions,
(e) transform the way you do any demonstrations,
(f) increase students' retention of what you teach,
(g) test student understanding (formative assessment),
(h) make some kinds of grading and assessment easier,
(i) facilitate testing of conceptual understanding,
(j) facilitate discussion and peer instruction,
(k) increase class attendance.

Duncan (2005, 2006a, 2006b) discusses these topics in more detail. Clickers are the most promising "teaching technology" to come along in decades. They can change the dynamics of a lecture so that students become more engaged, active learners. They provide immediate feedback to the instructor and to each student about what other students are thinking. They are relatively easy to learn to use and inexpensive. However, like any technology, clickers can be misused, and it is important to practice and to explain their use to students before starting.

References

Duncan, D. K., 2005, *Clickers in the Classroom* (New York: Pearson/Addison Wesley).

Duncan, D. K., 2006a, *Clickers in the Astronomy Classroom* (New York: Pearson/Addison Wesley).

Duncan, D. K., 2006b, Clickers, a new teaching aid with exceptional promise, *Astronomy Education Review*, **5**, 70–88.

Hake, R. R., 1998, Interactive engagement vs. traditional methods: a six thousand student survey of mechanics test data for introductory physics courses, *Am. J. Phys.*, **66**, 64–74.

Hammer, D., 1994, Epistemological beliefs in introductory physics, *Cognition and Instruction*, **12**(2), 151–183.

Lightman, A. and Sadler, P., 1993, Teacher predictions versus actual student gains, *Physics Teacher*, **31**(3), 162–167.

Mazur, E., 1997, *Peer Instruction: a User's Manual* (Upper Saddle River, NJ: Prentice Hall).
Schneps, M. H. and Sadler, P. M., 1987, *A Private Universe*, Harvard-Smithsonian Center for Astrophysics, Science Education Department, Science Media Group. Video available from Annenberg/CPB (www.learner.org).
Weiman, C. and Perkins, K., 2005, Transforming physics education, *Physics Today*, **58**, 11.

Comment

Paulo S. Bretones: I think that it is the same thing that I do with my students. I go with them to the "picture of the day" asking, using a picture of Mercury, "What is this?" Almost everybody answers "The Moon." So I answer that it is Mercury. After this, because of the conflict, I said: "No, this is Mercury." Is this the same situation?

Douglas Duncan: Yes. Students need to understand that an idea of theirs is wrong before they are open to learning a new idea. This very important point is dramatically shown by Schneps and Sadler (1987) in their video, *A Private Universe*.

8

Educational opportunities in pro–am collaboration

Richard Tresch Fienberg and Robert Stencel

Sky & Telescope, Sky Publishing, 90 Sherman St., Cambridge, MA 02140, USA; University of Denver, Denver, CO 80208, USA

Abstract: While many backyard stargazers take up the hobby just for fun, many others are attracted to it because of their keen interest in learning more about the Universe. The best way to learn science is to do science. Happily, the technology available to today's amateur astronomers – including computer-controlled telescopes, CCD cameras, powerful astronomical software, and the Internet – gives them the potential to make real contributions to scientific research and to help support local educational objectives. Meanwhile, professional astronomers are losing access to small telescopes as funding is shifted to larger projects, including survey programs that will soon discover countless interesting objects needing follow-up observations. Clearly the field is ripe with opportunities for amateurs, professionals, and educators to collaborate. Amateurs will benefit from mentoring by expert professionals, pros will benefit from observations and data processing by increasingly knowledgeable amateurs, and educators will benefit from a larger pool of skilled talent to help them carry out astronomy-education initiatives. We look at some successful pro–am collaborations that have already borne fruit and examine areas where the need and/or potential for new partnerships is especially large. In keeping with the theme of this special session, we focus on how pro–am collaborations in astronomy can contribute to science education both inside and outside the classroom, not only for students of school age but also for adults who may not have enjoyed particularly good science education when they were younger. Because nighttime observations with sophisticated equipment are not always possible in formal educational settings, we will also mention other types of pro–am partnerships, including those involving remote observing, data mining, and/or distributed computing.

8.1 Capacity lost and gained

Historically, amateur astronomers have repeatedly made important contributions to the science. Examples include sunspot records and the discovery of comets, asteroids, and supernovae. During the twentieth century, advances in technology, such as computers and electronic detectors, initially were limited to larger telescopes, but as that technology has become widely accessible, amateurs again can make important contributions (see Percy and Wilson, 2000; Overbeek and da S. Campos, 1987).

At present, professional astronomers are losing access to mid-sized meter-class telescopes at good sites as funding is shifted to larger projects. Meanwhile, increasingly ambitious automated all-sky survey programs are discovering countless interesting objects needing follow-up observations.

The increasing availability to skilled amateur astronomers of 0.3- to 0.5-meter-aperture optics and digital imagers on computer-controlled mounts means the potential recovery of some lost capacity. One challenge to the professional community is harnessing and training this potential for greater productivity. Pro–am organizations can help achieve this.

Innovation in Astronomy Education, eds. Jay M. Pasachoff, Rosa M. Ros, and Naomi Pasachoff. Published by Cambridge University Press.

Small-aperture telescopes can excel in time-domain-intensive efforts, where monitoring is required. A stable system can perform with 0.05-mag precision, down to ~15th magnitude, with care and the right software.

Serious amateurs keep up with astronomy and space news but don't always know what opportunities exist to contribute to the science, though articles in *Sky and Telescope* magazine (SkyandTelescope.com) and on the websites of organizations such as the American Association of Variable Star Observers (www.aavso.org/), the Center for Backyard Astrophysics (cba.phys.columbia.edu), and the Society for Astronomical Sciences (www.socastrosci.org/) are helping to spread awareness.

8.2 AAS working group for pro–am collaboration

The American Astronomical Society Working Group for Professional–Amateur Collaboration (WGPAC), established in 1999 and made permanent in 2004, provides a forum for collaboration between amateur and professional observational astronomers. Its website is located at www.aas.org/wgpac.

Participants include:

- American Association of Variable Star Observers (AAVSO)
- Association of Lunar and Planetary Observers (ALPO)
- Center for Backyard Astrophysics (CBA)
- Global Network of Astronomy Telescopes (GNAT)
- International Dark-Sky Association (IDA)
- International Meteor Organization (IMO)
- International Occultation Timing Association (IOTA)
- International Small Telescope Cooperative (ISTeC)
- Society for Astronomical Sciences (SAS)

8.3 What amateurs are doing

Here is a small sampling of the areas in which amateurs are regularly making real contributions to astronomical science:

- meteor showers – minor showers, lunar impacts;
- occultations – lunar limb profile, asteroid shapes and sizes;
- variable stars – light curves, detection of outbursts;
- CCD photometry – UBVRI, asteroid rotation, exoplanet transits;
- CCD astrometry – asteroid and comet orbits, binary-star orbits;
- search and discovery – novae, supernovae, GRB counterparts, comets, asteroids.

8.4 A broad spectrum of educational opportunities

In addition to advancing science, pro–am collaboration advances education on many scales and at many levels. On the micro-scale, amateurs with backyard telescopes can establish relationships with local primary and secondary schools, where they can help with classroom instruction or extracurricular activities such as science clubs. On the meso-scale, amateurs as well as professionals outside of academic settings can establish relationships with local colleges and universities, enabling undergraduate students to become involved in research projects. On the macro-scale, amateurs, professionals, and educators can link up across the globe through organizations such as those mentioned above, enabling cooperative research spanning multiple cultures and time zones.

8.5 Internet astronomy

Internet telescopes such as those provided by the Tzec Maun Foundation (tzecmaun.org) and Slooh (www.slooh.com) enable local daytime classroom astronomy, and not just for the "haves" with direct access to their own equipment. In the near future, daytime classrooms widely distributed in longitude will have their pick of faraway Internet telescopes with clear, dark nighttime skies. For a good list of remote-access telescopes, see the website www.phy.duke.edu/~kolena/imagepro.html#t.

More generally, amateurs and students will also play a key role as data miners and number crunchers through efforts such as the International Virtual Observatory (www.ivoa.net). Professionals generate 100% of the data but typically pick only the 1% of "low-hanging fruit" among data that becomes publicly accessible for their own research, leaving the other 99% available to everyone else.

Two noteworthy developments are Sky in Google Earth, whose creators modestly call it "the next frontier in astronomical data discovery and visualization" (arxiv.org/abs/0709.0752), and Microsoft Research's World Wide Telescope.

Amateurs are learning a lot about the scientific process by participating in two kinds of distributed computing: passive, such as SETI@home (setiathome.berkeley.edu), where unused CPU time is devoted to signal processing, and active, such as Systemic (www.oklo.org/), where the user downloads stellar radial-velocity data and manually adjusts parameters to fit exoplanet orbits. The SETI@home concept has proliferated into many other projects, not all of them astronomical.

Other roles played by amateurs, both online and offline, include education and outreach, innovation in both hardware and software design, and support for research through communication with elected officials.

8.6 Summary

The technology available to today's amateur astronomers, including computer-controlled telescopes, CCD cameras, powerful astronomical software, and the Internet, offers the potential to make real contributions to a variety of endeavors:

- scientific research,
- support of local educational objectives,
- participation in global educational objectives.

Through their participation, amateurs benefit from mentoring by expert professionals, pros benefit from observations and data processing by increasingly skilled and knowledgeable amateurs, and educators benefit from a larger pool of skilled talent to draw upon for astronomy/physics/math educational goals.

References

Overbeek, M. D. and da S. Campos, J., 1987, IAU Colloquium 98: the contribution of amateur astronomers to astronomy, *Monthly Notes of the Astron. Soc. Southern Africa*, **46**, 117.

Percy, John R. and Wilson, Joseph B., 2000, *Amateur–Professional Partnerships in Astronomy*, ASP Conference Proceedings, Vol. 220 (San Francisco: Astronomical Society of the Pacific).

9

Teaching history of astronomy to second-year engineering students at the University of Chile

José Maza

Departamento de Astronomia, Universidad de Chile, Casilla 36D, Santiago, Chile

Abstract: For over 20 years, the University of Chile has offered a course called "History of Astronomy" to second-year engineering students. The course partly fills the two-course "Humanistic Studies" requirement that each engineering graduate must complete. As a result of the course, men and an increasing number of women who will go on to work at and often become senior executives at major Chilean companies are exposed to the basics of astronomy and to its development over history. The first part of the course, "a tour to the scientific revolution," begins with the ancient civilizations, leads up to Newton's construction of modern science, and ends with the contributions to celestial mechanics of Euler, Clairaut, Lagrange, D'Alembert, and Laplace. The second part begins with William Herschel and the discovery of the Milky Way and proceeds over several weeks to a discussion of the big bang, the cosmic background radiation, and dark energy, before culminating with a lecture on the history of astronomy in Chile.

9.1 Who takes the course and why

Since 1980 I have been teaching a course in the history of astronomy to engineering students at the Faculty of Physical Sciences and Mathematics of the Universidad de Chile, at Santiago, Chile. The School of Engineering and Science admits some 550 students every year, selected from among the top students in the country. About 90% of the students who complete the five-semester common curriculum qualify as civil engineers, electrical engineers, industrial engineers, engineers in computer sciences, etc. Others earn a B.Sc. in physics, astronomy, mathematics, or geophysics. As part of the requirements for either the engineering or the B.Sc. degree, every student has to take two one-semester courses in "Humanistic Studies," choosing among fields including philosophy, poetry, literature, world history, sociology, history of Chile, history of art, and history of astronomy. The offerings vary from year to year, but typically students can select from among ten different subjects. Every subject is offered in a two-part curriculum.

"History of Astronomy I" and "History of Astronomy II" are presented in 30 lectures each, over a period of 15 weeks or one semester. Part I of the course begins with the Ancient Greeks and ends with Newton and the development of celestial mechanics. Part II starts with William Herschel and finishes with present developments in cosmology.

The main objective of this two-part course is to broaden the students' view by putting scientific advances into a historical perspective. Nowadays students entering our school have a weak command of language. They don't read or write with skill, and even their oral language skills leave much to be desired. The course in history of astronomy emphasizes reading and understanding. Students with a strong background in physics or math do not

Innovation in Astronomy Education, eds. Jay M. Pasachoff, Rosa M. Ros, and Naomi Pasachoff. Published by Cambridge University Press.

necessarily understand how most scientific developments came into being. In our lectures we present the connections among astronomy, physics, and mathematics, and the interplay among technological developments, theoretical advances, and scientific discoveries. By putting scientific discoveries into a broader context we emphasize the collective nature of science.

The history of astronomy class has been quite popular among the students; some 90 out of a total number of 450 second-year students take the courses every year. (Roughly 100 students fail at the first-year level.) Thus, on the order of 20% of the students in the School of Engineering and Science at the University of Chile take history of astronomy. In addition, some 80 students per year take a general astronomy course. That means almost 40% of the engineers trained at the University of Chile (the major school of engineering in the country) are exposed to some astronomy. Since a large fraction of our engineers becomes high executives in the most important companies in Chile, we consider it strategically important to expose these individuals to the basics of our science.

9.2 Some relevant statistics

For over a century the School of Engineering at the University of Chile has been strongly male-dominated. During the 27 years that I have been teaching these courses, the number of women entering the School of Engineering has grown, rising from 5% to 20%, and continuing to rise. There still are some refractory specialties, including mining and geology, but in other specialties, including computer sciences and industrial engineering (business management), the percentage of women keeps rising annually. The percentage of women taking the history of astronomy class is not different from the general one. In summary: the gender-balance situation is bad but improving.

The total number of professional astronomers (with a Ph.D.) working in Chilean institutions with a permanent contract is approximately 40. Only 15% right now are women. If we add the non-tenured jobs, the total number rises to 60. Two new astronomers per year are needed in order to maintain those numbers. The total number of students graduating in astronomy in Chile every year is approaching 20. Half of them continue with graduate work in astronomy; a significant fraction goes to a graduate school abroad. On the order of five students per year earn a B.Sc. in astronomy at our university. Our graduate program is relatively new – only six years old – but the total number of graduate students enrolled is approaching 20.

9.3 Outlines of the history of astronomy courses

9.3.1 History of Astronomy, Part I

1. Egyptians and Babylonians, Stonehenge
2. Classical Greek Astronomy
3. Ptolemy: the geocentric Universe
4. The Middle Ages: Arabic astronomy
5. Nicholas Copernicus: the heliocentric Universe
6. Tycho Brahe
7. Johannes Kepler
8. Galileo Galilei
9. Francis Bacon and René Descartes: the scientific method
10. Scientific societies; the foundation of the observatories of Paris and Greenwich
11. Christian Huygens

12. Olaf Roemer
13. Giovani Domenico Cassini
14. John Flamsteed
15. Isaac Newton: universal gravitation and the laws of mechanics
16. Edmond Halley
17. James Bradley: aberration of light and nutation
18. Nicholas Lacaille and the southern hemisphere
19. The 1761 and 1769 transits of Venus
20. Celestial mechanics: Euler, Clairaut, Lagrange, D'Alembert, and Laplace

9.3.2 History of astronomy, part II

1. William Herschel and the discovery of the Milky Way
2. Piazzi and the discovery of Ceres
3. Leverrier and Adams and the discovery of Neptune
4. The first measurements of stellar parallax: Bessell, Struve, and Henderson
5. Spectral analysis: Fraunhofer, Bunsen, and Kirchhoff
6. Spectral classification: Secci and Pickering
7. Astronomical photography
8. The size and shape of the Milky Way: Kapteyn and Shapley
9. Galaxies: Shapley, Curtis, and Hubble
10. Olbers's paradox
11. Cosmology: Einstein, de Sitter, and Lemaître
12. The expansion of the Universe: Hubble and Humason
13. Basic properties of stars: Hertzsprung and Russell
14. Stellar structure: Eddington and Chandrasekhar
15. Stellar evolution, supernovae, pulsars, black holes
16. The Big Bang: Lemaître and Gamow
17. The beginning of radio astronomy
18. Clusters of galaxies, radio galaxies, and quasars
19. Cosmic microwave background: Penzias and Wilson; big bang: the first few minutes; acceleration of the expansion; dark energy
20. History of astronomy in Chile

9.4 Contrasting Parts I and II of the course

Part 1 presents a tour of the history of science leading up to the scientific revolution and the construction of modern science by Isaac Newton. For that reason many topics like Chinese, Mayan, and Incan astronomy are omitted – not because they are uninteresting but because they don't contribute to this "road to Newton." An astronomer teaches these history of astronomy courses to engineering students. For that reason the focus is on science rather than history or the history of ideas. A large fraction of the first part of the curriculum is devoted to Greek astronomy, Ptolemy, Copernicus, Tycho, Kepler, Galileo, and Newton.

Part II, which begins with Herschel and the construction of big telescopes, proceeds through the rise of astrophysics in the nineteenth century and the main advances of the twentieth century, including stellar structure and evolution, the birth of radio astronomy, Olbers's paradox, and the big bang, and culminating in dark energy. Compared to Part I, Part II contains more astronomical concepts and less biographical information about key astronomers.

9.5 Future goals and acknowledgments

My lecture notes in Spanish are posted (as pdf files) on my personal web page: www.das.uchile.cl/~jose. Each topic in the syllabus averages 15 pages of notes. The students thus have to read some 300 pages of notes every semester. Sometime in the future I plan to print these notes as a book.

I would be glad to interchange ideas and share my experience from 27 years of teaching this course with anyone interested. Four decades ago I took a course on the history and philosophy of science at the University of Chile with Professor Desiderio Papp, whose inspiring lectures were the starting point of my interest in the history of astronomy. I want to acknowledge Professor Papp for stimulating my interest in learning and later teaching my own courses.

I would also like to express my gratitude to my colleague Diego Mardones for interesting discussions on issues relating to the history of astronomy. I acknowledge financial support from project *Centro de Astrofísica* FONDAP 15010003.

Comments

Rosa M. Ros: Engineers love to apply what they are learning. Do you have any experience in projects connecting astronomy and other science aspects that they are learning?

José Maza: The courses have been attended by more than 200 students. It is not possible to organize practical projects with them. They only go for 1 session to the observatory 20 km away at Santiago de Chile.

10

Teaching the evolution of stellar and Milky Way concepts through the ages: a tool for the construction of a scientific culture using astrophysics

G. Theureau and L. Klein

CNRS-LPCE, 3A av. de la Recherche Scientifique, 45171 Orléans, Cedex 02, France, and Paris Observatory, 61 av. de l'Observatoire, 75014 Paris, France

Abstract: We report on a two-semester experience at Orléans University (France) with a course in the history and epistemology of the concepts of stars and galaxies from antiquity to the early twentieth century. The framework is a *module d'ouverture* of the new Licence–Master–Doctorat reform of French universities, i.e., a transversal course aimed at providing scientific culture to a heterogeneous group of students from various fields mostly outside science. Because of the number of students and the heterogeneity of their backgrounds, the decision was made to offer the course in 22 lectures, divided into ten lessons plus one planetarium session. Special attention was paid to working with original historical texts. The final evaluation was based on students' reading notes and their commentaries on a historical text chosen from a list of various references covering the whole period under consideration. Each student, sometimes as part of a group of two or three, presented a report.

10.1 General

We report here on our experience teaching a course on the history and epistemology of the concepts of stars and galaxies from antiquity to the early twentieth century. The course, first offered at the University of Orléans in 2004–2005, was offered annually through 2007–2008.

The 22-hour course, which meets in 11 sessions of two hours each, is divided into four main parts, corresponding to four main concepts :

- world systems from the pre-Socratics to the philosophers of the Middle Ages (4 hours);
- concepts of mechanics and planetary motion from Aristotle to Newton (and a little beyond) (4 hours);
- spectroscopy and the nature of stars (6 hours);
- concepts of the Milky Way and the nature of the nebulae (8 hours).

(The full list of chapters is given in Table 10.1.)

Rather than a historical and time-oriented path, we chose to lay emphasis on an epistemological point of view, explaining the mechanisms of knowledge.

Our first goal was to present the evolution of ideas on the Universe from antiquity to the the twentieth century, showing that scientific endeavor relies not only on a critical combination of observation, experiment, and theory, but also on various kinds of preconceptions and thoughts shaped by the intellectual environment of a given epoch. Over the course of history, scientific views of the Universe have emerged from a variety of very different fields of thought, including:

Innovation in Astronomy Education, eds. Jay M. Pasachoff, Rosa M. Ros, and Naomi Pasachoff. Published by Cambridge University Press.

Table 10.1 *List of chapters*

(1)	World systems and migration of astronomical knowledge in the ancient world
(2)	Measuring the Earth and the Solar System
	Planetarium: from geocentric to heliocentric Universe, nebulous objects
(3)	From a closed to an infinite Universe
(4)	Elements of mechanics from Aristotle to Newton (including the notion of relativity)
(5)	The nature of the Sun and the stars. Spectroscopy, birth of astrophysics
(6)	The stellar source of energy and stellar evolution
(7)	The Milky Way: mythology and surmises
(8)	A stellar system: structure and models from Wright to Kapteyn
(9)	The nature of the nebulae: analogical and cosmogonical models
(10)	Galactic rotation, interstellar extinction and distances: final synthesis

- mythology (Indo–European, Greek, Semitic, Chinese, . . .),
- theology (Wright, Newton, . . .),
- philosophy (pre-Socratic, Plato, Kant, Comte, . . .),
- metaphysics (Pythagorean school, Kepler, . . .),
- physics (Aristotle, Galileo, Newton, Laplace, . . .),
- mathematics (Alexandrian school, Arabic astronomy),
- instrumentation (Al-Khwarizmi, Brahe, Lord Ross, Huggins, Barnard, . . .).

In this scheme, our goal was not only to present astronomy as a field in its own right, but also to discuss the evolution of ideas about the Universe as a part of human history and culture, emphasizing the importance of an open-minded approach to a large variety of cultures and viewpoints.

Our second aim was then to demonstrate to the students that science belongs to the patrimony of humanity and that it has no frontiers. Europe existed well before it was created as a political entity. Centuries before the Internet or the concept of globalization emerged, astronomers communicated throughout the world on the scale of continents (or more). For example, Babylonian, Greek, and Indian science traveled through primitive, Arabic, and Persian cultures before finally being taught at European universities.

Finally, and more pragmatically, this course became a way to gather on the same benches of the university students from both the humanities and the sciences, demonstrating that astronomy, with its long time scales and its involvement in an existential quest, requires both fields of knowledge. This kind of encounter is something out of the ordinary in France, where pupils choose specialties very early (at the age of 15) and where up to now almost no bridge has existed between universities.

10.2 Pedagogical context

How did the course actually go? It was proposed as an "unité d'ouverture," which literally would mean "open mind course" or "transversal course," and is characterized by a mixed audience:

- mix of backgrounds (law, language, pedagogy, history, biology, physics, mathematics, sports, . . .),
- mix of levels and ages (first, second, and third year of university).

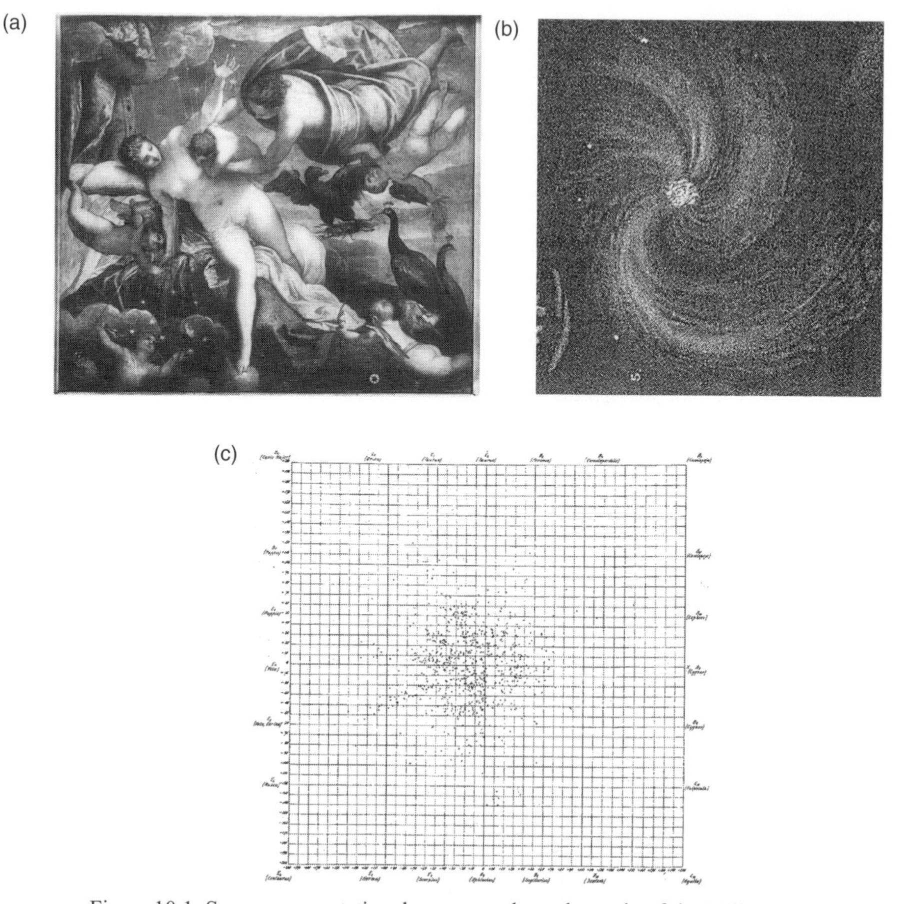

Figure 10.1 Some representative documents about the topic of the Milky Way. Left, by Tintoretto (1518–1594) *The Orgin of the Milky Way*, illustrating the myth of the Milky Way: Heracles was brought by Hermes to Hera's breast to get immortality. Note the stars escaping from the breast, illustrating the ambiguity of the nature of the Milky Way. Alinari Art Resource, NY, National Gallery, London, Great Britain. Two completely different representations of the Milky Way: (a) by Alexander (b) (1852), using the analogy with spiral nebulae (in this case, M99); (c) by Charlier (1916), from stellar material (counts and distribution of B stars). Figures from Chaberlot (2003).

This heterogenity meant that we had to use very few equations and if possible avoid overly technical descriptions.

We chose to favor document analysis rather than linear lectures based on sequences of dates or events. So the way we worked was as follows:

- we showed original documents (see Figures 10.1 and 10.2, for example),
- we presented original texts (see list in Table 10.2),
- we commented on them and placed them in their historical context.

All lectures were prepared in PowerPoint, and we edited a booklet with a copy of the assigned texts. We also built a dedicated website (lpce.cnrs-orleans/theureau/histoire.html) where we made available the following material:

Table 10.2 *List of texts or excerpts used*

Aristotle	*Physics*, *Meteorology*, *Treatise on the Heavens* (322 BCE)
Macrobe	*Commentary of Scipion's Dream by Cicero*, *The Republic VI* (fifth century)
Abou'l-Faradj	*Treatise on Astronomy and Cosmography* (1280)
Oresme	*Traité de la sphère*, *Traité du ciel et du monde* (1377)
Brahe	*Astronomiae Instauratae Progymnasmata* (1602)
Galileo	*Sidereus Nuncius* (1610), *Dialogo* (1632)
Wright	*An Original Theory of the Universe* (1750)
Kant	*Allgemeine Naturgeschichte und Theorie des Himmels* (1755)
W. Herschel	*Construction of the Heavens* (1785)
Laplace	*Exposition du Système du Monde* (1796)
Comte	*Cours de philosophie positive* (1839)
Secchi	*Le Stelle* (1877)
Clerke	*The System of the Stars* (1905)
Barnard	*Photographs of the Milky Way and Comets* (1913)
Eddington	*Stellar Movements and the Structure of the Universe* (1914)
Shapley	*Contributions from the Mount Wilson Solar Observatory* (1917)

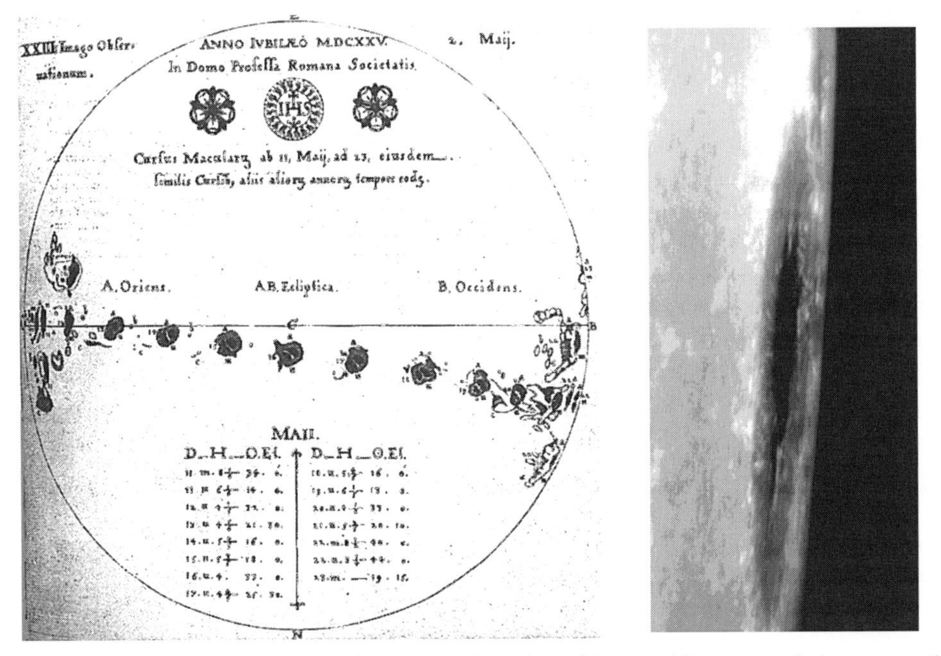

Figure 10.2 The nature of sunspots and rotation of the sun: (a) a sunspot during consecutive days (drawing by Scheiner, 1612, courtesy of the library of l'Observatoire de Paris), (b) sunspot at the limb (l'Observatoire de Pic du Midi, France).

- pdf version of the courses,
- bibliography,
- Internet links,
- general information about courses,
- instructions for examinations.

The website has been both a meeting point for the students and a way to deliver high-quality colored documents in electronic form. We proposed also an exchange of e-mails for anything concerning the course or the preparation of the evaluation.

At the end of the first part, in order to illustrate the debate between geocentric and heliocentric world systems, and to give a first feeling of how puzzling the nature of planetary motion and nebular objects can be, we held a dedicated planetarium session (Nançay Radio Observatory Astronomy Museum and Visitor Center, France). This "observing" session was a very good way to create "team" spirit among the students and, since most of them had never made scientific observations of the sky, it was a very good way for them to get a sense of the questions of interest to past generations of philosophers and astronomers.

10.3 Main difficulties

The main difficulty we faced in presenting the course was the heterogeneity of the audience: three different levels, different fields of interest and backgrounds, students spread over several departments of the university.

The project itself was ambitious, firstly, because it deals with a lot of information and concepts and requires a wide bibliography (with very few of the materials in French); secondly, because it requires real maturity and personal involvement on the part of the students, most of whom were unfamiliar with that kind of lecture; and, thirdly, because it was our first experience of this kind as teachers.

The greatest challenge, however, was to find a way to evaluate the students. The usual and easiest way consists of an examination where questions are asked on the course content. A more challenging choice is to demand of the students greater personal effort and more original thought. We adopted the latter approach.

The first year, we left the choice to the students between either an essay on a given topic or a report based on their reading of a historical text. No instructions were given. The result was a complete failure: 90% of the students submitted work that was cut and pasted from the Web, lacking any personal input. The graders spent most of their effort searching in Google Research for the sources the students had plagiarized.

The second year, thanks to this initial experience, we restricted the assessment to student reports on selections from a historical text or some letters, but with precise guiding instructions for reading and writing. The results were much better, still varied, but there were some very interesting results. (A typical list of books or essays we used is given in Table 10.3.)

Table 10.3 *List of monographs and essays used for the evaluation*

J. Buridan and N. Oresme	*Commentaries on the Mechanics of Ptolemy and Copernicus* (14th century)
Kepler	*Mysterium Cosmographicum* (1596)
Galileo	*The Copernican System: Set of Letters by Galileo* (1616–1624)
Galileo	*Dialogo*, 1st day (1632)
Fontenelle	*Entretiens sur la pluralité des mondes* (1686)
Comte	*Cours de Philosophie Positive*, Lessons 1 and 2 (1839)
P. Richet	*Solar Energy and the Age of the World in the 19th Century* (1999)
Humboldt	Kosmos (*Stellar Astronomy*, 1859)
Secchi	*Les Etoiles* (1895)
Arrhénius	*Le Destin des Etoiles*, extraits (1921)
Belot	*L'origine dualiste des mondes* (1924)

Note that on both occasions we used for the second evaluative session a more traditional test based on a list of questions concerning the course content and a commentary on a short historical text.

10.4 Preliminary conclusion

For the instructors, it has been a very nice and enriching experience, at least for the rich culture we acquired ourselves from the extensive bibliography we had to survey. As science teachers we also experienced a new pedagogical direction, closer to the one practiced by our colleagues in literature, philosophy, and history. We never overcame the challenge of working with such a heterogeneous audience of students who are increasingly less used to going through dense and difficult texts.

Probably our approach could be improved by reducing the content and increasing interactivity, e.g., by starting each session by reading and commenting on a couple of texts representative of the lecture.

A future possibility involves adapting this series of lectures into a course for distance learning (with the National Center of Distance Education – CNED and Université Paris VI).

10.5 Appendix: Examples of texts read with the students

10.5.1 Aristotle, Meteorology, *350 BCE*

Let us recall our fundamental principle [about the Milky Way] and then explain our views. We have already laid down that the outermost part of what is called the air is potentially fire and that therefore when the air is dissolved by motion, there is separated off a kind of matter – and of this matter we assert that comets consist. We must suppose that what happens is the same as in the case of the comets when the matter does not form independently but is formed by one of the fixed stars or the planets. Then these stars appear to be fringed, because matter of this kind follows their course. In the same way, a certain kind of matter follows the sun, and we explain the halo as a reflection from it when the air is of the right constitution. Now we must assume that what happens in the case of the stars severally happens in the case of the whole of the heavens and all the upper motion. For it is natural to suppose that, if the motion of a single star excites a flame, that of all the stars should have a similar result, and especially in that region in which the stars are biggest and most numerous and nearest to one another. [. . .] But this circle in which the Milky Way appears to our sight is the greatest circle, and its position is such that it extends far outside the tropic circles. Besides the region is full of the biggest and brightest constellations and also of what are called 'scattered' stars (you have only to look to see this clearly). So for these reasons all this matter is continually and ceaselessly collecting there.

10.5.2 Nicolas Oresme, Le livre du ciel et du monde, *about 1380*

Objections against Ptolemy's arguments against the Earth's rotation

(1) (Ptolemy): "The sky undergoes motion, because one sees some stars rise and set, and others turning around the northern pole." Oresme: "I say that if of the two parts of the world mentioned before, the upper one had a daily motion today and not the lower one, and if tomorrow the opposite were the case, the lower one undergoing motion and not

the upper one, i.e., the sky with its stars, we would have no means to notice this change, and everything would appear to us as being the same, today and tomorrow. We would always have the impression that that part of the world where we are is fixed, and that the other undergoes motion. Exactly as it appears to a man on a moving ship that the trees outside are moving."

(2) (Ptolemy): "If the Earth underwent rotation, strong eastern winds would be expected, whereas we actually notice that western winds are dominant." Oresme : "The answer to this objection is that it is not the (solid) Earth alone that undergoes this motion, but that the water and the air move together with the Earth, although both water and air have their own different motions under the influence of winds or other causes. It is the same as if air is confined in a ship: it will appear to a person embedded within this volume of air that it does not move."

(3) (Ptolemy): "If Earth underwent rotation from west to east, an arrow launched vertically at a given point should fall back to the Earth westward of that point." Oresme: "Concerning this experiment, one might say that the arrow conducted upward by the launch is rapidly transported eastward with the air that it traverses, and with the entire mass of the inferior part of the world as defined above, which undergoes a diurnal motion. This is why the arrow falls down at that place of the Earth where it had been launched."

10.5.3 W. Herschel, On the Construction of the Heavens*, 1785*

The nebula we inhabit might be said to be one that has fewer marks of profound antiquity upon it than the rest. To explain this idea perhaps more clearly, we should recollect that the condensation of clusters of stars has been ascribed to a gradual approach . . . When a nebulous stratum consists chiefly of nebulae of the first and second form, it probably owes its origin to what may be called the decay of a great compound nebula of the third form . . . In like manner our system, after numbers of ages, may very possibly become divided so as to give rise to a stratum of two or three hundred nebulae . . . This view of the present subject throws considerable light upon the appearance of that remarkable collection of many hundreds of nebulae which are seen in what I have called the nebulous stratum of Coma Berenices . . . There might originally be another very large joining branch [of our nebula], which in time became separated by the condensation of the stars.

10.5.4 Agnes Mary Clerke, in Problems in Astrophysics *(London: A. C. Black, 1903)*

The Milky Way is an integral part of the great sidereal system. It marks the equatorial girdle of a sphere containing stars and nebulae variously scattered and aggregated. The whole material creation is, to our apprehension, enclosed within this sphere. We know nothing of what may lie beyond . . . These materials consist of gaseous and white nebulae in all their varieties; of star clusters, globular and irregular; and of the sundry species of stars.

The great majority of white nebulae might be called globular clusters in disguise. They present a round surface, condensed centrally by gradations testifying to their true spherical form. The only obvious distinction between them and "balls of stars" is that they are irresolvable by any telescopic powers . . . This quality depends wholly upon distance – that round nebulae are neither more or less than remote globular clusters. [Anyway,] The space-relations of the two classes of objects are very different. Clusters frequent the Milky Way; white nebulae avoid it. The discrepancy may be capable of reconcilement, but by a somewhat elaborate artifice of speculation.

References

Alexander, S., 1852, On the origin of the forms and the present condition of some of the clusters of stars, and several of the nebulae, *Astronomical Journal*, **2**, Fig. 1.
Charlier, C. V. L., 1916, *Studies in stellar statistics, III: The Distances and the Distribution of the Stars of the Spectral Type B*, Meddelanden fran Lunds Astronomiska Observatorium, series II, number 14, Fig. 2.

Further reading

Belkora, B. L., 2003, *Minding the Heavens* (Bristol and Philadelphia: Institute of Physics Publishing).
Chaberlot, F., 2003, *La Voie Lactée*, (Paris: CNRS histoire des sciences).
Cotardière, P. de la (ed.), 2004, *Histoire des sciences* (Paris: Tallandier).
Duhem, P., 1997, *L'aube du Savoir, Epithomée du Système du Monde*, Collection Histoire de la Pensée (Paris: Hermann).
Hoskin, M. A., 1963, *William Herschel and the Construction of the Heavens* (London: Oldbourne History of Science Library).
Jaki, S., 1972, *The Milky Way, an Elusive Road for Science Science* (New York: History Publications).
Merleau-Ponty, J., 1983, *La Science de l'univers à l'âge du Positivisme* (Paris: VRIN).
Rashed, R., 1997, *Histoire des Sciences Arabes, Astronomie théorique et appliquée* (Paris: Seuil).
Smith, R., 1982, *The Expanding Universe* (Cambridge, UK: Cambridge University Press).

Comments

Julieta Fierro: I believe it is a great idea to have students from various disciplines because they learn more from each other, especially when they explain items from their own fields.

11

International Astronomical Union – education programs

Jay M. Pasachoff

Williams College – Hopkins Observatory, Williamstown, MA 01267, USA

Abstract: I describe the education programs of the International Astronomical Union's Commission on Education Development. Its work is carried out through its ten Program Groups on the Worldwide Development of Astronomy, Teaching for Astronomy Development, International Schools for Young Astronomers, Exchange of Astronomers, National Liaisons on Astronomy Education, National Liaison on Astronomy Education, Collaborative Programs, Commission Newsletter, Public Information at the Times of Solar Eclipses, and Exchanges of Books, Journals, Materials. All are described at the Commission's website at www.astronomyeducation.org.

11.1 The IAU and Commission 46

The International Astronomical Union (IAU) was founded in 1922 to "promote and safeguard astronomy . . . and to develop it through international co-operation." There are currently 9114 individual members in 67 countries. The IAU is funded through the adhering countries. Almost all of the funds supplied from the adhering countries' dues are used for the development of astronomy.

One of the 40 IAU "commissions," or interest groups, is Commission 46, formerly called *The Teaching of Astronomy* and more recently, at the 2000 General Assembly, merged with the group that dealt with the diffusion of astronomy to developing countries around the world and thus renamed *Astronomy Education and Development*. It is the only commission that deals exclusively with astronomy education; a previous Commission 38 (Exchange of Astronomers), which allocated travel grants to astronomers who need them, and a Working Group on the Worldwide Development of Astronomy, have been absorbed by Commission 46. The Commission's mandate is "to further the development and improvement of astronomy education at all levels, throughout the world."

In general, the Commission works with other scientific and educational organizations to promote astronomy education and development; through the National Liaisons to the Commission, it promotes astronomy education in the countries that adhere to the IAU; and it encourages all programs and projects that can help to fulfill its mandate. The Commission holds business sessions at each IAU General Assembly. Within the format of the IAU General Assemblies, the Commission organizes or co-sponsors major sessions on education-related topics, such as a Special Session held at the 2003 General Assembly in Sydney, Australia, on which a book was published: *Teaching and Learning Astronomy: Effective Strategies for Educators Worldwide*, edited by Jay M. Pasachoff and John R. Percy (Cambridge University Press, 2005).

Innovation in Astronomy Education, eds. Jay M. Pasachoff, Rosa M. Ros, and Naomi Pasachoff. Published by Cambridge University Press.

The Commission has also organized two major conferences on astronomy education – in the US in 1988 (Pasachoff and Percy, 1990), and in the UK in 1996 (Gougenheim, McNally, and Percy, 1998).

Other organizations also sometimes have education conferences. (See, for example, Percy, 1996, Fraknoi and Waller, 2004). And note the astronomy-education online journal *Astronomy Education Review* (Wolff and Fraknoi, 2002–6), aer.noao.edu.

The Commission has usually sponsored one-day workshops for local schoolteachers as part of the IAU General Assemblies and as part of several IAU regional meetings.

Immediately after the conference that is described in the recent book, a very successful teachers' workshop was held in Sydney, organized by Nicholas Lomb, Sydney Observatory. A teachers' workshop after the Prague IAU was held in Czech and did not involve as much international participation.

Until recently, Commission 46 was concerned primarily with tertiary (university-level) education and beyond, but several of its activities have an impact on school-level and public education.

At the Prague General Assembly, we had two days of Special Session, reports from which are included in this volume. Rosa Ros (Spain), the Spanish National Liaison, has joined me in organizing this Special Session on *Innovation in Effective Teaching/Learning Astronomy*, in order to further the goals of the IAU Resolution on education passed in Sydney. John Hearnshaw, head of our Program Group for the Worldwide Development of Astronomy, has organized a Special Session on *Astronomy in the Developing World*; the proceedings of that meeting are in a separate book, also published by Cambridge University Press.

You can find out more about IAU Commission 46 on Education and Development by visiting our website at physics.open.ac.uk/IAU46/, which is directly linked to www.astronomyeducation.org.

We are grateful to our vice-president (2003–6), Barrie Jones, for arranging the website at the Open University and for maintaining it with the major assistance of T. J. Moore and additional assistance from Darren Dawes. For the 2006–9 triennium, Magda Stavinschi (Romania) has succeeded me as President of the Commission and Rosa Ros (Spain) has become Vice-President.

Program Groups of IAU *Commission 46 on Education and Development*

- PG for the Worldwide Development of Astronomy
- PG for Teaching for Astronomy Development
- PG for International Schools for Young Astronomers
- PG for Exchange of Astronomers
- PG for National Liaisons on Astronomy Education
- PG for Collaborative Programs
- PG for Commission Newsletter
- PG for Public Information at the Times of Solar Eclipses
- PG for the Interchange of Books, Journals, and Materials.

Current officers and members and updated information is available at www.astronomyeducation.org.

11.2 Program Group for the Worldwide Development of Astronomy

The role of this PG is to visit countries with some astronomical expertise at tertiary (i.e., post high school) level, which are probably not IAU member states, but which would welcome some development of their capabilities in teaching and/or research in astronomy. For example, as a result of a visit in this program, Mongolia has become an interim member of the IAU and has received advice on broadening astronomy programs there.

11.3 Program Group for Teaching for Astronomy Development

TAD is intended to assist a country, currently with little astronomy, that wants to make a significant improvement in astronomy education. TAD operates on the basis of a proposal from a professional astronomy organization or on the basis of a contract between the IAU and an academic institution, usually a university.

The capabilities of the TAD program are limited to assistance with university-level activities, such as

(1) the creation of university-level astronomy/astrophysics courses and the faculty training and equipment associated with the development and first offering of such courses,
(2) a basic, largely educationally oriented research capability for faculty and students,
(3) travel (i.e., transportation) costs of foreign visiting lecturers and of students invited for study at foreign universities, and
(4) professional preparations needed as a prerequisite for plans to offer astronomy in schools and for the public.

TAD can provide advice about education of school teachers, but not financial support. The training of school teachers and the actual performance of school teaching and public outreach is considered to be part of the national resources.

11.4 Program Group for International Schools for Young Astronomers

ISYA seeks the participation of young astronomers primarily, but not exclusively, from astronomically developing countries. Participants should generally have finished first-degree studies.

ISYA seeks to broaden the participants' perspective on astronomy through lectures from an international faculty on selected topics of astronomy, seminars, practical exercises and observations, and exchange of experiences.

The most recent ISYAs as of the Prague meeting were in Morocco in 2004 and in Puebla, Mexico, in 2005. Since the Prague meeting, there was an ISYA in Malaysia in March/April 2007.

11.5 Program Group for Exchange of Astronomers

The program group makes travel grants, usually for durations of over three months, to qualified individuals in order to enable them to visit institutions abroad where they may interact with the intellectual life and participate in the research of the host institution. It is the objective of the program that astronomy in the home country be enriched after the applicant returns. The program group publishes, both on the IAU website and in IAU Information Bulletins, all the information needed to apply for a grant under the IAU Exchange of Astronomers program.

11.6 Program Group for National Liaisons on Astronomy Education

The main duty of each National Liaison on Astronomy Education is

(1) to write the triennial national report, to make it a valuable resource for countries wishing to enhance their astronomy education, and
(2) to transmit to the educators of his/her own country the insights that they might glean from the reports and conferences.

11.7 Program Group for Collaborative Programs

This Program Group works on activities co-sponsored by UNESCO, COSPAR, the United Nations (UN), the International Council of Scientific Unions (ICSU), etc., and carries out interactions with other international organizations.

11.8 Program Group for the Commission Newsletter

The Newsletter is published twice a year, and is available (including back issues) at www.astronomyeducation.org.

11.9 Program Group for Public Information at the Times of Solar Eclipses

Timely advice for countries that will experience a solar eclipse. We maintain a website at www.eclipses.info. We consult with local astronomers and with newspapers.

We were also active for the transit of Venus, and we maintain a website at www.transitofvenus.info.

11.10 Program Group for the Interchange of Books, Journals, and Materials

We are restudying the role of this program group in the context of new electronic document possibilities, but we can still link people needing written material with those for whom the material is surplus. Further, this Program Group has taken over the IAU responsibility of liaison with Cambridge University Press to maintain the list of 20 or so institutions around the world that receive gratis copies of all IAU publications.

11.11 New Program Groups

Discussions at Prague have led to the formation of new Program Groups, including one to deal with kindergarten-to-high-school (K-12) education. Our Commission may also have some responsibilities in liaison with the expected 2009 International Year of Astronomy.

References

Fraknoi, A. and Waller, B. (eds), 2004, *Cosmos in the Classroom 2004*. San Francisco, CA: Astronomical Society of the Pacific. Handouts, papers, and resource sheets from a major national meeting on teaching into astronomy for non-majors. Includes many innovative teaching ideas. www.astrosociety.org/events/cosmos/cosmos04/cosmos.html.

Gouguenheim, L., McNally, D., and Percy, J. R. (eds.), 1998, *New Trends in Astronomy Teaching*, Proceedings of IAU Colloquium 162 (London), (Cambridge: Cambridge University Press).

Pasachoff, J. M. and Percy, J. R. (eds.), 1990, *The Teaching of Astronomy*, Proceedings of IAU Colloquium 105 (Williamstown), (Cambridge: Cambridge University Press). adswww.harvard.edu.

Pasachoff, J. M. and Percy, J. R. (eds.), 2005, *Teaching and Learning Astronomy: Effective Strategies for Educators Worldwide*, (Cambridge: Cambridge University Press). www.cambridge.org/9780521842624.

Percy, J. (ed.), 1996, *Astronomy Education: Current Developments, Future Coordination* (San Francisco: Astronomical Society of the Pacific Conference Series, vol. 89).

Wolff, S. C. and Fraknoi, A., 2002–2006, *Astronomy Education Review*, aer.noao.edu.

Comments

Stewart Eyres: Comment on PG for distribution of books, etc. Many third-world countries have communites with no electricity, let alone Web access – perhaps the disbanding of this group is premature.

Jay Pasachoff: The undertaking is difficult to organize and expensive.

Further comments by the current PG secretary: It is difficult to match needs to availability and to organize distribution.

12

Astronomy in culture

Magda Stavinschi
Astronomical Institute of the Romanian Academy, Bucharest, Romania

Abstract: Astronomy is, by definition, the sum of the material and spiritual values created by mankind and of the institutions necessary to communicate these values. Consequently, astronomy belongs to the culture of each society, and its scientific progress does nothing but underline its role in culture. It is interesting that there is even a European society bearing the name "Astronomy for Culture" (SEAC). Its main goal is "the study of calendric and astronomical aspects of culture." Owning ancient evidence of astronomical knowledge, dating from the dawn of the first millennium, Romania is interested in this topic. But astronomy has a much deeper role in culture and civilization. There are many aspects that deserve to be discussed. Examples include the progress of astronomy in a certain society, in connection with its evolution; the place held by astronomy in literature and, generally, in art; the role of science fiction in the epoch of mass media; astronomy and belief; astronomy and astrology in the modern society, and so forth. These are problems that can be of interest for the IAU; but most important might be the educational role astronomy can play in the formation of the culture of the new generation, in educating the population about the need to protect our planet, and in the ensuring of a high level of spiritual development of society in the present epoch.

12.1 Astronomy's ubiquitous role in culture

Which of these statements makes the point most clearly: "astronomy in culture," "astronomy and culture," or "culture without astronomy"? These are only a few variants, each with its own sense. Perhaps the last question is the most pertinent. Does culture really exist without astronomy? The history and evolution of human civilization demand a negative response.

12.2 What we mean by culture

When we think of a *culture* (the *Hellenistic* one, for instance), we mean a set of customs, which may be expressed artistically, religiously, or intellectually, that differentiate one group or society from another. On the other hand, we often use the notion of culture in a different sense: shared beliefs, ways of regarding and doing, which orient more or less consciously the behavior of an individual or group. An example would be the *laic culture*. Moreover, the body of knowledge acquired in one or several domains also constitutes a *culture*, such as the *scientific culture* of an individual or group. Finally, the totality of the *set of human cultures* amounts to nothing less than *civilization*. Now, if we come back in time to the history of civilization, we find a constant component in every culture: astronomy. Not only is astronomy present in every culture but it has also often played a decisive part in its evolution.

Innovation in Astronomy Education, eds. Jay M. Pasachoff, Rosa M. Ros, and Naomi Pasachoff. Published by Cambridge University Press.

12.3 Astronomy and archaeology (archaeoastronomy)

A study of prehistorical civilizations and societies reveals evidence of astronomical knowledge even this far back. We should not be surprised that this is so. The necessities of life at the time made it useful to be able to predict seasons, to orient oneself by the stars, and so on. Archaeoastronomers can testify to the ubiquity of such evidence. In my own country, Romania, for example, sanctuaries dating back to the dawn of the Christian Era prove that our ancestors, the Dacians, were very familiar with the celestial vault.

Such knowledge also entailed interpretations and raised questions, some of which are deeply philosophical. What is mankind's place in the Universe? Who created the sky and the Earth? Are we the only rational beings who populate this Cosmos? These are among the questions people have always asked and have not yet answered.

12.4 Astronomy and history

With the birth of historiography, people began to record the most significant events they experienced: wars, epidemics, fires, natural catastrophes (earthquakes, floods, drought), the birth or death of a leader, etc. As calendars were imprecise, chroniclers often linked events to terrestrial and, more often, cosmic events. For example, the claim might be made that such-and-such an event happened before the darkening of the Sun or in the days when a tailed star came across the sky.

12.4.1 Comets as historical markers

In ancient history comets were thought to disrupt the cosmic order. This idea persisted for centuries. In his *Natural History*, Pliny wrote: "The comet is in general a dreadful star, and cannot be turned into derision." This observation is the starting point for the connection of comets to various doomful events. For example, the comet that appeared in Nero's time was regarded as the cause of the atrocities he perpetrated. Another comet was thought to have announced Vespasian's death. And in the year 400 a brilliant comet came across the sky at a time when a series of calamities befell Constantinople. Today such "fatal" appearances of comets give clues to their identification. Comets were said to have announced the deaths of Emperor Constantin (336); Attila (453); Muhammad (632); Casimir, King of Poland (1058); Richard the Lion-Hearted, King of England (1199); and Philip the Fair, King of France (1314). Such a list could fill many pages, as could the titles of the books that describe comets as companions to the deaths of kings. If the appearance of a comet did not actually coincide with a leader's death, it was simply invented, such as the "comet" said to have appeared at the time of Charlemagne's death, in 814. What is important is that historical events were fixed in time with the aid of "heavenly signs," and, conversely, that comets were identified on the basis of known moments in history.

12.4.2 Astronomy and the birth of Jesus

Even the birth of Jesus is established according to such astronomical landmarks. Romanians take pride in the fact that it was Dionysius Exiguus – born in Tomis, today's Constanta, a Romanian port at the Black Sea – who had the idea of counting years from the birth of Jesus. By the time Dionysius came up with this idea, over five centuries had elapsed since the birth of Jesus. The fact that Dionysius made an error of some four to seven years is, however, insignificant for that time, all the more as even today we are not able to establish accurately when this event of crucial importance for humanity occurred.

The Star of Bethlehem also is associated with a cosmic phenomenon. There are two categories of astronomical phenomena identifiable with it: a comet or a special planetary configuration. Fascinated by the spectacular close approach of Jupiter and Saturn on the night of 17 December 1603, Kepler had the idea that the Christmas Star could be just such a phenomenon. It might be the triple conjunction of the year –7, when Jupiter and Saturn approached one another three times in Pisces: on 29 May, on 6 October, and on 1 December. What led the Mesopotamian astrologers to Palestine? They could forecast this phenomenon, and associated Pisces with the Jewish people and with Israel, whereas Jupiter was considered a royal planet, and Saturn, the protector of Israel. Accordingly, that sign could be interpreted as an announcement of the coming (Saturn's protection) of (Jupiter) the King of (Pisces) the Jews.

The first conjunction (May 29) convinced the Magi to set out. The duration of their trip (approximately 1000 km) was about four months, so they reached Jerusalem at the moment of the second sign. Herod, the foreign usurper installed on the throne by the Romans, became anxious when the Magi arrived; he saw his reign threatened by the King of the Jews. After the Magi met Herod, the third sign occurred: the conjunction of 1 December, the one that showed them the way to Bethlehem. The "prophetic" star, assimilated to the third conjunction, could be seen southwards, in the dusk. The Magi went in that direction and reached Bethlehem, 8 km south of Jerusalem.

Is there another possible variant? The conjunction only drew the attention of the Magi to Israel; they waited for a sign, which appeared as the nova of March, –5. They arrived in Jerusalem, and only the appearance of the second nova (April, –4) brought them to Bethlehem.

12.4.3 Eclipses as historical markers

Eclipses, both solar and lunar, constitute another set of phenomena reported in the chronicles of all cultures. Such a record dates as far back as the year 3450 BCE. The oldest example of an historical event correlated with an eclipse, however, dates back to the battle between the Medes and the Lydians in 585 BCE. Frightened by the darkening of the Sun, both camps hastened to make peace. The birth of Muhammad (570) is also associated with a solar eclipse visible from Mecca. A lunar eclipse figures in the story of Columbus, who procured meals from the aborigines of Jamaica as a donation for the lunar eclipse he "prophesied" for the night of 29 February 1504.

12.5 Astronomy and politics

The relationship between astronomy and politics goes beyond the linking of important political events to astronomical phenomena. In fairly recent history, for example, one need only look to the aftermath of World War I. In 1919, the year after the bloody war's conclusion, astronomers became the first scientists to establish an international scientific society with the creation of the International Astronomical Union. Another little-known example dating from that time also involves an eclipse.

Albert Einstein's general theory of relativity, completed between 1913 and 1916, during the war, needed confirmation. An opportunity to confirm the theory arose with the total solar eclipse of August 21, 1914, visible over southern Russia. But because of the outbreak of war, which pitted Russia against Germany, the German team that went to Russia to perform the observations was imprisoned (though they were soon exchanged for a number of high-ranking Russian prisoners of war and returned safely to Berlin). The privilege of confirming Einstein's theory thus fell to English astronomer Arthur Eddington at the eclipse of May 29, 1919. Britain and Germany had also been fierce foes during the war, but now a British test corroborated a German theory, thus proving also the internationalism of scientific endeavors. The British demonstration of the validity of a German theory helped overcome the prejudice scientists from the victorious Allied countries continued to feel toward their colleagues from the defeated Central Powers after the war.

It is interesting to note, too, the interest some political figures took in astronomy. Thomas Jefferson, for example, who had recently completed his two terms as president of the United States, observed the annular solar eclipse of September 17, 1811. Over a century later King Baudouin I of Belgium (1930–1993), a defender of scientific and cultural institutions in general, was particularly devoted to astronomy. Perhaps uniquely for a sovereign, he understood the role played by precession in the difference between the motion of the Earth's poles and that of the celestial poles. Baudouin made regular use of the telescope he had installed in his summer residence in southern Spain. In homage to his love for astronomy, a statue of Baudouin was erected in front of the Royal Observatory of Belgium, which the King often visited.

12.6 The relationship between astronomy and other sciences

Everything that concerns astronomy also concerns *geography*. First of all, Earth is a planet, hence a body belonging to the Solar System. The better we know the surrounding world, the better we know the planet that hosts us. Terrestrial reference frames, orientation, time, mean-time zones – all these prove the link between astronomy and geography. It should be emphasized that the need to determine the latitudes at sea contributed appreciably to the further development of astronomy.

The link between astronomy and *mathematics* is similarly self-evident. Not only do all theories and problems of celestial mechanics and astrophysics depend on math, but the astronomical uses of mathematics also helped the science of mathematics advance.

With regard to *physics*, the Universe is the greatest physical laboratory. *Chemistry*? Where are we discovering elements, whose existence was just intuited on the Earth, if not in the Universe? *Meteorology*? How can we save our own planet without knowing what happens on Venus or on other similar bodies? *Technology*? The same space missions made possible by technology have also made important contributions to the further development of technology. *Biology*? What can be more important than studying how life evolves in conditions different from terrestrial ones, or investigating why the dinosaurs disappeared, or searching for life in the Universe? *Pharmacology*? More and more sophisticated space experiments tackle the production of medications or life-saving substances in cosmic conditions.

12.7 Astronomy and art

This fairly new field is a particularly rich source for researchers interested in the role of astronomy in culture. As early as prehistoric times, painters on cave walls displayed considerable knowledge of the constellations and recorded celestial phenomena. Cross-culturally, folk artists incorporate symbolic images of the Sun, Moon, and stars. Cosmic imagery is a common motif in Romanian peasant art, whether on platters, clothes, or even homes. The most spectacular phenomena, comets and eclipses, are ubiquitous in art. Among the most celebrated representations of Halley's comet is the stylized image of its 1066 apparition

Figure 12.1 Halley's Comet, detail from the Bayeux Tapestry, eleventh century. (By special permission of the City of Bayeux.)

woven into the Bayeux tapestry, on display in Bayeux Castle, in Normandy. Commissioned by Queen Mathilde, wife of William the Conqueror, the tapestry, woven between 1073 and 1083, illustrates the victory of her husband over Harold III of Norway at the battle of Hastings (Figure 12.1).

The earliest painting of this most famous comet was done by the Florentine Giotto di Bondone. When a brilliant comet adorned the night sky late in 1301, Giotto painted it in detail on the fresco of the Arena chapel in Padua as a representation of the star of Bethlehem.

Countless other painters have incorporated their awe of the Cosmos into their work. Van Gogh's *Starry Night* is only one such masterpiece.

12.8 Astronomy and music

Kepler's interest in music is famous. He found *musical* harmonies between polyhedrons and planetary orbits. It is also well known that for William Herschel music and astronomy represented two complementary professions.

Numerous musical creations, including Beethoven's *Moonlight Sonata* and Debussy's *Clair de lune*, have been inspired by the Cosmos. What about the *music of spheres*? I am convinced that anyone who has spent at least one night alone with the stars has heard their unearthly music. What is impressive today is that cosmic acoustic waves are a serious topic. Was Kepler right? Was Strauss (whose Blue Danube waltz is so recognizable)?

12.9 Astronomy and heraldry

Heraldry is the science that studies blazons (coats of arms) and their art (herald means harbinger). Although the first blazons appeared only in the tenth century, this fact can be accepted only *stricto sensu*, because surely the ancient Egyptian depiction of the Sun's disk on the head of the Sun God is nothing if not a primitive blazon. Even if we limit ourselves to "official" blazons, we find a lot of cosmic images: the Sun, the new Moon, stars, comets. There are studies dedicated to the interpretation of each heraldic figure. As the blazon was the highest symbol of a family, it is not surprising that cosmic iconography was so prevalent.

12.10 Astronomy and folklore

Cosmic themes are plentiful in the art of all peoples, but my focus here is the customs every cosmic phenomenon generates. I think of eclipses, not only because they were the most important astral events for Europe, and especially for Romania, now, at the turn of millennia, but also because an eclipse is such an impressive show, so despite their rarity, there are no cultures without customs related to eclipse phenomena. For example, in a Romanian chronicle dating to 1764, one finds the following advice: "In the case of a solar eclipse, do not water cattle that day." Some cultures mark eclipses by lighting fires, others by fasting, etc.

From the long list of astronomical influences on folk customs, one recently discussed is the usage of an African tribe (the Dogon, in Mali) that celebrates Sirius every 50 years. What is strange here? Today we know that Sirius is a visual binary with an orbital period of 50 years, but Sirius B was first detected optically in 1862. How did these African tribespeople identify a periodic motion "discovered" only in the nineteenth century? It was briefly thought that a possible explanation is that millennia ago, the luminosity of the companion might have exceeded that of the primary star. It is now widely accepted that the knowledge of Sirius B was brought to the Dogon by visitors from Europe or that the anthropologists who reported the story misinterpreted what they heard from the Dogon.

12.11 Astronomy and literature: *belles lettres* and science fiction

The Cosmos is a recurrent theme in poetry. I am astonished by a question asked with more and more obstinacy: if astronomers unravel the Moon's mysteries, will poets will still find it an intriguing subject for their verses? Will moonlight continue to enchant lovers once the Moon is better understood? From my point of view, just the opposite will prove true. As we learn more and more about the Universe, we are confronted with many new questions that we could not even ask before. We realize that the Universe is infinitely mysterious, that the Cosmos will remain eternally unfathomable. Some people think that science and technology undermine human emotion. I don't agree – although, to be honest, sometimes a glimpse of a contemporary painting or an earful of the music young people adore tempts me to say otherwise! Every generation experiences shock in the face of innovation, but eventually what endures is not the cutting-edge fad but rather the perceptiveness of the human sensibility, shaped, of course, by its host culture.

Literature has always recorded the thrill of the new, the beautiful, the impressive – in general, of everything that touches the soul. I refer here not to the miracle of sunlight or the charm of moonlight but rather to the concrete astronomical phenomena immortalized in the celebrated pages of literature. One need think only, for example, about the eclipses mentioned by Homer (*The Odyssey*, 1178 BCE), Aristophanes (*The Clouds*, the eclipse of 424 BCE), Cicero (about Quintus Ennius, who mentions the eclipse of 400 BCE), Plutarch (71 CE), or, much later, John Milton (*Paradise Lost*, the eclipse of 1667), Mark Twain (*A Connecticut Yankee at King Arthur's Court*), Thomas Hardy (the poem *At a Lunar Eclipse*, 1903).

But there is another literary genre that is even more interesting for many of us: science fiction – SF – literature (or movies). Here the author's fantasy tends to involve us completely. Of course, an author like Jules Verne will always be famous for his imagination and his talent for literature, as well as for the creation of the SF novel. Many other authors, including Isaac Asimov, Stanislaw Lem, Ray Bradbury, Kurt Vonnegut, and Arthur Clarke, knew how to mine scientific knowledge for ideas that inspire the literary creation of utopias and dystopias.

Unfortunately, sometimes the rush for sensation (or big bucks) can lead to the appearance of books and movies based on scientific fallacies. Then, who can differentiate between fantasy and aberration? Who can evaluate the literary value of such productions? In Romania, for instance, young graduates of a humanistic secondary school lack even the most elementary astronomical notions, public scientific conferences are considered the "regrettable heritage of an atheistic–scientific education," and the content of the mass media responds to commercial motives rather than educational ones. Astronomers have their work cut out for them to deal with this problem.

12.12 Astronomy and education

Science fiction is hardly the only or the worst problem. The dissemination of false "scientific information" often leads the public to panic for no reason at all (let us recall, for example, the scare about the "alignment" of planets on May 5, 2000). The press, radio, TV, and the Internet all bombard the public incessantly with astrological nonsense. People have no idea whether the heavenly bodies actually do or do not have any influence on human destiny. The burden is on us to educate the public in the purest sense of the word. Whoever understands the workings of the Universe realizes its significance for humanity, realizes that we must protect our planet, realizes that if the atomic race continues, the moment will arrive when we will actually destroy our planet, much before its natural end.

12.13 Astronomy and mass media

The rapid evolution of the mass media has affected astronomy, especially with regard to astronomical education. On one hand, newspapers and especially TV have exposed the public to specific advances in space science and to unique cosmic phenomena. Those of us of a certain age, for example, will never forget the excitement we experienced in July 1969, while watching Neil Armstrong's first step on lunar soil. In 1999, Romanian television offered a world première: the broadcast of the lunar shadow, from the MIR orbital station, during the total solar eclipse of 11 August. We have been privileged to watch various space missions, orbital rendezvous, astronauts passing from one spaceship to another; we have experienced the excitement of *Apollo* 13 and such tragedies as the explosion of the space shuttle *Challenger.*

The mass media continue to offer unforgettable hours spent alongside the most renowned specialists, moments in which the public has the chance to understand what happens in the Universe. The famous TV series hosted by Carl Sagan and the educational movies on the TV channels Discovery History are only two examples of the positive influence the mass media can have on astronomy education.

Of course, there is another side of the picture: the broadcast of misinformation about astronomical events. Usually misinformation is not countered by a follow-up correction, as, for instance, the announcement concerning the start of the third millennium on January 1, 2000. Even worse, there are dramatic "prophecies," like the prediction that so-called "alignment" of planets on May 5, 2000, would bring great earthquakes and other natural catastrophes. Recall particularly the 1999 total solar eclipse. Nostradamus's vague predictions had been interpreted to apply. The famous astrologer Madame Teissier announced a great catastrophe for that day. The no-less celebrated Paco Rabanne said that the MIR station would fall down over Paris and promised to give up his career if 11 August did not usher in the end of the world. All over the world, newspapers, instead of announcing that the last total solar eclipse of this millennium would have the maximum in Romania, alleged that the Romanians were waiting for the return of Dracula. They wrote that the Romanians, terrified by the apocalypse, took refuge in forests. *Le Figaro* reported a genuine crisis of mysticism in Romania; it alleged that in Bucharest the eclipse would cause epidemics, floods, and catastrophes, but not the apocalypse – what a kindness! The same publication announced that a Colombian, Alonso Manzano, killed his wife before killing himself. He left a short letter, explaining that this was his attempt to outrun the end of the world that would occur at the eclipse. Note that Colombia was not even within the eclipse path!

In Mexico, the ecclesiastical authorities had to intervene in order to lay to rest the rumor that the eclipse would last for three days and would bring the end of the world. Under the influence of such prophecies, many people sold all their possessions and were planning to commit suicide.

To a certain extent, astronomers can take advantage of such situations. Since the public is in a state of anxious alert, they have the opportunity to give scientific explanations of what is really going on.

12.14 Astronomy and astrology

Many people ask: which came first, astrology or astronomy? Given that the earliest human societies harbored a belief in magic, astrology can be considered a step in the right direction. The Olympian gods of the ancient Greeks were very close and also very volatile. Once people realized that the heavens have other dimensions and that the course of the celestial bodies is immutable, they leaped to the conclusion that divine justice, associated with the perfectly regular motion of these bodies, must be rigorous and must influence the Earth. Evidence for the influence lay close to hand in the alternation of day and night, the seasons, and the tides.

Aside from the diurnal and seasonal rhythms, the early sages who observed the skies tried to discover other possible links between Heaven and Earth. Such a study naturally generated *astronomy*. But some of the links they supposedly found and the casual coincidences they promoted as rules led to the development of *astrology*.

The two fields ceased to be synonymous only in the first years of the Christian Era. The Chaldeans and Egyptians, the incontestable masters of astrology, turned astrological doctrines, which began as a caste privilege, into a common asset. The Arabs raised the art of elaborating horoscopes to the rank of a genuine industry.

In the fourteenth century kings had official astrologers; for instance, Nostradamus was the astrologer of Catherine de Medicis and King Charles IX. Even in the twentieth century, some countries' leaders consulted astrologers.

Famous astronomers like Tycho Brahe, Kepler, and Galileo elaborated horoscopes. This fact was often used as an argument in favor of astrology. But Kepler himself recognized that the one who used "the foolish and depraved daughter (astrology) to nourish the wise but poor mother (astronomy)" must not be convicted. The same Kepler asked: "Without this naive hope to read the future in the sky, were you so wise to study astronomy for itself?" Indeed, the necessities of the occult science – astrology – sometimes helped the development of the real science – astronomy. As an example, the need of knowing the exact instant of the birth of significant individuals contributed to a quick spread of the new pendulum clocks. Even the ancient methods of predicting eclipses and establishing accurate positions of stars were related to the needs of astrology.

So Kepler was right. And he was right, too, when he wrote in 1627, referring to Tycho Brahe: "His reason, so sure, knew how to distinguish the *general* effects of the stars from those that would affect the *individual* life of the man. Most of the mortals, believing in miraculous prophecies, do not understand this."

But we are far from that epoch. Today we can say that we have sufficient arguments to prove the quackery of astrology. The outcome is still far from decisive. For many people astrology constitutes only an amusement. But there also are many people who believe in it, especially when astrologers make use of scientific terminology and generate computer-based horoscopes. In a bewildered world, the influence of certain "occult sciences" can be catastrophic. Even if we remain silent about spiritism, fortune-telling, etc., it is our moral duty as astronomers to debunk astrology.

12.15 Astronomy and religion

The association between astronomy and astrology is not so far from the relationship between astronomy and religion. The problem is extremely intricate and constitutes the subject of interesting debates, at least over the course of the last several years. I myself participated in such a meeting in April 2006, in Paris, where astronomers, biologists, theologians, adherents of different religions, even atheists, tried to answer questions human beings have posed since the dawn of time. To tackle the problem of the possible relationships between religion and science – conflict? contrast? contact? confirmation? – we have to define religion as exactly as possible. My own opinion is that there is no conflict between science and religion because they have nothing in common. They are merely two distinctly different ways of considering the world.

The problem is much too complex to be discussed here. As a member of a society where for half a century every religion was taboo, I realize that I lack the religious education necessary to discuss such a problem. This realization doesn't keep me from asking and trying to answer questions about the origins of the Universe, its evolution, as well as the possibility of the existence of other rational beings in the Universe. I shall pose one question that simply shocked me: if there are other rational beings in the Universe, and if God does exist, is God their God also?

12.16 Astronomy and philosophy

Of course, the relationship between astronomy and religion is very close to another connection, perhaps the most important, that is, between astronomy and philosophy. I don't believe it is possible to defend a philosophical doctrine in which the Universe has no place. Having begun my talk with definitions, I will end with definitions, too. What, essentially, is philosophy? It is a cultural domain based on a set of interrogations, meditations, and rational researches on the being, causes, values, etc., which makes use of many different approaches to investigate the relation between human beings and their world.

Philosophy began as a scientific way of thinking about nature and the causes that make possible the existence of the Universe, life, human beings, society. I think it is useless to try to explain the relationship between astronomy and philosophy. Everything that defines philosophy intimately contains the Universe and especially human existence in the Universe.

12.17 Cultures cannot exist without astronomy

The idea that no culture can exist without astronomy has preoccupied me for many years. I also believe that, from this standpoint, astronomy is much more than a science. I am convinced that enumerating the possible links between astronomy and all the other domains of knowledge is far from being exhausted. For us, the most important thing is to understand astronomy in this context, and to try to prove that this oldest science means not only scientific knowledge, but also represents an essential factor for the progress of our society. This aspect is crucial now, in the early years of a new millennium, when it is incumbent on humanity to stop for a moment and ponder not only the steps and missteps it has taken up to now but also, and more importantly, the steps it must take from now on.

Comments

Mary Kadooka: Could we have a copy of your PowerPoint?
Magda Stavinschi: It can be downloaded from my website at www.astro.ro/ASTRONOMY&CULTURE.ppt.
Further comments: This might be put on CD to accompany the proceedings. Permissions, etc. will need to be investigated.
Malcolm Smith: There are matters of copyright and permissions with related expenses, which must be looked into carefully.

Jay Pasachoff: You included so much excellent information that it will be very good to have it in the proceedings, where we can study it at leisure. The Dogon question was looked into carefully a couple of dozen years ago. It was shown that the Dogon had contact with Europeans, so their knowledge of Sirius B was certainly by diffusion from Europe or the whole story resulted from a misinterpretation.

In one of the Harry Potter books, he took an exam in which he studied Orion in June from the Hogwarts Astronomy Tower. Such a sighting is, of course, impossible. Perhaps this accounts for the grade of “B” Harry got in that course, as we learned in the next book.

13

Light pollution: a tool for astronomy education

Margarita Metaxa
Philekpaideutiki Etaircia, 63 Eth. Antistaseos, 15231 Athens, Greece

Abstract: The current educational crisis reflects a general cultural, political, and moral or "spiritual" crisis. Within this environment the role of astronomy education in schools is crucial. After identifying the problems we face while teaching astronomy, I offer a number of proposals for teaching astronomy in a modern classroom that build on the problem of light pollution. This problem exists nearly everywhere and is still growing rapidly. Maintaining dark skies not only at prime astronomical locations but also elsewhere depends on the awareness of the public, particularly of key decision-makers, including lighting engineers. Continual promotion of awareness of light pollution and its effects is necessary. If we wish to preserve the astronomical environment, we must promote effective education with respect to this problem.

13.1 The educational crisis and the importance of astronomy education

At the heart of the teaching crisis lies the problem of unmotivated students. From being in the field of education as a student, tutor, lecturer, and teacher, and after analyzing the large number of learning theories that have unfolded, I believe that an education should provide a balanced foundation, which should be achieved regardless of the specific area of study. This foundation should lead any individual learner towards achieving his/her fullest potential. Such a foundation, for example, would enable a graduate student to succeed in an environment that many educational specialists around the world call a "rapidly changing environment." Fundamentally, this set-up treats the teacher as the mentor and coach of the student who is actively engaged in the educational process. In an educational model that views students as individuals at the beginning of a life-long intellectual adventure within a constantly changing society, students become the primary actors in the educational process rather than the dutiful audience of the teachers.

Astronomy education can easily focus on preparing a graduate for a "rapidly changing environment" for a "life-long learning" journey. Because of its practical applications and its philosophical implications, astronomy education is deeply rooted in the history of virtually every society. It provides a window for introducing scientific discipline in general to students. Astronomy education occurs not only inside the formal classroom but also when people visit planetaria and museums, read relevant newspaper articles or popular books on astronomy, watch television shows or hear radio programs on related topics, visit astronomical sites on the Internet, or view the night sky while on an overnight hike. Students may, in fact, be more influenced by such "informal" education than by what teachers impart in a formal classroom setting.

Innovation in Astronomy Education, eds. Jay M. Pasachoff, Rosa M. Ros, and Naomi Pasachoff. Published by Cambridge University Press.

13.2 The problems of astronomy education and the learning process

The problems of astronomy education permeate all levels of education in all countries of the world. We thus have to:

- change the traditional way of teaching astronomy,
- help teachers with the materials/ideas available.

Students rarely have the foundation that we expect; they hold misconceptions about the physical world that actually inhibit the learning of scientific concepts. Additionally, students are not able to remember as much as we think they can, and they lack a general framework in which to keep the knowledge they do glean. Today's students are not younger versions of ourselves. They plan to pursue other careers and do not share the the same passions, ways of thinking, and interests that we have. In addition the environment in which they are growing up is completely different from the one in which we were reared. If we fail to adjust our teaching methods accordingly, we do our students a grave disservice.

13.3 Other ways of teaching astronomy: using light pollution as a tool

The problem of light pollution offers an excellent educational opportunity to introduce the topic to our students, and through them also to lighting professionals, government officials and functionaries, as well as the public at large. We present here the model that Commissions 46 and 50 and the International Dark Sky Association (IDA) have found to be an effective way of teaching astronomy.

The model is a collaborative one, involving teachers, researchers, and students, with each partner working in his/her area of specialty. A framework is established where scientists can contribute their unique assets with potential large impact and multiplier effect, taking advantage of an existing end-user support structure.

In order to optimize student learning and build understanding and knowledge that outlasts the final exams, our criteria and goals should be realistic. We should teach prerequisite notions prior to more exotic ones.

13.3.1 The program

The program:

- covers all levels of education, formal and informal,
- is multi-disciplinary (involves: physics + astronomy + technology + environment),
- depends on collaboration to manage scientific work, and
- stresses both scientific and social components.

Through the program students, the public, etc.:

- familiarize themselves with the problems of light pollution through astronomy, physics, and computer science,
- consider the cultural and social dimensions of the impact of light pollution, and
- appreciate the preservation of the heritage and environment throughout their country.

13.3.2 Activities connected with light pollution: an introduction to astronomy education

Through the various activities that we propose, we intend our students and the general public to be encouraged to become familiar with the night sky and to treat the Universe as a whole as our home, the environment we belong to in the broadest sense. Additionally, we intend to increase awareness of the effects of light and air pollution and to attempt to influence planning authorities to produce efficient and effective lighting schemes. Well-designed lights will not only cut down light pollution but will also save energy.

The activities proposed include astronomy projects that familiarize students with the night sky (through visual or photographic exercises) as well as an introduction to stellar spectra. The potentially adverse effects of lighting projects, technology projects, and environmental projects are evaluated. These activities can be found on the IDA's web-pages: www.darksky.org.

In order for the program to be effective, students should be divided into groups based on their preferences and abilities. For students to decide exactly what the focus of their area of study will be, they will need to engage in brainstorming. To complete the proposed activities and to understand the status of light pollution in their local area, they will need to work collaboratively. They will thus find themselves engaged in the scientific way of working!

Students and teachers who participate in the program's activities should make their results on the light pollution problem available to the public by informing and communicating with local authorities, environmental associations, and scientific societies. By doing so, they contribute to the formation of the "critical mass" that will finally influence planning authorities to produce efficient and effective lighting schemes. This approach has already been effective in many cases.

By engaging in these activities, each student discovers how the intellect, emotions, body, desires, intuition, and imagination need to interact in order to have effective motivation and learning.

13.3.3 Adaptation to school curricula

By incorporating physics, astronomy, technology, and environmental studies, this program makes it easier:

(a) for students to increase their understanding in astronomy and develop useful skills,
(b) for teachers to implement astronomy projects using the environment as a generalizing background that can then be connected to different subjects.

Table 13.1 below presents the key concepts that exist in almost every curriculum, according to the level and the subject.

Table 13.1 *Physics and astronomy in natural science complex*

Educational level	Physics level	Astronomy level	Environmental level
Elementary school	Introduction to experiments, observations	Time, seasons	Surrounding world
Middle school	Introduction to interactions, motion, waves	Solar System, place of the Earth	Natural sciences
High school	Interactions, e/m radiation, light sources	Introduction in astrophysics	Ecosystem

13.4 Theoretical framework

Effective teaching in the contemporary world demands a connection with events in our everyday lives. This connection is successfully provided by the UNESCO model for environmental studies, which requires that each project be classified as **natural, historical, social**, or **technological**. The goal of this model is "the development of citizens/people with knowledge, sensitivity, imagination, and an understanding of their relationship with their physical and human environment, ready to suggest solutions and participate in decision making and implementation." The model fits well with a light-pollution project, and we strongly recommend it.

13.5 Current projects – accomplishments

Our educational efforts took place all over the world.

(a) This model was first implemented by the Greek light pollution educational program, which had been arranged through the Greek Ministry of Education and Religion with support and finance from the EU. The two-year program, which ran from 1997 to 1999, was a proposal of the Astrolaboratory of the Arsakeio of Athens.

In November 2000, people from all over Europe took part in the continent's biggest educational and cultural event on the Web – the "netd@ys project." Both the Greek Educational Project (selected as one of the three "labelled projects" for Greece) and the Internet Forum on Light Pollution have been connected to the netd@ys project. An international UNESCO-backed conference on "Youth and Light Pollution" was held in Athens in November 2003.

(b) Educational efforts in Chile, led by Dr. Malcolm G. Smith, Director of CTIO/NOAO, in collaboration with the active group RedLaSer (**Red de Estudiantes de La Serena**), were very impressive. Various groups (schools network, University of La Serena, Mamalluca Municipal Observatory, US National Science Foundation, AURA Observatories) collaborated to bring the Gemini portable planetarium program to over 85 000 Chilean schoolchildren from over 100 localities in the first four years of operation. The program includes a dramatic demonstration of the effects of light pollution on the visibility of stars in the night sky. Teachers in Hawaii and in La Serena visited each other's schools in 2003 under the Gemini/RedLaSer/"Sister-city" program "StarTeachers"; Internet2 connections are being used for a variety of international videoconferences related to light pollution now that AURA's observatory in Chile is connected. The Astronomical Society of the Pacific's "Project ASTRO," through the excellent outreach team at the US National Optical Astronomy Observatories in Tucson is developing programs over this link, including "ASTRO-Chile," involving Spanish-language videoconference exchanges between teachers in Tucson, Arizona, and La Serena, Chile.

In addition, the Office for the Protection of the Quality of the Sky (OPCC) was given the task of making pragmatic improvements to at least one significant lighting project in each of the IInd (Paranal), IIIrd (Las Campanas) and IVth regions (La Silla, Tololo and Pachon) of Chile. The OPCC is a joint project of AURA, ESO, CARSO, and CONAMA (the Chilean National Environment Commission).

(c) An extraordinary Internet 2006 Star-Hunting Party, involving over 18 000 people from 96 countries on all continents, reporting 4591 nighttime observations, took place at www.globe.gov/GaN/index.html. The program was sponsored by the National Optical Astronomy Observatories, Windows to the Universe, CADIAS and ESRI.

communication of this sort can increase the perceived isolation of the students, and reduce the tutor's ability to establish a rapport with individuals.

On the other hand, we are developing our use of the discussion forums in particular. In principle these can allow students to interact more naturally, and enable peer learning and more exploration of the subject matter. Tutors have gradually developed their own skills, realizing that there is a fine line between guiding and dictating discussion. Tutors are seen as experts in the field (as well they should be), and an untimely or inappropriate intervention can stifle a useful discussion. The experience with e-mail (which came into general use before the discussion forums) has trained the tutor into giving a full and in-depth answer, an approach that needs to be unlearned for the forum setting.

14.5 The student body

The diversity of our student body reflects our success in providing a flexible yet challenging suite of modules. There is no typical student in our courses, but a selection of types gives an idea of the students we attract:

- retired industrial professional with a PhD in Chemical Engineering,
- English teacher with a keen interest in astronomy,
- employee of an examinations board with responsibility for schools astronomy curriculum,
- high-school student studying for University entrance exams,
- primary-school classroom assistant working towards a degree,
- an "empty nester" parent,
- sales person travelling the world on business,
- leading light of a local amateur astronomy society,
- employee of a science center looking to gain IT qualification with relevance to their interests.

Alongside this diversity of students is a corresponding diversity of needs and aims. Clearly all the students are initially attracted to these courses by an interest in astronomy. However, some are wanting to engage their minds, while others have specific applications in mind. These might be to enhance their teaching, to change careers, or to improve their prospects in university entrance competition. Understanding and addressing these differences may well be key to the future development of the programme and our ability to adapt to the changing market.

14.6 Future directions

Currently we are looking at how we might further develop our programs and take advantage of our experience. This evaluation also includes the systems we have established to support the distance-learning process, which are largely transferable to other courses.

14.6.1 Full degree course in the UK context

The UK higher education community is experiencing a new regime, with the progressive introduction of competitive tuition fees. While still highly regulated, higher education is widely perceived as becoming more expensive for the student, and many are looking to minimize these costs. Two areas that we might be able to take advantage of are in the desire to remain at home and the lower price of part-time programmes under the UCLan fees policy.

Thus with the launch of a full BSc (Honours) programme in astronomy, we are exploring the opportunities to attract students from the traditional 18 to 21 year old UK degree market.

14.6.2 University entrance competition

Against the background of a new fees regime, competition for places at the most prestigious institutions is increasing. In addition there is a perceived crisis in the national university entrance exams, in that it is very difficult to differentiate the most able students. Consequently many students studying for university entrance are looking at ways to make them stand out from their peers. Our entry-point modules provide one mechanism by which students might demonstrate their suitability for particular university courses. After initial experience in a single school, we are now looking at developing our presence in the post-compulsory school and college sector.

14.6.3 Distance-learning courses

Further distance-learning courses beyond the astronomy area are also being considered. Situated as we are in a Department of Physics, Astronomy and Mathematics, we are looking at using our experience to enter the teacher-training market, initially filling a shortage of mathematics and physics high-school teachers. These programmes would be complementary to more traditional locally run courses, and cater to teachers who for whatever reason cannot attend face-to-face programmes.

14.6.4 Curriculum developments

Our curriculum development is driven primarily by student demand. In particular we have identified a decline in the take-up of our entry point modules. With this in mind, we are developing an astrobiology thread, drawing on a broader range of interests than the purely astronomical.

14.6.5 Delivery and support

As virtual learning environments develop, there will be more opportunity for a richer learning experience. Future developments envisage more interactivity both with learning object and with fellow students. Tutors will be able to provide more tailored support, and also encourage student interaction. With the development of Virtual Observatory it will also be easier to expose our distance learning students to contemporary astrophysics and genuine research work. All these possibilities are made possible by the wide spread penetration of the Internet, and we will be experimenting in a developing landscape of communications options.

14.7 Conclusion

The astronomy by distance-learning programs at UCLan have developed organically over a decade. They have taken student preferences and needs into account, while at the same time maintaining a clear and strong academic lead. The result is a suite of modules that can cater to both short-term learners and those determined to acquire higher awards. The programmes are both flexible and rigorous, and able to meet a wide range of students. We have learned lessons about learning materials, student support and course management systems that are invaluable to running such a program effectively, and satisfying the students needs.

S. Eyres was supported in presenting this paper by a Newly Appointed Lecturer Award from the Nuffield Foundation.

Comments

Andre Lipand: What are the requirements for enrolling into the courses and what are the associated costs?

Stewart Eyres: To enroll, one must have GCSE level (school leaving, approximately age 16) physics/math. For people from the UK, the cost is £205 per module; for people from outside the UK, the cost is £230 per module for EU students and £340 for others.

15

Edible astronomy demonstrations

Donald Lubowich

Department of Physics and Astronomy, Hofstra University, Hempstead, NY 11549, USA

Abstract: By using astronomy demonstrations with edible ingredients, I have been able to increase student interest and knowledge of astronomical concepts. This approach has been successful with all age groups from elementary school through college students, and the students remember these demonstrations after they are presented. In this paper I describe edible/drinkable demonstrations and present three edible demonstrations I have created to simulate differentiation during the formation of the Earth (and planets); the expansion of the Universe, and radioactivity. Sometimes the students eat the results of the astronomical demonstrations. These demonstrations are an effective teaching tool and can be adapted for cultural, culinary, and ethnic differences among the students.

15.1 Introduction

In this paper I will present some examples of how to use edible ingredients or materials in astronomy and space science demonstrations. I describe some edible demonstrations to illustrate the following concepts: differentiation; plate tectonics; convection; mud flows on Mars; formation of the galactic disk; formation of spiral arms; curvature of space; expansion of the Universe; radioactivity and radioactive dating. Some of the demonstrations require prior cooking, baking, or other preparations.

I will describe in detail three demonstrations that I have performed in my introductory astronomy classes and with elementary school children using chocolate or chocolate milk to demonstrate density and differentiation during the formation of the terrestrial planets; using a cookie to show the expansion of the Universe; and using popcorn to simulate radioactivity.

Other edible demonstrations include simulating mud flows on Mars with melted chocolate on cake (angel food cake, which is porous and white, works well); simulating the formation of craters with a round candy or M&Ms falling into frosting; simulating the collapse and formation of the galactic disk by tossing and rotating a pizza (also discussed by Charles Liu in the April 2003 issue of *Natural History*, American Museum of Natural History, www.findarticles.com/p/articles/mi_m1134/is_3_112/ai_99818082); and simulating plate tectonics and earthquakes by colliding crackers with layers of peanut butter, jelly, or cream cheese to show the collision of plates and how one layer can go underneath another layer (suggested by Hillary Olsen). A chocolate volcano (molten chocolate cake) can also be used to show volcanic activity (see recipe from the Chef2Chef website forums.chef2chef.net/showflat.php?Cat=&Number=436798). A pepperoni pizza or a fried egg can also be used as a model of the Milky Way. The formation of spiral arms can be illustrated with cream slowly poured into stirred coffee or hot chocolate. (Make sure to pour in a different colored liquid from the coffee or hot chocolate.)

Innovation in Astronomy Education, eds. Jay M. Pasachoff, Rosa M. Ros, and Naomi Pasachoff. Published by Cambridge University Press.

Figure 15.1 Differentiation with floating cereal and sinking M&Ms (photo by Jay M. Pasachoff).

15.2 Differentiation and density

I demonstrate differentiation and density by using melted chocolate, chocolate milk, or custard as a base. I then drop in two foods with different densities so that one floats on the top and one sinks to the bottom of a glass bowl, beaker, or pan so that the density differences are easily visible to the students. In more advanced classes one can measure the density. I use three different densities. For this demonstration I melt chocolate and stir in nuts or candy (anything that will sink to the bottom; M&M peanuts or almonds work fine because the colored candy coating is easily visible) and marshmallows (or anything that will float; candy sprinkles or stars will also work). Sometimes I add a little cream or butter to the chocolate to make it less dense and more fudgey. It is often easier to cut and serve fudge than thick chocolate. After it cools and hardens I show the separation into different layers and describe the method by which the Earth forms to explain why the Earth has a nickel-iron core but sand at the surface. I also discuss the concept of density and how when rocks are heated they can become liquid and molten. I have also done this demonstration with chocolate milk or milk with a cereal that floats on top (Figure 15.1).

I have also melted and refrozen ice cream with different added ingredients to show differentiation. Rocky road ice cream works well because of the nuts and marshmallows, but any ice cream with at least two added ingredients of different densities should work.

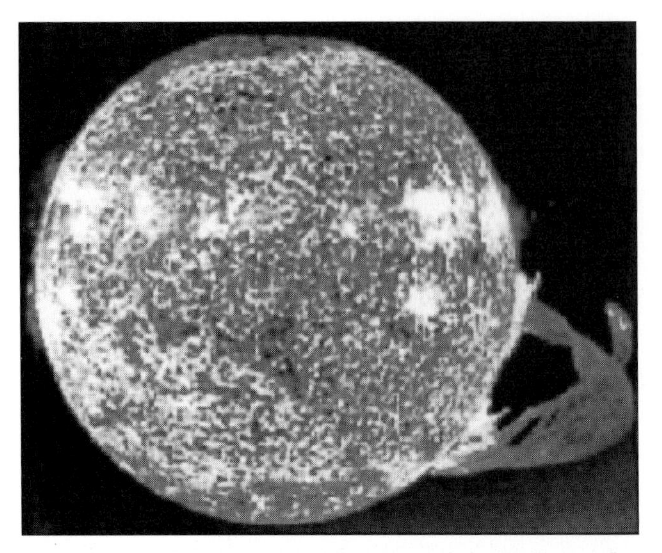

Figure 15.2 Solar prominence and supergranulation UV image from Skylab space station, Dec. 1973, starchild.gsfc.nasa.gov/Images/StarChild/solar_system_level2/sun_skylab.gif.

Although more difficult to make and prepare, custard or pudding can be used in a similar demonstration of differentiation. Because custard is made with eggs, milk or cream, and sugar, it can be quickly and easily made in a classroom or laboratory as part of the demonstration. Custards or puddings such as flan, zabaglione (or sabayon), crème brûlée (French), crema catalana (Spanish), or burnt cream (English) can be used with fruit, berries, nuts, or candy pieces, plus whipped cream or marshmallows floating on top, to show the multiple layers and to illustrate differentiation. Candy-coated chocolate candies such as M&Ms work well because the bright-colored coating is easy to see. The top layer of sugar forms a crust with a high-temperature source such as a hand-held (propane) torch. The top crusty layer can also be used to show earthquakes or starquakes. Crème brûlée is shown in www.cookingforengineers.com/hello/259/958/640/IMG_3335_sharp.jpg.

15.3 Convection

Convection is an important phenomenon in geology, atmospheric physics, solar physics, and stellar interiors. Important examples of convection include plate tectonics, storms and wind flows, solar granulation (Figure 15.2), and stellar interiors (Figure 15.3). One of the best edible/drinkable demonstrations of convection and plate tectonics is the Hot Chocolate Mantle Convection Demo by Philip Medina, using Swiss Miss powdered-milk chocolate (www.mrsciguy.com/convection.htm), shown in Figures 15.4 and 15.5. One can also stir the cocoa and serve hot chocolate after the demonstration is completed.

One can also simulate convection in a container using miso soup (soybean soup), as described in the Jet Propulsion Laboratory Educator's Guide to Convection (education.jpl.nasa.gov/educators/convection.html); using mica flakes in boiling water (www.earth.northwestern.edu/people/seth/Demos/CONVECT/convect.html), described by Seth Stein and John DeLaughter; and cold cream slowly poured into a cup of hot chocolate, tea, or coffee (also demonstrating upwelling), described by Brien Park (www.nps.gov/archive/brca/Geodetect/Plate%20Tectonics/convection%20currents.htm). Bubbling in any thick liquid,

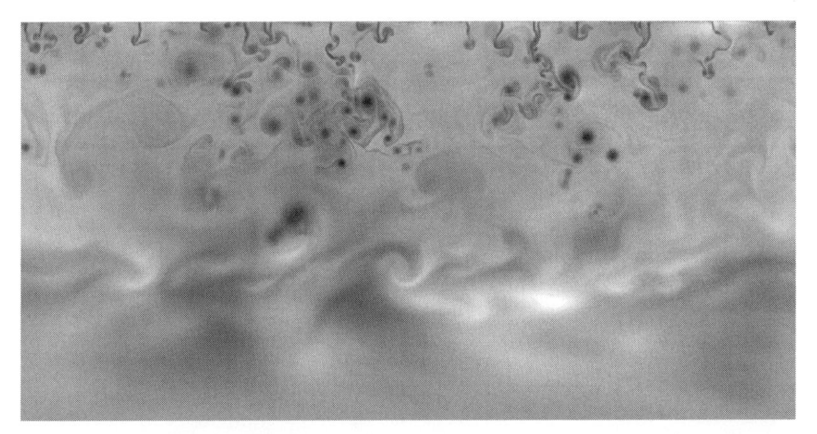

Figure 15.3 Turbulent mixing. Convective penetration of unstable gas into a layer of stable gas in stellar interiors, astro.uchicago.edu:80/rranch/Computing (includes a movie).

Figure 15.4 Starting the hot chocolate convection demonstration.

such as oatmeal or hot fudge, can also be used to demonstrate convection and solar granulation.

15.4 Curvature of the Universe

The curvature of the Universe can be described as open, flat, or closed, as shown in Figure 15.6. Flat or closed Universe models are more easily understood by students than a negative curvature (open Universe) model, which is a model of an accelerating Universe. Pringle potato chips (www.pringles.com/pages/index.shtml and encyclopedia.thefreedictionary.com/Potato%20chips) or crisps (suggested by Dr. Steve Lawrence), are an edible example of a negative curvature.

15.5 Expansion of the Universe

The traditional demonstration of the expansion of the Universe involves blowing up a balloon, with "galaxies" glued onto the surface or drawn onto the surface of the balloon. The second traditional example of the expanding Universe is baking raisin bread. Often a diagram, figure, picture, or a computer animation of expanding raisin bread is used, as shown

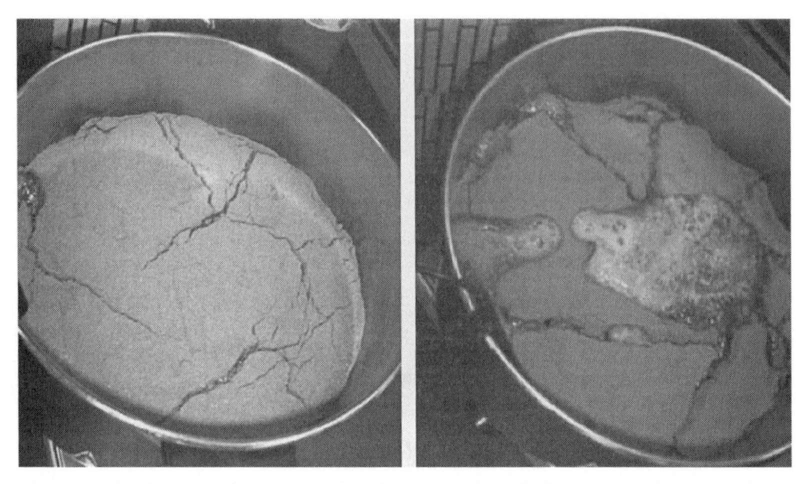

Figure 15.5 Hot chocolate convection demonstration. Left: mantle, plates, cracks. Right: plates, cracks, and "lava."

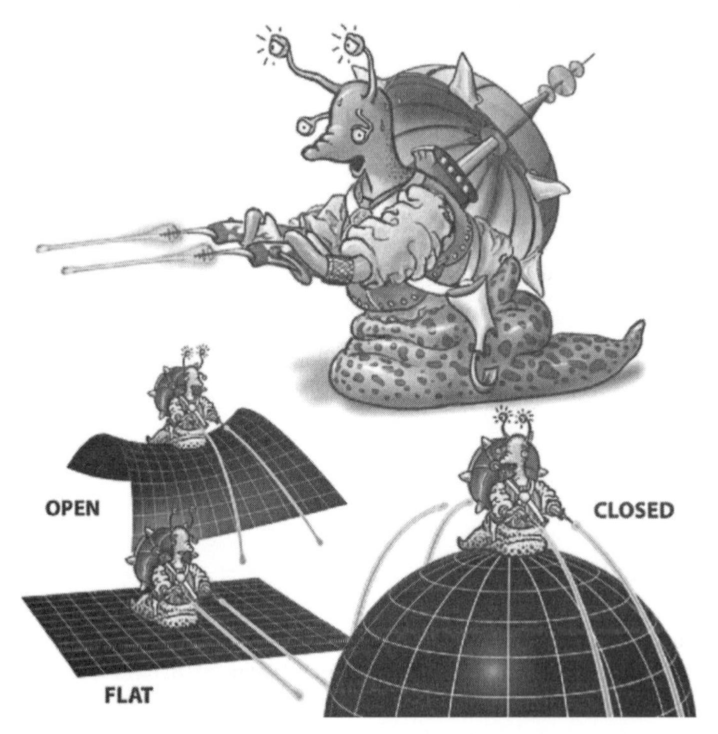

Figure 15.6 The geometry of space-time in the Universe, map.gsfc.nasa.gov/m_mm/mr_content.html (WMAP site). (Courtesy of Wilkinson Microwave Astronomy Probe, NASA's Goddard Space Flight Center.)

in (wmap.gsfc.nasa.gov). Instead of raisin bread, I bake a very large cookie, 25 cm in diameter, with nuts, candy pieces (sprinkles, stars, or hearts) or chocolate chips/chunks/M&Ms placed in rows on the surface. I compare the baked cookie to unbaked cookie dough containing similarly spaced nuts, candies, or chocolate chips. Although not three-dimensional like

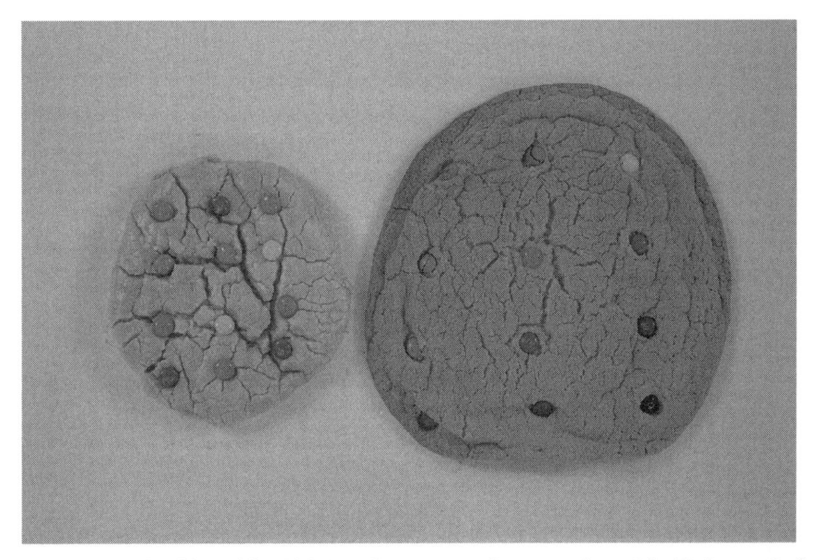

Figure 15.7 Cookies with M&Ms to demonstrate the expansion of the Universe. Left: raw cookie dough with M&Ms separated by 3.5 cm. Right: baked cookie with the separation of the M&Ms 7.0 cm. (Photo by Donald A. Lubowich.)

bread, the cookie spreads out, enabling students to see the expansion effects (Figure 15.7). I have used both homemade and store bought cookie dough. I use cookies because they can be baked in a short time in a microwave oven. I have also baked bread or cake with chocolate chips for this demonstration. One can also bake "big-bang" brownies. (Use a more cake-like recipe, which includes baking soda or baking powder, so that the dough expands more.)

15.6 Popcorn radioactivity

I use popcorn as an example of radioactivity and show how one can measure a "popcorn half-life" and determine how long the popcorn was popped. I then use this result to explain radioactive dating and how it is used to determine the age of rocks (the oldest Earth and Moon rocks ~4.4 to 4.5 Gyr). Recently the age of stars has been measured from the detection of stellar uranium and thorium abundances. (The age determined from the U/Th ratio is 13–15 Gyr.)

One can use a pot or a commercial popcorn maker for the demonstration. I use a Toastmaster air popcorn popper, made by Salton, available at many department stores, kitchen and bath stores, and Amazon.com, in the $12–$15 price range. I suggest an air popcorn maker that does not use any oil. It is cleaner and easier to use to determine the number of popped and unpopped kernels. In this model the popcorn comes out of the top and is caught in a container to be analyzed. Other types of popcorn poppers make the popcorn in a closed container that is part of the popcorn maker. Although this is easier to use if you are making popcorn to eat, the open system is better for a classroom demonstration. In order to better calibrate your results, it is important to use a popcorn popper with an on-off switch, such as the Toastmaster air popcorn maker. Some more expensive popcorn poppers have a lever to stir the popcorn for more uniform heating. My procedure is as follows.

Table 15.1 *Radioactivity simulation with popcorn*

time popped	number popped	number unpopped	total number of kernels
0 s	0	100	100
10 s	35	65	100
20 s	85	15	100
30 s	95	5	100

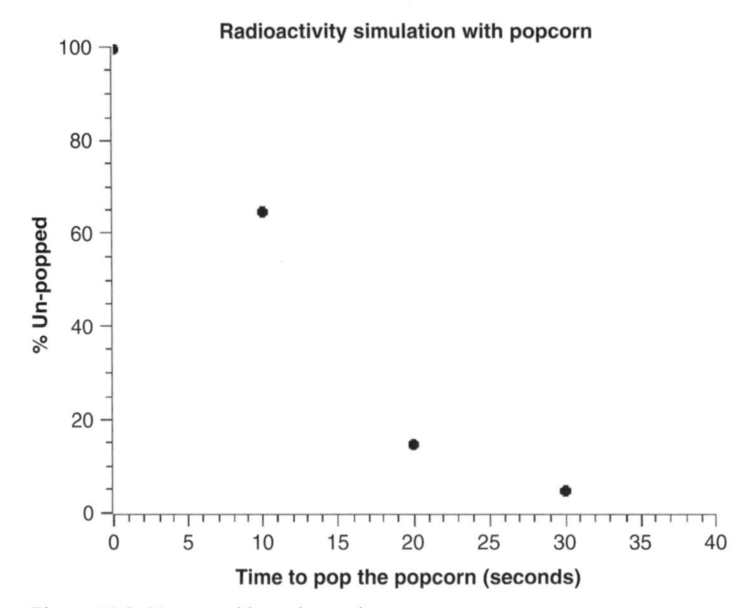

Figure 15.8 Unpopped kernels vs. time.

(1) Show the students one bag of 100 fully popped kernels and another bag of 100 partially-popped kernels. Ask them if they can determine which bag has been popped longer. The students will then understand that if the popcorn is popped for a longer time, then more kernels will be popped. Ask them if they can estimate how long the partially popped bag of popcorn was popped.
(2) Have students count out groups of 100 kernels. In a longer class I use groups of 300 kernels to obtain better statistics and a longer half-life.
(3) Turn on the popcorn maker and wait until it gets hot. Always start the timing when first kernel pops. Ask if the students can predict which kernel will pop first.
(4) Have students count the popped kernels and unpopped kernels. This is not as easy as it seems because there will be some unpopped and some partially popped kernels. One must make a judgment if a specific kernel has popped.
(5) Plot the percentage of unpopped kernels as a function of time. (One can also plot the percentage of unpopped kernels.)
(6) Estimate a popcorn half-life.
(7) Test this prediction by popping 100 kernels for the estimated half-life and determine measured half-life.

(8) Pop another 100 kernels for a time known to the instructor but not the students. Have the students determine the number of popped and unpopped kernels and then estimate the time that the popcorn was popped.
(9) Compare this time estimate with the actual time for popping this group of 100 kernels. Then explain that this method is similar to the method scientists use to determine the age of rocks or stars.
(10) Make popcorn for the class. Bring salt, paper plates, and butter-flavored salt.

The data for one demonstration are shown in Table 15.1 and the results are plotted in Figure 15.8. I estimate that the popcorn half-life is approximately 11 seconds. The short popping times are because I used only 100 kernels. In a longer class I would use groups of 300 kernels and obtain more data points. I then compare Figure 15.8 with the exponential decay curve for a radioactive element.

I explain that since no one can determine which kernel will pop first (or the order of popping), the popping is a random statistical process, and one can only obtain information from a large number of kernels (atoms). I then explain that radioactive decay is a spontaneous random statistical process whereby an atom emits a particle and changes into a different atom at a known rate. The rate described by the half-life ($\tau_{1/2}$) is the time for half of the atoms to decay into another atom. For example, ^{235}U decays to ^{207}Pb with $\tau(1/2) \sim 700$ Myr and ^{238}U decays to ^{206}Pb with $\tau_{1/2} \sim 4.5$ Gyr.

For the more advanced classes, I discuss different isotopes and radioactive decay pathways (mother/daughter), the number of disintegrations/s $\sim$ activity; the exponential decay law $N(t) = N_O(t) \exp(-\lambda t)$; and carbon dating.

15.7 Conclusion

Edible astronomy demonstrations provide a unique way to interest students and to help them remember astronomical and physical concepts. These demonstrations can be adapted for different cultures and ethnic groups. These demonstration have been successful with all age groups, and are an effective teaching tool. Be creative, create your own edible demonstrations, and have fun teaching astronomy.

Comment

Bill MacIntyre: Just a reminder that the use of food is *taboo* with *Pasifika* students, as it does go contrary to the Samoan, Tongan, etc., cultures.

From the Chandra X-ray Center's blog

At the Chandra booth in the exhibit hall [at the January 2008 American Astronomical Society meeting in Austin, Texas], we put out a 1.5 gallon jar and filled it with 5100 jelly beans. The trick was we didn't just fill the jar with randomly color jelly beans. Instead, we made 96% of them black (that's a lot of black licorice) and 4% of them multicolored. Why, you might ask? Well, the black jelly beans represented the "dark Universe" that consists of 70% dark energy and 26% dark matter, with 4% being "normal" matter that includes galaxies, stars, planets – everything else. We held a contest to see who could calculate (or guess) the number. The one who had the closest to the actual number won the jelly beans and a Chandra calendar. We had over 80 entries – largely from the younger set at the meeting – so we think it was a success. Whether the winner will actually eat all those black licorice beans is another question for another blog. See a picture, read more, and find a mathematical exercise on the jelly bean Universe, chandra.harvard.edu/blog/node/47.

16

Amateur astronomers as public outreach partners

Michael A. Bennett

Executive Director (retired), Astronomical Society of the Pacific, 390 Ashton Avenue, San Francisco, CA 94112, USA

Abstract: Amateur astronomers involved in public outreach represent a huge, largely untapped source of energy and enthusiasm to help astronomers reach the general public. Even though many astronomy educators already work with amateur astronomers, the potential educational impact of amateur astronomers as public outreach ambassadors remains largely unrealized. Surveys and other work by the Astronomical Society of the Pacific in the US show that more than 20% of astronomy club members routinely participate in public engagement and educational events, such as public star parties, classroom visits, work with youth and community groups, etc. Amateur astronomers who participate in public outreach events are knowledgeable about astronomy and passionate about sharing their hobby with other people. They are very willing to work with astronomers and astronomy educators. They want useful materials, support, and training. In the USA, the ASP operates "The Night Sky Network," funded by NASA. We have developed specialized materials and training, tested and used by amateur astronomers. This project works with nearly 200 local astronomy clubs in 50 states to help them conduct more effective public outreach events. In just two years it has resulted in nearly 3600 outreach events, reaching nearly 300 000 people, In this presentation we examine key success factors, lessons learned, and suggest how astronomers outside the US can recruit and work with "outreach amateur astronomers" in their own countries.

16.1 Introduction

During the 1990s, as the Astronomical Society of the Pacific's nationwide Project ASTRO began to take shape, we discovered an interesting unanticipated result. Originally, we assumed that most of the astronomers who would volunteer to serve as partners for teachers would be professional astronomers. Instead, nearly 50% were amateur astronomers. As partners, they were as successful as professionals, and often more so. In general, they were very knowledgeable about astronomy, and very passionate and committed as well. We began to realize that amateur astronomers might be a very important force as volunteer science educators.

In 2002 the ASP began a US-wide program to help amateur astronomers conduct more effective, more organized public outreach programs. The program has been very successful and continues to grow. Although our experience and data are from the United States, we believe our conclusions may apply to amateur astronomers in many different countries. We believe that amateur astronomers represent a huge, largely untapped source of "outreach energy." We encourage astronomers and astronomy educators around the world to consider more formal cooperation with amateurs to increase the quantity and quality of public outreach programs conducted by amateurs.

Innovation in Astronomy Education, eds. Jay M. Pasachoff, Rosa M. Ros, and Naomi Pasachoff. Published by Cambridge University Press.

16.2 The amateur astronomy community in the United States

How many amateur astronomers are there in the US? The answer depends in large part on how one defines "amateur astronomer." The combined unduplicated circulation of the two large monthly astronomy magazines, *Sky and Telescope* and *Astronomy*, is approximately 300 000. But it is safe to say that not everyone who reads a monthly magazine can be called an amateur astronomer. Instead, we chose a somewhat limited definition of amateur astronomer – one who has joined an amateur astronomy club. There are about 675 known amateur astronomy clubs in the US, with an average membership of 75. Thus total club membership in the US probably exceeds 50 000. We refer to these individuals as "affiliated amateur astronomers." Amateur astronomers as a group are quite knowledgeable about astronomy (Berendsen, 2005).

16.3 Public outreach activities of amateur astronomers

In 2002 the ASP conducted a web-based survey of US amateur astronomers, funded by a planning grant from the National Science Foundation (Storksdieck *et al.*, 2002). The survey was available from March through May. It was well-advertised in a variety of publications normally seen by amateur astronomers. Following the survey, we conducted six follow-up focus groups of amateur astronomers to gain further insight into the results. There were 716 completed surveys, and a total of 38 people participated in the six focus groups. We emphasize that this study is not statistically valid and is almost certainly affected by self-selection effects. Nevertheless, it produced some suggestive conclusions.

In the US, the average size of an astronomy club is 75 people. Approximately 25% of the respondents say that they participate in some sort of club-sponsored public outreach activity. Public outreach is defined as a program or event involving individuals who are not members of the club. Of the "outreach amateur astronomers," 17% say that they are involved with a public outreach event four or more times per month; 20% said two times a month; 34% said one time per month; and 30% said one to seven times per year. So, of those doing any outreach at all, more than 70% participate in at least one event per month. Conservatively, US affiliated amateur astronomers participate in some 50 000 public outreach events per year. Using other data suggesting an average attendance at all public events of approximately 100, we estimate that amateur astronomers reach some 500 000 members of the general public every year.

16.4 Types of public outreach audiences and venues

The most common public outreach venue is public star parties organized by the club (mentioned by 82% of respondents), closely followed by classroom visits (81%). Other common venues are other types of public events (such as community fairs) and museum/science center events. The most common audience types mentioned were the general public (71%), school children (70%), families (50%), and youth/community groups (48%). More events are held during daylight hours than during darkness, and about half of all events (both daytime and nighttime) involve actually looking through a telescope.

The content covered by amateur astronomers covers virtually the entire gamut of astronomy. Not surprisingly, sky watching, sky tours, and telescope use rank very high. General astronomy, the Solar System and space exploration, current events in astronomy and space exploration, cosmology, dark sky issues, and constellation stories were most frequently mentioned.

16.5 What do "outreach amateur astronomers" want and need?

In both the survey and the follow-up focus groups, we asked amateurs what support or help would be most useful in assisting their public outreach efforts. Their most consistent concern was the time it takes to search for new and useful information. They felt that tested, preselected materials, visual aids, and simple activities would be most useful to them. They also wanted more content knowledge, and help with presentation skills.

16.6 A first step – the NASA Night Sky Network

Conceived and implemented so far by the ASP, the NASA Night Sky Network (NSN) is a project designed to help amateur astronomers conduct more effective public outreach activities of all kinds. With initial funding from the Navigator Public Engagement Program at NASA/JPL, NSN was publicly launched in March 2004. It is currently funded by a coalition of NASA SMD missions and EPO forums.

The NSN provides amateurs with tested Outreach ToolKits on specific topics that can be used in a wide variety of ways with many different types of audiences. It also provides training in their use. At present, the program is aimed at amateur clubs rather than individuals, in order to maximize efficient use of the Outreach ToolKits. All materials and training are free. Clubs must apply and meet certain minimum requirements in order to become a member of the NSN. Individual club members are authorized to use the materials and log events. In order to maintain its membership, a club must log at least five public outreach events per year in which NSN materials are used.

So far, five Outreach ToolKits have been produced and distributed. They are:

- PlanetQuest (search for extra-solar planets),
- Our Galaxy: Our Universe (size/distance/time scales),
- Black Hole Survival Kit,
- Shadows and Silhouettes (eclipses and transits),
- Telescopes.

Each kit contains short demonstrations and manipulatives, activities, games, and handouts, plus background information and a training CD. Many of the activities are designed for use at or near telescopes so they can be used during a star party. Whenever possible, the topics and materials are related to objects that can be observed with amateur telescopes. Finally, all the materials are rigorously pilot-tested *by amateurs* to ensure their acceptance by the amateurs.

Since the network's inception, some 200 US amateur astronomy clubs have joined the NSN, with a total membership of approximately 20 000. There is at least one NSN club in each of the 50 United States, plus a member club in Puerto Rico. Over 2000 club members are registered with NSN. To date, NSN participants have logged over 4500 public outreach events reaching nearly 390 000 people.

The NASA funders consider the NSN a very successful public outreach program, and funding is expected to continue. Future Outreach ToolKits built around the Solar System, a topic of particular interest to amateurs, are currently being developed and tested.

16.7 Conclusions

The ASP's experience with amateur astronomy clubs in the US makes it very clear that a significant fraction of amateur astronomers enthusiastically participate in a wide variety of public outreach activities, and that they respond very well to support for their outreach efforts. It is also clear, however, that the content and materials must be tailored to their specific needs.

References

Berendsen, M., 2005. Conceptual astronomy knowledge among amateur astronomers. *Astronomy Education Review* **4**(4). (Washington, DC: Association for Universities for Research in Astronomy, Inc.) aer.noao.edu/AERArticle.php?issue=7§ioin=2&article=1

Storksdieck, M., Dierking, L. D., and Wadman, M., 2002. *Amateur Astronomers as Outreach Ambassadors*: results of an online survey prepared for the ASP under NSF Planning Grant ESI-0002694. (Annapolis, MD: Institute for Learning Innovation). www.astrosociety.org/education/resources/AAISASurveyResults.pdf

17

Does the Sun rotate around Earth or does Earth rotate around the Sun? An important aspect of science education

Syuzo Isobe

The Japan Spaceguard Association, 1-60-7, 2F, Sasazuka, Shibuya, Tokyo ISI-0073, Japan

Abstract: The majority of astronomy teachers, especially active ones, tend to teach concepts based on the newest theories. One typical example is the concept that Earth rotates around the Sun. For school students, any concept may not be obvious, but is usually obtained from their daily lives. We would like to stress, therefore, that not only different astronomical items but also scientific ones should be properly taught in school, depending on students' grade, level of ability, level of interest in the topic, and expected future jobs. Teachers should not teach the same concept on the same level to all students.

17.1 Earth rotates around the Sun

A good starting point for children is to watch the night sky. If they are patient, they may notice the motions of stars on the celestial sphere. At first most children believe that the Sun rotates around Earth. In addition they realize that different planets move with different speeds through the constellations.

In the sixteenth century, Copernicus proposed that Earth rotates around the Sun. However, there are no obvious differences of planetary positions that depend on whether the Sun rotates around Earth or vice versa.

Now, of course, schoolteachers know that Earth rotates around the Sun under the inverse-square Newtonian force of gravity, and they intend to teach this concept to their students.This "fact," however, does not convey a completely true picture of the Solar System. The Solar System bodies rotate around a center of gravity that is slightly offset from the center of the Sun. If schoolteachers try to teach the Solar System in too much detail, both cases shown in the title of this paper become wrong. See Figure 17.1.

A reason for this kind of problem is that the most thorough schoolteachers do not always take account of the students' grade, individual ability, level of interest in the topic, and expected future jobs.

17.2 What is the shape of Earth?

Here, we will show a much clearer case, in which schoolteachers sometimes confuse about the shape of Earth.

Usually, schoolteachers say Earth is a sphere. It is the right answer for most students. However, if the students' interest level is much higher or their future jobs will require further knowledge, the teacher should say a figure of Earth is an ellipsoid or geoid, as shown in Figure 17.2. Instead of Earth's mere radius as a sphere, we have equatorial and polar axes of 6378 km and 6357 km, respectively, for the ellipsoid, and a 21-km difference for the case of a geoid.

Innovation in Astronomy Education, eds. Jay M. Pasachoff, Rosa M. Ros, and Naomi Pasachoff. Published by Cambridge University

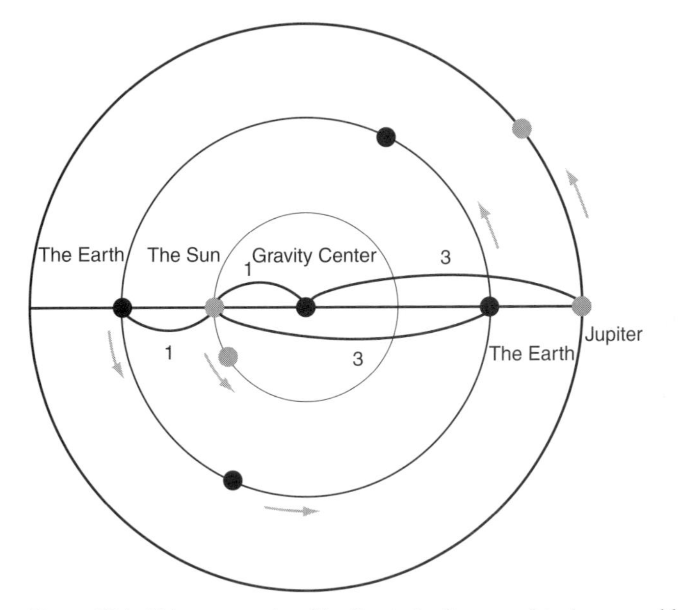

Figure 17.1 If the mass ratio of the Sun to Jupiter were 3 to 1, one could not obviously say that Earth rotates around the Sun. In any mass ratio of the Sun and Jupiter, this phenomenon occurs to some degree.

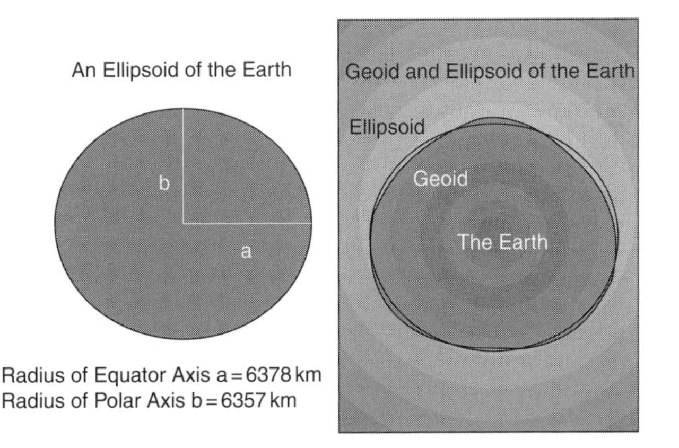

Figure 17.2 Differences of shape between a sphere, ellipsoid, and geoid.

17.3 Conclusion

It may be good for school students to study all the scientific concepts in full. However, since students are given too many things to study in school, they cannot cover the all the items. Moreover, if all the students succeeded in mastering everything by rote, the students' personalities could disappear, making it difficult for them to develop individual human identities.

In conclusion, we stress again that scientific items should be taught in school with an awareness of the students' grade, individual abilities, levels of interest in a given topic, and their expected future jobs.

The author would like to express his sincere thanks to all the IAU Commission 46 members for supporting his education activities.

Comment

Silvia Torres-Peimbert: Of course there are more important problems than teaching astronomy. But our job is to try to teach the best astronomy possible depending on the age, interest, and future work of the child.

18

Using sounds and sonifications for astronomy outreach

Fernando J. Ballesteros and Bartolo Luque

Observatorio Astronómico, Universidad de Valencia, Edificio Institutos de Investigación, Pol. La Coma, 46980 Paterna, Valencia, Spain; Departamento de Matemática Aplicada y Estadística, Escuela Superior de Ingenieros Aeronáuticos, Universidad Politécnica de Madrid, Plaza Cardenal Cisneros 3, 28040, Madrid, Spain

Abstract: Good astronomy pictures, like those of the HST, play an important and well-known role in astronomy outreach, triggering curiosity and interest. This same aim can also be achieved by means of sounds. Here we present the use of astronomy-related sounds and data sonifications to be used in astronomy outreach. These sounds, which people are unlikely to hear in the normal course of things, are a good tool for stimulating interest when teaching astronomy. In our case, sounds are successfully used in "The sounds of science," a weekend science-dissemination program heard on the principal national radio station, Radio Nacional de España (RNE). But teachers can also easily make use of these sounds in the classroom, since only a simple cassette player is needed.

18.1 Introduction

This paper presents neither a teaching methodology nor a study program but simply a very simple tool easily used in the classroom. Because it is so easy to implement, its results are strikingly effective for provoking student interest. The tool involves the use of sounds and sonifications that are either of astronomical origin or related in some way to astronomy.

The power of beautiful astronomical images to trigger our curiosity and interest is well known. Attractive images provide good reinforcement for explanations of astronomical concepts. When people look at them, they find them beautiful, strange, curious, rare, spectacular – something they are unlikely to encounter in daily life. Images, therefore, are a good hook to attract people to astronomy in particular and to science in general. When looking at these images, many students, ask, "Is this really out there?" Several even think, "I would like to see that with my own eyes." A few may even say to themselves, "Wow, I would like to do something like that myself," and eventually become professional astronomers. In short, astronomical images are a very good outreach tool.

It turns out that sounds can play the same outreach role as astronomical images. Sounds of astronomical origin can be used very effectively for astronomy outreach. They are also a good tool to stimulate interest when teaching astronomy. These sounds are attractive in some cases because of their intrinsic beauty; in others, because of their exotic origin; and in general because they are sounds that people are unlikely ever to hear in the normal course of events.

On the other hand, people are growing so accustomed to – almost over-saturated by – pictures that they are at risk of finding them somewhat dull. Sounds thus provide an excellent alternative to images. In some cases, in fact, sounds are superior to images. For example, in the case of pulsars, the images are not very spectacular but the sounds are strangely attractive.

Innovation in Astronomy Education, eds. Jay M. Pasachoff, Rosa M. Ros, and Naomi Pasachoff. Published by Cambridge University Press.

Finally, although some astronomical images are newly available in Braille, in general blind people can be exposed to the wonders of astronomy more effectively through sound than through visuals.

18.2 Use in the classroom

Clearly, sounds are very easy to use in the classroom, as they require only a simple infrastructure: just a cassette player or, more recently, an mp3 player. What might be more difficult is for the teacher to obtain these sounds. Fortunately, in the Internet age, there are several sources. Both Altavista and Alltheweb have search engines for sounds. If you are patient, it is possible to find exactly what you are looking for. Some research centers are also making compilations of sounds related to their work. The excellent page "Space Audio," at the University of Iowa, lists radio recordings of sounds of many atmospheric and astronomical phenomena. As the trend continues to grow, more scientific and astronomical sounds will become available over the Internet.

An interesting alternative is to sonificate your own data – that is, transform your own research results into an audible format. Doing so will give the data a new and interesting dimension that can help you disseminate your research. Nowadays there are several programs available to facilitate this task. "Sonification Sandbox," for example, is a commercial Java multi-platform software, usable in almost any operative system. "Sounds of Space," software developed by the University of California at Berkeley, for Windows XP and Mac, can even help you to detect through audio some structures that would pass unperceived in a visual exam. The creators of the software "Sounds of Space" took data from satellites Helios-1 and 2 that measured the intensity of solar emission, passing frequency to musical tone and emission intensity to sound intensity, resulting in a very interesting piece of sound. And given that there were two satellites, the sound was in stereo!

Once you have the sounds, what do you do with them? There are different ways to use sounds during class time. The most obvious way is to use them for emphasizing or highlighting what you are explaining. For example, if you explain what an aurora is, you can conclude your exposition by saying, "And it sounds like this," which undoubtedly will grab the students' attention. It might be better, in fact, to start with the sound, and only begin your explanation once you have the attention of the class. Or you can just play a game of "guess what this sound is." The limit is the imagination of the teacher.

18.3 Astronomical sources of sound

Using astronomical sounds is so surprising because we all know, including many students, that in space *there is no sound*. How, then, can we be talking about astronomical sounds? Where do they come from? This component of surprise is a good teaching tool in itself, which should be taken into account for its challenge to expectation. In many cases, as the two previous examples of pulsars and aurorae showed, the sounds will be radio signals passed to sound. Radio transmission is one of the most productive sources of sounds, and you can find also radio emissions from black holes, lightning storms on Saturn, or ionization tracks from shooting stars, for instance.

But there are cases where the sounds are real. Sometimes there can be a direct record of sound, as when a shooting star crosses the sound barrier, and sometimes inaudible but real sounds can reconstructed, as in the case of sound waves crossing the solar surface: because of the vacuum of space, this sound does not reach Earth, but can be indirectly "recorded" by instruments, like the SOHO satellite, and reconstructed afterwards.

Other effective sounds for classroom use are not of direct astronomical origin but nonetheless support the explanation of astronomical concepts. The sound of a train passing, for example, can help explain the Doppler effect and its relationship with astronomy. The music from Kepler's *Harmonices Mundi* represents the different speeds of orbital movement as notes, with each planet having an associated melody. A very fruitful source of sounds comes from the world of astronautics, including Neil Armstrong's remarks, audio from the Voyager disk, telemetry of Sputnik, etc.

18.4 Conclusion: it works!

The most important lesson we can give about the use of this tool is that we know it works. The authors of this paper have a successful radio program based on this tool, called "Los Sonidos de la Ciencia" – the sounds of science. It is a 10-minute-long program, aired at weekends for the past three years on the principal national radio station, Radio Nacional de España (RNE). The show has been a great success, earning first place among "best morning radio programs" in the Spanish version of the popular *ciao* listing (ciao.es), as well as first honorable mention for science dissemination in the national phase of "Science on Stage." Listeners write often to the program asking for more information about the concepts explained on the show. From their input we know that one of the most important components of the program is that they enjoy listening to those sounds that they would otherwise never hear. That is the key.

Acknowledgments

We extend our thanks to Radio Nacional de España, RNE-1, and mainly to the team of "No es un día cualquiera," for their support, help, and friendship.

Sources

Alltheweb (www.alltheweb.com/?cat=mp3), Yahoo! Inc., 701 First Avenue, Sunnyvale, CA 94089, USA.

Altavista (www.altavista.com/audio), Overture Services, Inc., 74 North Pasadena Avenue, 3rd Floor, Pasadena, California 91103, USA.

Los Sonidos de la Ciencia (matap.dmae.upm.es/WebpersonalBartolo/radio.html), with records of all the emitted programs.

Radio Nacional de España, RNE (www.rne.es), RadioTelevisión Española S. A., Casa de la Radio, Avda. Radio Televisión, 4 28223. Pozuelo de Alarcon, Madrid, Spain.

SOHO (soi.stanford.edu/results/sounds.html), SOHO MDI, SOI group, Stanford, CA 94305, USA.

Sonification Sandbox (sonify.psych.gatech.edu/research/sonification_sandbox/sandbox.html), Sonification Lab, School of Psychology, J. S. Coon Building, 654 Cherry Street, Atlanta, GA 30332-0170, USA.

Sounds of Space (cse.ssl.berkeley.edu/impact/sounds_apps.html), University of California, Berkeley, Space Sciences Laboratory, 7 Gauss Way # 7450, Berkeley, CA 94720-7450, USA.

Space Audio (www-pw.physics.uiowa.edu/space-audio/), Department of Physics and Astronomy, University of Iowa, Iowa City, Iowa 52242, USA.

Comments

Unrecorded questioner: Do you tell your audience the frequency of the sounds they have heard?

Fernando J. Ballesteros: Yes, and I recommend that others who use sound for astronomical teaching or outreach do so.

Daniel Fischer: The scientists turning plasma wave disks from the Voyagers into sound didn't do it for fun – they (esp. PI Dave Gurnett) actually *listen* to their disks to better understand the processes of plasma flowing around planets (in a similar pattern to the turbulent flow of water).

19

Teaching astronomy and the crisis in science education

Nick Lomb and Toner Stevenson
Sydney Observatory, PO Box K346, Hay market, NSW 1238, Australia

Abstract: In Australia, as in many other countries, the fraction of high school students voluntarily choosing to study the core sciences such as physics and chemistry has dropped in recent decades. The reasons for this worrying trend include the perception that these sciences are difficult subjects that lack relevance to the lives of the students. Family influence to choose courses that are believed to be more likely to lead to highly paid careers is also a major factor. Astronomy, by contrast, has a broad public appeal and seems untainted by the negative feelings associated with most other scientific fields. As a result, astronomy can be a useful tool to stimulate students' scientific interest, not only in formal college-level courses but also in informal education centers. Investment in public facilities and the provision of resources for astronomy outreach can be highly beneficial by engaging the imagination of the public. We will discuss activities offered at Sydney Observatory, where school student attendances have increased by 50% in the last five years.

19.1 Introduction

In Australia, as in a number of other countries, there is concern about the dropping number of students at high school and at tertiary levels enrolling in core scientific subjects like physics, chemistry, and mathematics. The reduced number of students suggests that in the near future there may not be enough people with science, engineering, and technology (SET) skills to satisfy the demand from research and industry.

In this paper we look at the extent of the problem in Australia and comparable countries and examine its possible causes. We also consider whether the teaching of astronomy – in particular, astronomy at informal education centers such as planetaria and public observatories – can have a role in reducing the shift away from science.

19.2 Concern over the anticipated shortage of future scientists

The Australian Government recently released an "Audit of Science, Engineering and Technology Skills" (Australian Government, Department of Education, Science and Training, 2006a). The report emphasizes that productivity and growth in the country's economy is dependent on the availability of skilled scientists and engineers to meet industry and research needs. With shortages of skilled people in some areas, there has been strong recent growth in demand for many SET skills. The annual demand for scientists over the next ten years is estimated at five per cent. This is mainly to replace the large numbers of "baby boomers" – people born between 1946 and 1959 – expected to retire over that period.

Since it is difficult to predict what will be the skill needs of future industries, however, the report considers that the formal estimates could under-represent the growth in demand. Future

Innovation in Astronomy Education, eds. Jay M. Pasachoff, Rosa M. Ros, and Naomi Pasachoff. Published by Cambridge University Press.

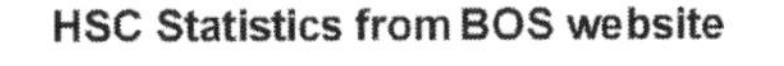

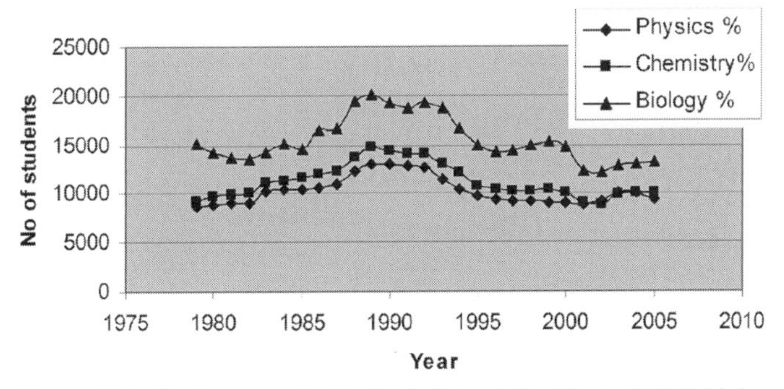

Figure 19.1 Students enrolled in High School Certificate (HSC) biology, chemistry, and physics in the Australian state of New South Wales.

industries are likely, though, to be science-based, including nanotechnology and biotechnology, new materials and IT (information technology), which would clearly need workers with SET skills.

While there is an anticipated increasing future need for people with SET skills, the audit reports a drop in high school enrolments in science subjects. It states that Australia-wide from 1993 to 2003 the percentage of final year high school students studying physics dropped from 4.1% to 3.4%, chemistry from 4.5% to 3.6%, and biology from 6.9% to 4.6%.

There has also been a fall in the number of students studying mathematics at advanced or intermediate levels. A report from the Australian Mathematical Sciences Institute (Barrington, 2006) states that Australia-wide from 1995 to 2004 the number of advanced mathematics students dropped from 14.1% to 11.7% and the number of intermediate mathematics students from 27.2% to 22.6%.

There are large variations in student enrollments in the different science subjects among the different Australian states. Figure 19.1 indicates that in the state of New South Wales (Board of Studies NSW, 2006) there was a build-up of final year High School Certificate (HSC) student numbers in science subjects to 1990 and then a drop-off to 1995. Since then, with curriculum changes and other actions, it has been possible to keep student numbers steady, especially in physics and chemistry.

Figure 19.2 looks at the fraction of students enrolled in the three main science subjects for the HSC. From the 1980s to 2005 there has been a steady growth in the number of students retained for the final years of high school, with total enrollments doubling during that period. This led to a steeper decline in science enrollments expressed in percentage terms, but again, for the last decade, the fraction of students enrolled in science subjects, especially physics and chemistry, has held steady.

Australia is not the only country where there has been a drop in science enrollments. Figure 19.3, using figures from the Institution of Physics (2006), gives the number of mathematics and science entries as a percentage of the total entries to the UK A-level examinations. From 1990 to 2006 the fraction of mathematics students dropped from 11.7% to 7.5%, while the number of physics students dropped from 6.6% to 3.6%. Surprisingly, as can be seen from the figure, the fractions of biology and chemistry students stayed approximately constant.

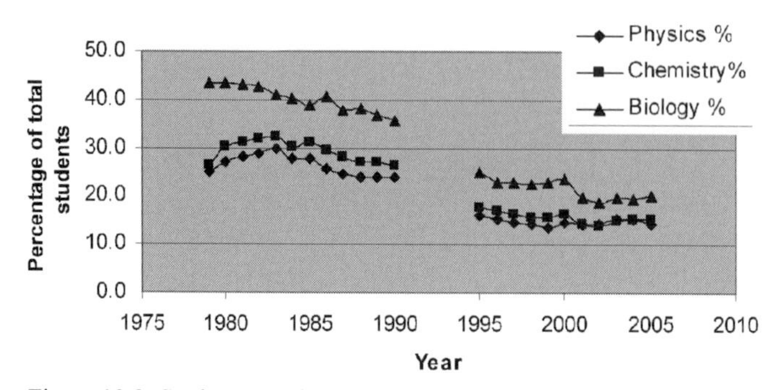

Figure 19.2 Students enrolled in High School Certificate (HSC) biology, chemistry, and physics in the Australian state of New South Wales as a percentage of total enrolments. There is a gap in the data as the total number of students enrolled for the HSC for the years 1991 to 1994 is unavailable.

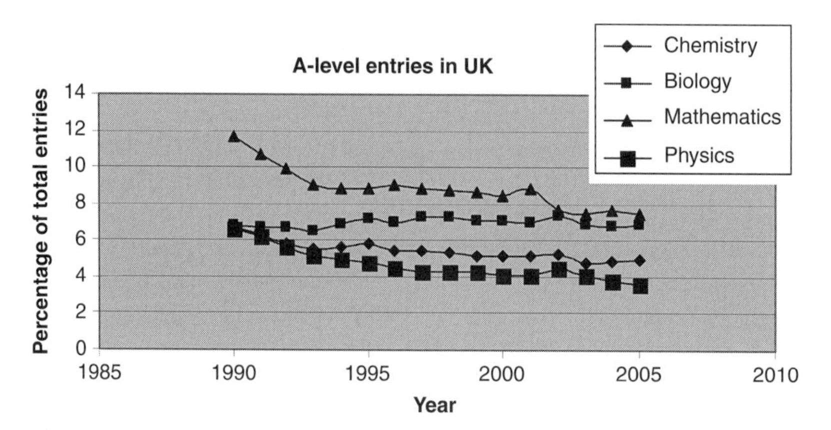

Figure 19.3 The number of mathematics and science entries as percentage of the total entries to the UK A-level examinations.

After analyzing data from 19 countries, Global Science Forum (2006) of the Organization for Economic Co-operation and Development also expressed concern about science enrollments. Their report states that though absolute numbers of science and technology students increased in most countries during the ten years from 1993, the relative share of S&T students among the total student population has dropped. In many countries a downwards trend is seen even in the absolute numbers of students for physical sciences and mathematics.

19.3 Why are students turning away from science?

Lyons (2004) examined why students chose to do or chose not to do science for their last two years of high school. His findings were based on a survey of 196 students and detailed interviews with 37. Though these numbers are too small to have statistical validity they are highly indicative. Among the findings are the following.

- Surprisingly, all students regard physical science courses as irrelevant and boring, whether they have chosen the subjects or not. They consider that the subjects are based on the presentation of facts without context and that in the classroom they are just the passive recipients of these facts.

- Students consider physics and chemistry as the most difficult of science subjects, and hence only students confident in their abilities chose to do them.
- Those who choose to do physics or chemistry do so for strategic reasons, in the sense that those subjects are believed to improve the students' opportunities in the final high school examinations and at university.
- Parental influence, attitudes, or example are crucial to the choice of science subjects. Students only choose science subjects if there is parental support; it also helps if a parent or other family member demonstrates an interest in science by reading science books and magazines or watching science television documentaries.

A larger Youth Attitudes Survey (Australian Government, Department of Education, Science and Training, 2006b) of 1830 high school students and recent school leavers obtained similar indications. They found that students chose science courses because of

- the perceived usefulness of the courses for future prospects, including improving university entrance scores,
- the students' aptitude for the subject,
- encouragement of the students by their teachers and their parents.

On the other hand students did not choose science courses because

- they consider that science is boring,
- they believe that science does not appear useful for their future,
- they lack aptitude for the subjects,
- they receive little parental encouragement.

19.4 Stopping the decline in enrolments: the role of astronomy

Unlike most physical sciences, astronomy has a wide popular appeal. There are, for instance, tens of thousands of amateur astronomers around the world, while there are few, if any, amateur physicists or chemists. In the USA large numbers of non-science major students choose to do Astronomy 101, the introductory college-level astronomy course. In recent years, physics departments at Australian universities have also introduced introductory astronomy courses to boost otherwise rapidly falling student numbers.

Astronomy, thus, can be used to combat what students, fairly or unfairly, regard as a "boring" curriculum and to stimulate the interest of high school students in science. Informal education facilities, such as planetaria and public observatories like Sydney Observatory, have a major role to play in improving the attitudes of students to science.

Schools can visit Sydney Observatory on any morning during school term. The decision on whether, and when, to bring students is up to the individual school or teacher. Obviously teachers consider a visit to the Observatory has educational value for their students for, as indicated by Figure 19.4, the numbers of student visits, especially from primary schools, has greatly increased in recent years.

Each school visit takes about one and a half hours and concentrates on the astronomical topics chosen by the teacher from a range of options. Unlike the perception of students as passive receptacles in school science classes, visits to the Observatory are designed so that the students are active participants in the learning process (Lomb, 2005). During each visit

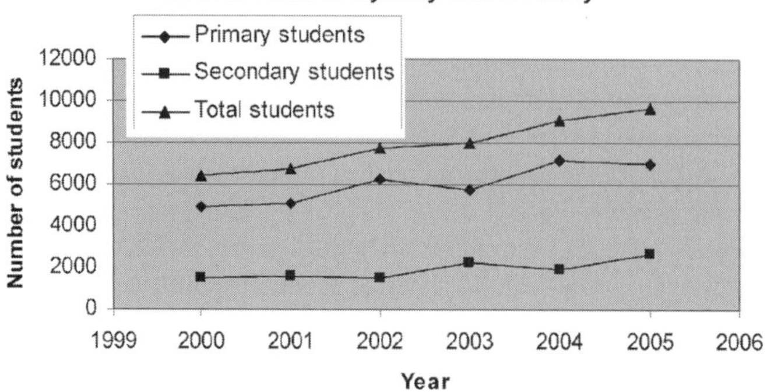

Figure 19.4 Visits by school students to Sydney Observatory.

students experience a 3D presentation in the Space Theatre; visit the exhibition with its variety of interactive displays; discover the southern stars in the small planetarium; and go up to the telescopes where, on clear days, they can see the Sun in hydrogen alpha light or see a bright planet or star.

A visit to Sydney Observatory, like visits to other informal astronomy education facilities, can successfully fulfill what museum researchers Falk and Dierking (1992) call the "Interactive Experience Model." This model is a dynamic process occurring at the intersection of three overlapping contexts – personal, social, and physical. Within the Sydney Observatory setting, students have the personal experience of encountering astronomy through 3D images and films in the Space Theatre, the social interaction in the small planetarium and while exploring the exhibits with other students, and the physical experiences of looking through a real telescope and of being in a unique environment.

19.5 Conclusions

In many countries in the last ten years or so, there has been a strong tendency for a decreasing number of students to choose to study science subjects, especially physics and chemistry. The students are turning away from science as they see it as difficult, boring, and irrelevant to their lives.

Astronomy, with its proven capability to interest students, can be used to stimulate their interest in science. The need to encourage students to study science creates an important role for informal science centers such as planetaria, astronomical visitor centers, and public observatories like Sydney Observatory.

References

Australian Government, Department of Education, Science and Training, 2006a, Audit of science, engineering and technology skills, summary report, www.dest.gov.au/sectors/science_innovation/policy_issues_reviews/key_issues/setsa/report.htm.

Australian Government, Department of Education, Science and Training, 2006b, Youth attitudes survey: population study on the perceptions of science, mathematics and technology study at school and career decision making, www.dest.gov.au/NR/rdonlyres/5D23C185-9031-4881-9FC9-89514F935ACB/13057/Youth_Attitudes_Survey.pdf.

Barrington, F., 2006, Participation in Year 12 Mathematics across Australia 1995–2004, International Centre of Excellence in Mathematics and the Australian Mathematical Sciences Institute, www.ice-em.org.au/pdfs/Participation_in_Yr12_Maths.pdf.

Board of Studies NSW, 2006, Statistic for Higher School Certificates, boardofstudies.nsw.edu.au/bos_stats/.
Falk, J. and Dierking, L., 1992, *The Museum Experience*, (Washington, DC: Whalesback Books).
Institute of Physics, 2006, The number of entries to A-level examinations in sciences and mathematics 1985–2006, www.iop.org/Our_Activities/Science_Policy/Statistics/file_7579.doc.
Lomb, N., 2005, The role of science centres and planetariums, in Pasachoff, J. M. and Percy, J. R. (eds), *Teaching and Learning Astronomy, Effective Strategies for Educators Worldwide*, (New York: Cambridge University Press), 221.
Lyons, T., 2004, Choosing physical science courses: the importance of cultural and social capital in the enrolment decisions of high achieving students, www-ra.phys.utas.edu.au/IOSTE_XI_Lyons.doc
Organisation for Economic Co-operation and Development, Global Science Forum 2006, Evolution of student interest in science and technology studies, policy report, www.oecd.org/dataoecd/16/30/36645825.pdf.

Comment

Silvia Torres-Peimbert: The same situation is also present in Mexico, where students are inspired by astronomy. We are thus likely to become interested in science and engineering.

20

Astronomy for all as part of a general education

J. E. F. Baruch, D. G. Hedges, J. Machell, K. Norris, and C. J. Tallon

University of Bradford, Bradford BD7 1DP, UK; Bradford College, School of Education, Gt Horton Road, Bradford BD7 1AY, UK

Abstract: This paper (1) evaluates a new initiative in support of the aim of Commission 46 of the IAU to develop and improve astronomy education at all levels throughout the world, (2) describes a free facility to support education programs that include basic astronomy and are delivered to students who have access to the Internet on www.telescope.org/, (3) discusses the role of robotic telescopes (generally not truly autonomous robots but remotely driven telescopes) in supporting both students and their teachers, (4) shows that although robotic telescopes have been around for some time almost all of them are designed to cater for a tiny percentage of students, (5) shows how truly autonomous robots offer the possibility of delivering a learning experience for all students in their general education, (6) discusses the Bradford Robotic telescope, a facility available free of charge, which is on track to deliver the initial levels of astronomy education to all school students in the UK, (7) describes problems of delivering a web-based education program to very large numbers of students, including delivering such a programme with teachers who have little confidence working with IT and little knowledge of basic astronomy, and (8) discusses practical solutions.

20.1 Background

The aims of Commission 46 of the International Astronomical Union (IAU) include the support and development of astronomy education,[1] since astronomy attracts people to education in science and technology. It is a problem for many developed countries that their young people are avoiding university courses in the physical sciences, engineering, technology, and computer sciences (House of Lords Science and Technology Committee, 2006; Gavaghan, 1999). In the past, young people got their first introduction to science traditionally through either astronomy or radio. The development of the Internet has led to a dramatic decline in the size of the Radio Ham community, leaving only astronomy as the classical accessible introduction to the physical sciences. But astronomy is not what it once was.

With the growth of light pollution, astronomy has ceased to be easily accessible, and it is no longer a natural route into careers in science technology and engineering. In addition to looking at the light-polluted night sky, young people can find many other activities to fill the time when they are not at school, including the Internet, computer games, and TV.

It is the well-documented experience of teachers both in the UK and in other developed nations that children under the age of 12 are enthusiastic about science and well motivated to study science (see POSTnote: Primary Science, 2003). As they grow into their teenage years and puberty, however, science becomes less and less attractive, with falling numbers of

[1] Physics.open.ac.uk/IAU46/guidelines.html

Innovation in Astronomy Education, eds. Jay M. Pasachoff, Rosa M. Ros, and Naomi Pasachoff. Published by Cambridge University Press.

students opting to take science post 16. In the UK, figures released recently by the CBI (Confederation of British Industry, 2006) showed that the number of students enrolling for a post-16 course in advanced-level physics has dropped by 56% in the past 20 years, and the similar A-level course in chemistry has fallen by 37%.

In many countries astronomy is not on the normal curriculum, but the basic ideas of astronomy still need to be taught, often by people who have little or no scientific training, to students in the early years of their education. Students want to know such things as why the Australians don't fall off the face of the Earth and why we have day and night, winters and summers, tides and eclipses. We know that students who lack an education that includes basic astronomy are left with a wide range of scientific misconceptions. Gallup polls in recent years, for example (2006a, b), showed that 72% of Americans believe in some form of creationism, with varying numbers around 45% saying that human beings were created less than 10 000 years ago and 35% unsure (November 2004 Gallup). By implication it can be presumed that they believe that the Earth, Sun, and Universe were created at the same time. Since the USA cannot be regarded as an underdeveloped country, one can only speculate about the wild ideas that would be found where education is much more difficult to obtain.

There is clearly a great need for astronomy education. Many people, especially young people, have an immediate interest in the possibility of extraterrestrial life and in the dangers of Near Earth Objects. Teaching these ideas from the point of view of science and not superstition can help learners grapple with the ideas and methods of science and so become economically valuable in their home countries.

In reality teachers and learners do grapple with these ideas. Robotic telescopes are a valuable resource for those who are making their first encounters with astronomy.

20.2 Education for all

The objective of delivering astronomy education as a part of a general education is included in the education programs of most countries. In the UK the National Curriculum (www.nc.ak.net) starts at the age of 5 and continues up to 16. Students learn the fundamentals of scientific enquiry based in contexts that include "life processes and living things," "materials and their properties," and "physical processes." In these contexts they learn about ideas and evidence in science, specifically that it is important to collect evidence by making observations and measurements when trying to answer a question in science. They develop a set of investigative skills that begin with asking questions – How? Why? What will happen if? – and decide how they might find answers to them. They are shown how to use first-hand experience and simple information sources to answer questions. If they experiment, they are encouraged to think about what might happen before deciding what to do. They develop the understanding to recognize when a test or comparison is unfair. They develop their skills in obtaining and presenting evidence and considering and evaluating evidence.

The National Curriculum is divided into key stages 1 to 4. At key stage 1, for children from ages 5 to 7 years, the only link with astronomy is that children are encouraged to ask questions and think of ways of answering them. They could clearly ask a lot of astronomically based questions. Why doesn't the Moon fall out of the sky? Why does the Sun travel across the sky? Why is day followed by night? Generally, however, their questions are much less stimulating than this.

At key stage 2, for children aged 7 to 11 years, the scientific enquiry program is continued, developing the concept that it is important to test ideas using evidence from observation and measurement; that it is also important to ask questions that can be investigated scientifically; that there is a wide range of methods of communicating information, including graphs and IT. In the section on the Earth and beyond, under physical processes, they are taught the shape of the major bodies (approximately spherical); the changing position of the Sun during the day and its relationship to shadows; the fact that the Moon orbits the Earth; the cause of day and night; and the astronomical basis of the year.

At key stage 3, for children aged 11 to 14 years, astronomy component section 7L – the Solar System and beyond – could require about nine hours of study. The section builds on the early ideas to include eclipses and seasons and extends into understanding planets, satellites, and the Sun as a star. Students consider how evidence about the Solar System has been collected and evaluate the strength of the evidence.

At key stage 4, for teenagers aged 14 to 16 years, the syllabus has changed for the section on Earth and beyond. It used to state: a knowledge of the relative sizes and positions of the planets, stars, and other bodies in the Universe; consider gravity and the evolution of stars, the ideas about the origin and evolution of the Universe and the search for life elsewhere. It now merely states that the Solar System is part of the Universe that has changed since its origin and continues to show long-term changes. All this is taken in the developing context of scientific enquiry.

A robot telescope would be an ideal component of such an education program, supporting the students' enquiry and learning with their own data, providing real access to primary data with observations that the students can specify according to their own investigations. It is necessary to specify exactly how the robot telescope can help. At key stage 1, it depends upon what questions the students ask. These are likely to be limited to the Sun, Moon and stars, and the answers are likely to require naked-eye observing or the support of a telescope to provide broad panoramas of the night sky to combat the effects of light pollution. At key stage 2, a key requirement is to use images of the Moon at different phases to show that it orbits the Earth. It is also useful to provide simulations of the rotation of the Earth, its path around the Sun and the path of the Moon around the Earth when images of constellations might be helpful.

At key stage 3, images of the planets and constellations are sufficient for pupils to understand the local Universe and that the Sun is a star. At key stage 4, the telescope can be used to take deep fields in which the color is a part of the dataset. The images can be used to ascertain galaxy types, check Hubble distances through the apparent size of similar objects with known redshifts, and look at the age of clusters through an understanding of the changing colors of stars as they age.

All the areas of the Science English National Curriculum associated with astronomy can be supported, with large field images of constellations, smaller images of clusters that can also produce the whole Moon, and deep-sky images that typically only cover a small fraction of the Moon. All these need to be supported by a range of color filters and ideally cameras that will follow the stars as they set or circle the pole.

With this basic course in astronomy it is possible to understand our models of the Universe that we inhabit and to follow public announcements of new discoveries and developments. It is also possible to understand that science can explain the basics of the world we live in whilst providing a host of interesting research questions.

20.3 Robotic telescopes in support of education

There are at least 200 telescopes in the world that have the term robotic ascribed to them. Using the terminology in Baruch (1993), most of these systems are either automatic or remote telescopes. The Bradford Robotic Telescope appears to be the only generally available system that is fully robotic in the Baruch (1993) sense of the term, which stipulates that the telescope self-schedules.

The remote telescopes typically include the Slooh (www.slooh.com) operations and the Faulkes Telescopes (www.faulkes-telescope.com). The normal mode of operation for these systems is for users to book time on the telescope and then insert the details of the object they wish to observe. They normally have a webcam, but these rarely work in the dark. When the image is returned, observers can optimize the settings for their observations. If there are problems such as intermittent cloud, observers have to work out what the problem is, first ascertaining that the camera is working and that the dome is open. The big advantage of these systems is that they can observe anything with any set of combinations of filters and multiple images to support the study of variable objects. Typically there are half-hour slots. An observing program may take two of these. For a good site like Tenerife, with 200 clear nights per year, such a system can support 2000 or so observing programs per year. This schedule could accommodate 2000 students or 2000 classes of students if they all require the same image. This is a small fraction of the young people and adults who would wish to study astronomy at any one time.

It is possible for lots of observers to watch, and eavesdrop, on the observations being taken by the single telescope user, but this is clearly a two-class society. Most of the eavesdroppers will want to drive the system. The real question is whether the learning experience will be as effective if most of the learners feel like second-class citizens.

Experience also shows that people are very slow when operating such robots. Tests with novice observers have shown that a robot can work between 4 and 20 times more quickly, depending on the type of observing program being followed. The key problem with this type of remote observing is that one telescope can service only a very limited number of users.

The alternative self-scheduling system also has its disadvantages. In absolute terms it can work more quickly than a remotely driven system, but it is still limited to making one observation at a time. The great advantage that it has is that if it aims to support only those people starting to understand the local Universe following a specified education program, then there is only a very limited number of objects that they will request. Direct experience over more than ten years shows us that only one in 10 000 requests is outside a list of fewer than 30 objects. Few of these are faint, and a robot telescope can take images of these 30 objects in less than an hour. It is this truth that enables a truly autonomous robot telescope to service many thousands of people learning about astronomy and deliver an inspirational learning experience to every child who is following the course of study.

This approach of concentrating on education is also supported by the scheduling, which is weighted according to the number of requests for an object. In this way a class of students asking for the Moon will always receive a higher priority than a single request. Steps are taken in the system to ensure it is not abused by individuals requesting the same observation many times over.

20.4 A robotic telescope for all

Servicing thousands of users is only part of the challenge set by Commission 46. It is not enough to supply the course on the web, although most developed countries have Internet access for their schools or they are in the process of acquiring it. It is also necessary to support the teachers both in the delivery of the course and in building their understanding to a level where they are confident to deliver the course.

From the Gallup data quoted above, it can be seen that even in the developed countries unscientific ideas can be widespread, as in the USA. It is therefore necessary to adopt a scientific approach, generating the concepts and opening up those concepts for questioning and experimental confirmation. It is only in this way that the profusion of false ideas about the Earth, Sun, Moon, and the rest of astronomy can be laid to rest.

Computers with simulations are invaluable in the classroom. They can show the Moon orbiting the Earth, illustrating how it keeps one face towards the Earth as it cycles through its phases. The alternative for teachers is to find a dark space with 30 or more children and move half-painted white balls around, representing the Earth and Moon. These generally produce little learning, merely the memory of the excitement of the class sitting in the semi-darkness for the lesson, whilst the teacher tried to show something about the Moon. Redshifts and Hubble's law are also much more easily demonstrated on a computer screen and backed up with carefully chosen observations using a robotic telescope.

Computers also have the advantage of providing the teacher with a managed learning environment to support the design and delivery of the lesson and a class management system to keep the teacher informed of the progress of individual learners. Both these are incorporated into the Bradford Robotic Telescope systems.

20.5 Inspiring teachers and students with science

The Bradford Robotic Telescope system consists of a range of cameras to take images of the night sky and produce videos of each night. The cameras use either a 1000-pixel-square low-noise astronomical CCD or a TV CCD with avalanche amplification. The astronomical CCDs are all associated with filter systems and produce a range of sky panoramas 40 degrees, 3 degrees, 20 arc minutes, and 12 arc minutes, although the 12 arc-minute system is mainly used for guiding with almost double the number of pixels. The TV cameras watch the stars setting over Mount Teide and the stars circling the pole star.

This array of instrumentation provides the data for the teacher to deliver basic astronomy. The instrumentation and its nightly data support the website to deliver understanding of the basic astronomy concepts that underpin a general education. The website includes a list of 34 basic astronomy modules, supported by 25 observing activities using the telescope. Each of the modules stands by itself or can be used with other modules and/or observing activities to deliver the area of understanding sought by the teacher. There are extensive teacher-support materials and extension activities for the brighter children. The organization of the scheduling of the telescope ensures that it can handle every child. The learning and class management systems support teachers and reduce their work load when they use the Bradford Robotic Telescope to deliver these basic ideas.

20.6 Conclusions

The Bradford Robotic Telescope strives to support the aims of Commission 46 in developing and improving basic astronomy as part of a general education for all who have access to the Internet and speak English. It has been shown that an autonomous self-scheduling telescope is capable of handling large numbers of users, but it must be supported by an extensive educational website to develop understanding of the ideas and concepts that underpin modern astronomy.

Comments

Silvia Torres-Peimbert: Congratulations on the work you have carried out! Having the opportunity to observe is very difficult, so I expect to be able to use this facility.

John Percy: Two speakers referred to the effect of "hormones" on young people's interest in science. The most receptive audience for astronomy is the population "beyond hormones," namely, later-life learners, or "seniors." And their population is increasing because of the "baby boom" effect. We should remember that there are large audiences for astronomy beyond school and university.

J. P. DeGreve: I like the project. But it has great opportunities for extensions. Are such extensions planned (different audiences, different applications of the concept)?

J. E. F. Baruch: Yes, we wish to extend the work to support independent adults; parents supporting their children's studies to include the different cultures of our ethnic minorities; and then the other languages, cultures, and curricula around the world.

References

Baruch, J. E. F., 1993, Robots in Astronomy. A review paper. *Vistas in Astronomy*, **35** 4, 399–438.

Confederation of British Industry, 2006, UK's world-class science base under threat as young people turn their back on science, www.cbi.org.uk/.

Gallup, 2006a, www.unl.edu/rhames/courses/current/creation/evol-poll.htm.

Gallup, 2006b, www.pollingreport.com/science.htm.

Gavaghan, H., 1999, Physics grapples with its image problem, *Nature* **398**, 265–268.

House of Lords Science and Technology Committee, 2006. *Science Teaching in Schools*, London: The Stationary Office, 9–16.

POSTnote: Primary Science, 2003. London: Parliamentary Office of Science and Technology (No. 202, September 2003), p. 2.

21

Cosmic deuterium and social networking software

Jay M. Pasachoff, Terry-Ann K. Suer, Donald A. Lubowich, and Tom Glaisyer

Williams College – Hopkins Observatory, 33 Lab Campus Drive, Williamstown, MA 01267, USA; Caltech, Pasadena, CA, USA; Department of Physics and Astronomy, Hofstra University, Hempstead, NY 11549, USA; Columbia University School of International and Public Affairs and Netcentric Campaign now at Columbia University School of Journalism, New York, NY 10027, USA

Abstract: We discuss social networking software and give examples of its applicability for bringing together references for the study of cosmic deuterium. Our website at www.cosmicdeuterium.info provides links to all papers in the field, which is relevant to cosmology since all the Universe's deuterium was formed in the first 1000 seconds after the Big Bang. Understanding cosmic deuterium is one of the pillars of modern cosmology. Studies of deuterium are also important for understanding galactic chemical evolution, astrochemistry, interstellar processes, and planetary formation. By 2006, social networking software had advanced with popular sites like facebook.com and MySpace.com; the Astrophysical Data System had even set up MyADS. Social tagging software sites like del.icio.us have made it easy to share sets of links to papers already available online. We have set up del.icio.us/deuterium to provide links to many of the papers on www.cosmicdeuterium.info. Links to a del.icio.us site are easily added, the prime advantage of such software. Use of keywords allows subsets to be displayed, though only papers already online can be linked without being separately scanned. The opportunity to expose knowledge and build an ecosystem of web pages that through its use by many people captures knowledge collaboratively is considerable. Setting up such a system marries one of the earliest stages of the Universe with the latest software technologies. The method of setting up a del.icio.us social-tagging site, which is so easy, is applicable to a wide range of educational purposes.

21.1 Introduction

For the education of newcomers to a scientific field and for the convenience of students and workers in the field, it is helpful to have all the basic scientific papers gathered. For the study of deuterium in the Universe, in 2004–5 we set up a website at www.cosmicdeuterium.info with clickable links to most of the historic and basic papers in the field and to many of the current papers.

Cosmic deuterium is especially important because all deuterium in the Universe was formed in the epoch of nucleosynthesis in the first 1000 seconds after the Big Bang, so study of its relative abundance (about 1 part in 100 000 compared with ordinary hydrogen) gives us information about those first minutes of the Universe's life. Thus the understanding of cosmic deuterium is one of the pillars of modern cosmology, joining the cosmic expansion, the three degree cosmic background radiation, and the ripples in that background radiation. On the website, some papers had to be scanned and posted while others are available through links to the Astrophysical Data System (adswww.harvard.edu) or to publishers' websites.

21.2 Social networking and social tagging software

By 2006, social networking software (tinyurl.com/zx5hk) had advanced with popular sites like facebook.com and MySpace.com; the Astrophysical Data System (adswww.harvard.edu)

Innovation in Astronomy Education, eds. Jay M. Pasachoff, Rosa M. Ros, and Naomi Pasachoff. Published by Cambridge University Press.

had even set up MyADS. Social tagging software sites like del.icio.us have made it easy to share sets of links to papers already available online and to make those links widely available. We have set up del.icio.us/deuterium to provide links to many of the papers on cosmicdeuterium.info, furthering previous del.icio.us work on del.icio.us/eclipses and del.icio.us/plutocharon. It is easy for the site owner to add links to a del.icio.us site; it takes merely clicking on a button on the browser screen once the site is opened and the desired link is viewed in a browser. Categorizing different topics by keywords allows subsets to be easily displayed.

So far, we have not subcategorized the papers on del.icio.us/deuterium. At present, on ADS, there are 910 papers with deuterium or deuterated in the title (523 refereed) and 2434 papers (1458 refereed) with deuterium or deuterated in the abstact. These results do not include astro-ph nor a full-text search for deuterium. So there are too many papers to incorporate all of them into our del.icio.us list on a regular basis. But still, del.icio.us/deuterium can be useful for organizing the basic papers and those ongoing papers that we feel should be singled out because of their importance or because they review results.

The opportunity to expose knowledge and build an ecosystem of web pages that use the functionality of a facebook.com type application to capture knowledge collaboratively is considerable. This is something that might be explored as the study of deuterium advances. It would represent an interesting marriage between one of the youngest elements in the Universe with the latest software technologies.

21.3 Information about www.cosmicdeuterium.info

Researchers and scientists collaborated via computer networks long before the Internet became public domain. But these scholars were unable to reach out to everyone who sought the knowledge that they possessed. Today, the Internet has evolved into a reliable network over which knowledge can be stored and shared by anyone who wishes access to it.

The cosmic deuterium website is an archive of articles from published science journals. Most of its literature explores the use of deuterium as a "cosmological baryometer" and the page is organized based on different approaches to the subject (e.g. Deuterium in the Sun, Deuterium in QSOs, Theoretical Nucleosynthesis). This structure anticipates the questions that researchers and students might ask and will hopefully stimulate thoughts and ideas which will further the growth of cosmology.

"Overview" includes articles concerning the general theories about deuterium as a cosmological marker. Deuterium is not produced in significant quantities by any known natural process. It is inferred that most of the present deuterium has been around since near the birth of the Universe. In addition, the deuterium abundance is thought to be indicative of the density of the Universe during its early evolution. Accurate detection and measurement of the deuterium/hydrogen (D/H) ratio is therefore incredibly important to cosmologists who seek to understand the baryonic density of the early Universe, its evolution and ultimate fate.

"Earliest Background" explores the hypotheses that inspired the foundational works in the search for deuterium and the measurement of the D/H ratio. The first attempts at detection occurred over 50 years ago. Since then several methods of detection have been developed and tested. Some of these methods and their results are discussed in the sections that follow.

"92-cm Spin Flip" comprises articles that investigate radio observations of the hyperfine line of DI at 91.6 cm (equivalent to the 21 cm line of hydrogen). This is a challenging technique for determining the D/H ratio and the few detections that have been made all have large errors in the measurement.

"Ultraviolet Deuterium" explores observations of the atomic transitions of deuterium and hydrogen of the Lyman series in the far ultraviolet. Several space telescopes have been used to detect these absorption lines, the FUSE (Far Ultraviolet Spectroscopic Explorer) was built with this purpose in mind.

"Deuterated Molecules" examines deuterium by tracking the interstellar abundances of molecules that contain it. There is large variation among the measured ratios between the deuterated molecule column density and its non-deuterated counterpart. These ratios are useful for study of the chemical evolution of the interstellar medium (ISM) but not as parameters for the Big Bang models. The heterogeneity of the ratios is thought to be from a process known as fractionation by which some regions of the ISM is enriched by deuterated molecules.

"Deuterium in QSOs" considers the detection of deuterium along the line of sight of distant quasars. The measurements obtained are highly red shifted and should give a more accurate value for deuterium's primordial abundance, and presumably a better estimate of the early baryonic density of the Universe. However, detection of red shifted deuterium is limited by the small number of suitable QSO absorption systems and by the high density of the hydrogen absorption lines in this regime.

"Deuterium in the Sun" discusses the creation and destruction of deuterium during solar flares. It is suggested that the deuterium produced in the atmospheres of active stars is ejected through flares into the ISM and contributes to the contamination of the primordial D/H ratio. It is important but difficult to distinguish between the recently created deuterium and that which existed in the early Universe.

"Theoretical Nucleosynthesis" compares observed abundances of light elements (deuterium, helium-3, helium-4 and lithium-7) with those predicted by theories of the formation of these elements shortly after the Big Bang. While the measured concentration of deuterium is roughly what is expected by the Big Bang Theory, it is higher than predicted if the Universe is made entirely of baryons. This has led to the assumption that the Big Bang produced non-baryonic matter (dark matter) as well.

"Deuterium in the Orion Nebula and Other Nebulae" discusses the various measurements that have been made of the D/H ratios in these mostly HII regions. The ratios fluctuate wildly across the Orion Nebula in particular, but several solid detections of the deuterium Balmer series have been made in that region.

The final section, "KNAC papers" comprises articles written by students of the KNAC (Keck Northeastern Astronomy Consortium) colleges. These students have observed and analyzed data in both radio and optical wavelength with Jay Pasachoff (Williams College) and Donald Lubowich (Hofstra University). Excellent results, including the identification of deuterium-alpha emission line in the Orion Nebula have been obtained through the student research program.

By organizing the most relevant articles on this subject we are helping students, researchers or anyone with an interest to better interact with the information, thus promoting the use of the Internet as a research and educational/knowledge based tool.

21.4 Comparative tagging

An advantage of the cosmicdeuterium.info website over del.icio.us/deuterium is that we include basic papers that we have scanned. A disadvantage is that it is relatively hard to keep up-to-date and to make changes when we want to add papers. But when we do make the changes/additions, we can add the new papers into categories with complete control. A minor advantage of cosmicdeuterium.info is that we can include papers not included in ADS (though we could submit them to ADS); for example, we include the student papers from the Keck Northeast Astronomy Consortium for work done at Williams College with Pasachoff and Lubowich.

In either case, how up-to-date the sites are depends on how much effort is placed in monitoring new deuterium papers by the keepers of the sites. Often, new papers are noticed by their appearance on ADS or astro-ph (xxx.lanl.gov/archive/astro-ph).

Poster highlights

Astronomy in the laboratory

Bunji Suzuki

Astronomy and Earth Science, Saitama Prefectural Kasukabe Girl's High School, 6-1-1 Kasukabe-Higashi, Kasukabe-shi, Saitama 344-8521, Japan

Abstract

It is not easy to practice astronomical observation in a high school. It is difficult to teach authentic astronomy because real-world conditions cannot be reproduced in the classroom. However, our ideas produced some interesting experiments. We produced emission spectra by using a gas burner and welding. The spectra in the experiment closely resembled those of a meteor. The black-drop phenomenon became important in the case of Venus's passage between Earth and the Sun in the nineteenth century. We tried to reproduce this phenomenon by using a small steel ball painted black, solar light, and an artificial illuminant. Of course, these experiments are in conditions that are very different from the actual physical conditions.

Artificial light, diffuser, and mask to mimic the transit of Venus.

Innovation in Astronomy Education, eds. Jay M. Pasachoff, Rosa M. Ros, and Naomi Pasachoff. Published by Cambridge University Press.

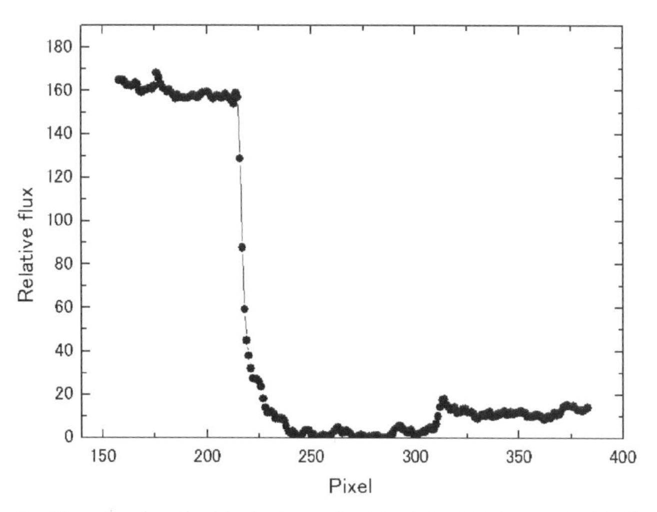

Profiles showing the black-drop effect (a) Solar light on a steel ball.

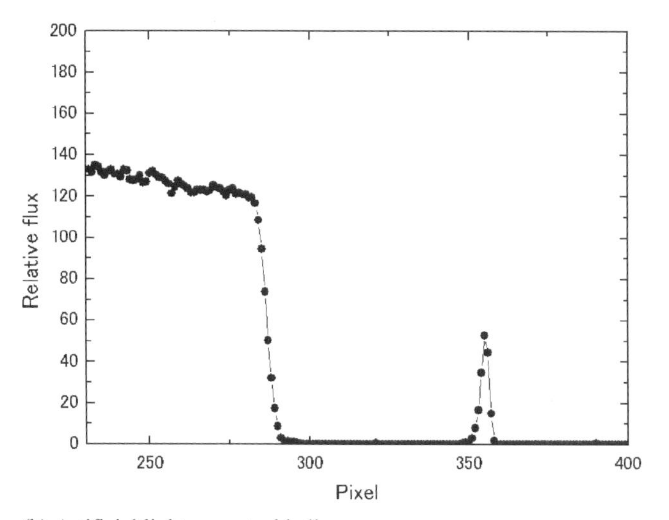

(b) Artificial light on a steel ball.

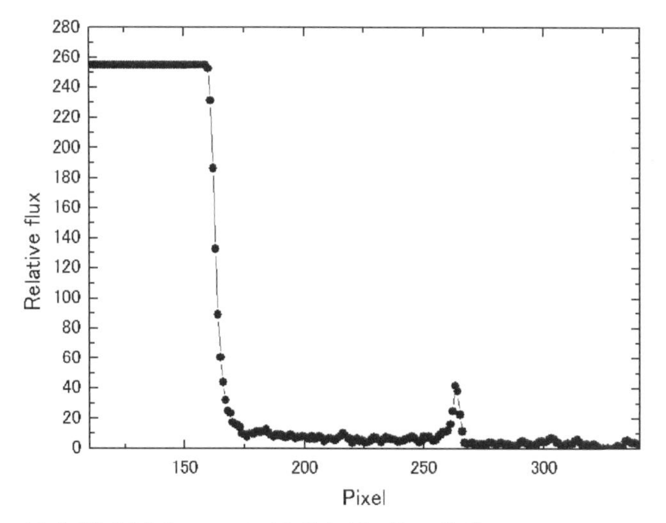

(c) Artificial light on a steel ball (with oil applied).

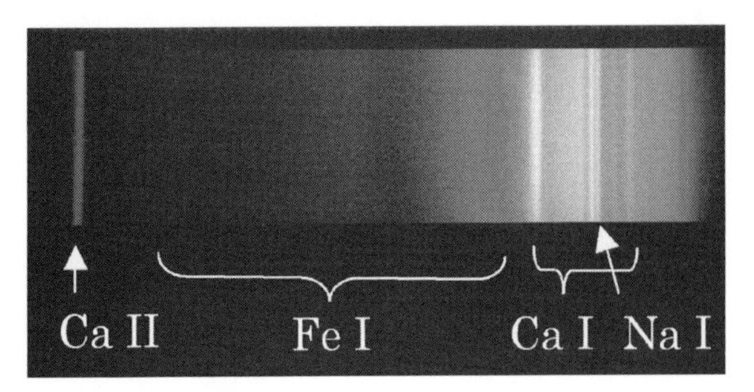

The meteor spectrum.

However, we think that they provide a very effective method for enhancing students' interest in astronomy. We are planning other experiments with similar themes.

Crayon-colored planets: using children's drawings as guides for improving astronomy teaching

Ana Beatriz de Mello, D. N. Epitácio Pereira, E. A. M. Gonzalez, R. V. Nader, and B. C. G. Lima

Observatório National, R. Gal. José Cristino 77, 20921-400 Rio de Janeiro/RJ, Brasil; Observatório do Valongo, Universidade Federal do Rio de Janeiro, Ladeira do Pedro Antônio 43, 20080-090 Rio de Janeiro/RJ, Brazil

Abstract

Many institutions around the world run programs of astronomy teaching for young children, but it is often very difficult to evaluate the efficiency of the methods and material employed. Interviews and written tests are usually unsuitable for kids. Our group of researchers and students has been teaching children up to the fifth grade, at the Observatório do Valongo (Brazil), notions about the Solar System through games and activities and has used their drawings to evaluate their previous knowledge and our pedagogic methods. The artistic expression of children can not be compared with that of adults. Usually children are not concerned in exactly reproducing the reality as the older ones do. They just represent what they find to be most interesting or amusing. This first representation, which represents their previous knowledge, is influenced by how the concept was formed, through experiences at home and notions learned at school. The left-hand figure shows one of these pre-excursion drawings, showing some common mistakes. The students, then, go on an excursion to the observatory, where they participate on a tour of the campus and take part in several

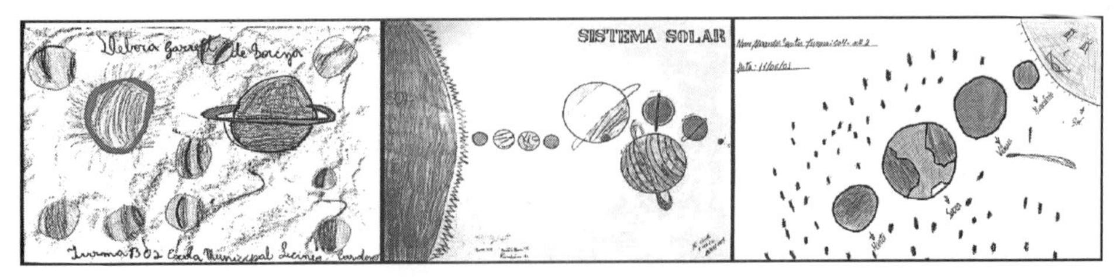

Drawings made before (left) and after (middle and right) the excursion.

workshops, two of which concern the Solar System: "Dimensions in our Solar System," that deals with the relative sizes of the celestial bodies, in which the kids build a model of them in scale; and "Solar System Bingo," an interactive lecture on our planetary system followed by a thematic Bingo. Additional activities include observation of sunspots and a portable planetary session. After the activities, they are asked to make a new drawing. This second one is affected by what they have just learned. Proportionality is a very simple concept; however, most young children are not much familiarized with the proportions of the Solar System bodies. Nevertheless, this perception was greatly improved through our activities, as shown in the middle and right figures. Their initial lack of knowledge about the Solar System components was unexpected. However, after our activities, the students demonstrated interest in all the planets and small bodies, and the existence of rings in all gaseous planets. The quantitative analysis of the 382 drawings, defined by some objective topics, statistical tests, and a more detailed discussion of the results will be presented in a forthcoming paper.

Challenges of astronomy: classification of eclipses

S. Vidojevic and S. Segan

Department of Astronomy, Faculty of Mathematics, Students lei trg 16, 11001 Belgrade, Republic of Serbia

Abstract

The usual common approach for classification of eclipses is via apparent distribution of objects. By using a computer-based environment we have realized a new approach. The idea behind this new approach is to form celestial triplets based on the spatial and electromagnetic characteristics. If we include observer position, this approach results in a very simple classification. All types of eclipses can be divided into two general groups: apparent and true eclipses.

Malargüe light pollution: a study carried out by measuring real cases

B. García, A. Risi, M. Santander, A. Cicero, A. Pattini, M. A. Cantón, L. Córica, C. Martínez, M. Endrizzi, and L. Ferrón

UTN Regional Mendoza y san Rafael-CONICET, Observatorio Pierre Auger-Malargue; LAHV, INCIHUSA, CRICYT-CONICET

Abstract

The use of artificial illumination to ensure human activities during the night contributes to so-called light pollution. This kind of contamination affects the view of the night sky and unnecessarily consumes a significant amount of energy. Since the 1980s, light contamination has been a subject of study; there are several cities in the world where the use of street lights is being regulated, especially for public illumination. In the city of Malargüe, in the south of the Mendoza province, a city regulation was passed on April 14, 2005, by which the sky is protected from this kind of contamination. As a consequence of this accomplishment, an interdisciplinary team will perform a series of measurements of the impact of traditional street lights and of those recommended for the control of light pollution in this city. This poster presents the general characteristics of this topic and the preliminary results of the measurements.

A 70-watt balloon light, mirrored on its upper half.

A mirrored balloon light in a Malargüe plaza.

A traditional 150-watt ballon light.

A traditional balloon light.

San Martin square, with its traditional balloon lights.

Sarmiento square, with its mirrored balloon lights. Both pictures were taken with the same settings and the same camera. Clearly, the light pollution into the sky and onto the surrounding trees is less with the mirrored balloon lights. The point is not about lighting in general but about providing lighting in the proper way.

Simple, joyful, instructive: ignite the joy for astronomy

Yasuharu Hanaoka and Shinpei Shibata
Astrophysics, Yamagata University, Japan

Abstract

This poster presents the basic program of a non-profit organization called the Society of Little Astronomers, founded in 1998 by Professor Dr. Shinpei Shibata, to foster the growth of science literacy. Participants attend a workshop where they spend approximately an hour constructing their own telescope. All participants are encouraged to emulate Galileo, whose

Children observing with the telescopes they have made.

story is told. Amateur astronomers help participants practice pointing and focusing. Each participant earns a certificate stating that he or she constructed a personal telescope and observed the Universe with it.

Successive innovative methods in introducing astronomy courses

Tapan K. Chatterjee
University of the American (Department of Physics), A.P.1316, Puebla. Mexico

Abstract
This poster describes an introduction to astronomy presented in several universities and institutes in Mexico, aimed primarily at revitalizing the interest of physics students in astronomy. The program began with a popular overview of the history of astronomy, focusing on Tycho, Kepler, Galileo, Newton, and Einstein. Students attended a weekly star party, where they learned how to make telescopic observations and take photgraphs. Some participants went on to construct their own telescopes or to buy them from the National Institute of Astrophysics, Optics, and Electronics of Mexico, in Puebla. After being exposed to the latest electronic information on topics on the frontiers of astronomy and physics, including dark matter and cosmology, some students became so fascinated by astronomy that they demanded a regular course in astrodynamics, which the authors immediately instituted. Many students were also inspired to join the Mexican society for the popularization and comprehension of science.

The 2005 annular eclipse: a classroom activity at EPLA

Herminia Filgaira-Alcalá
Colegio E. P. C. A. Ctra de Bétera s/n, 46100 Godella, Valencia, Spain

Abstract
On October 3, 2005, an annular solar eclipse was seen in part of Spain, including the city of Valencia and its surroundings. The last time something similar happened in Valencia was just a century ago. These unusual astronomical events are an excellent opportunity for students to learn astronomy, as they live it and feel part of it. In our school, "Colegio E. P. L. A.," at Godella, it was a wonderful opportunity, as the annular eclipse happened during playtime. For two weeks the school prepared several activities making use of the eclipse as the central item (what is called in the Spanish educational model a "transverse subject"), explaining to the students (according to their levels) what the eclipse was and preparing them for safe eclipse observation: i.e., the children fabricated during class time their own eclipse glasses with black polymer, and for the little ones, the window of a classroom was covered with black polymer for secure observation of the phenomenon. Thanks to the advance preparation, the children not only understood what they were observing but also observed safely. Some of them expressed the wish to become professional astronomers as a result of the experience. Surprisingly, the Valencian government's Education Council strongly recommended that children be locked inside the classroom during playtime, with the classroom window blinds closed, and recommended that they look at the eclipse only over the Internet! As a result of this misguided advice, many students missed out on this unique opportunity. We are proud to be one of the few Valencian schools where pupils could see and enjoy the eclipse.

The Armagh Observatory Human Orrery

M. E. Bailey, D. J. Asher, and A. A. Christou
Armagh Observatory, College Hill, Armagh BT61 9DG, Northern Ireland, UK

Abstract

The Armagh Observatory Human Orrery is a dynamic model of the Solar System in which *people* play the role of the moving planets. Stainless steel disks, or tiles, mark with a high level of precision the orbits of the classical planets, the "dwarf planet" Ceres, and two comets. The users' interactions with the model lead to greater awareness of their place in space and improved understanding of our planet's changing position with time. The Human Orrery is an innovative concept, being the first outdoor exhibit to show with precision the elliptical orbits and changing positions of the main bodies in the Solar System. It draws people into science and mathematics in an engaging way, and introduces a variety of key concepts in astronomy, mathematics, and space science. The Human Orrery lends itself to many different kinds of learning activities. Perhaps the simplest example, known as "walking the orrery," involves people moving along their assigned planetary orbits at a steady pace on the call or clap of a group leader. Another is to locate the positions of the planets at today's date. Standing on the Earth tile, one can then look towards the "Sun" to see whether Mercury and Venus may be visible as morning or evening stars, to the right or left (in the northern hemisphere) of the "Sun" (and, similarly, the real Sun), respectively. Looking away from the Sun tile shows which planets are visible at night, the result being easy to check in the real sky on the next clear night. These activities help people make the connection between what is seen in the model and what is seen in the sky. In our experience, people who experience the Human Orrery develop a deeper understanding of the Earth's motion and their position in space. Exposure to the Human Orrery helps, quite literally, to bring astronomy "down to Earth." See star.arm.ac.uk/orrery/ for more information.

View of the Armagh Observatory from the South, showing the Human Orrery in the foreground. (Image courtesy of Miruna Popescu.)

A University College Dublin lifelong-learning class visit to the Human Orrery, showing participants "Walking the Orrery." (Image courtesy of Miruna Popescu.)

Orrery demonstrators at the launch of the Human Orrery in November 2004, illustrating the development of an alignment of planets in the Solar System. (Image courtesy of Miruna Popescu.)

What mathematics is hidden behind the astronomical clock of Prague?

Michal Krizek, Alena Solcová, and Lawrence Somer

Institute of Mathematics, Academy of Sciences, Zitna 25, CZ-115 67 Prague 1, Czech Republic; Department of Mathematics, Faculty of Civil Engineering, Thakurova 7, CZ-166 29 Prague 6, Czech Republic; Department of Mathematics, Catholic University of America, Washington, DC 20064, USA

Abstract

We found that there is a surprising connection between the astronomical clock of Prague and triangular numbers. We noted the special properties of these numbers that make the regulation

The face of the astronomical clock of Prague. (Photo by Jay M. Pasachoff). See also the cover photo.

A gear of the astronomical clock of Prague. (Photo courtesy of Michal Krizek, Alena Solcová, and Lawrence Somer.)

of the bellworks more precise. In particular, we proved a necessary and sufficient condition guaranteeing that a periodic sequence is a Sindel sequence, named for the famous Prague mathematician and astronomer Jan Sindel. Sindel, who was the rector of Prague University in 1410, invented the mathematical model of the clock. Triangular numbers and Sindel sequences could be used for enhancing students' interest in mathematics and astronomy.

Solar System – Practical Exercises and *Astronomy – Practical Works* for secondary scholars

Aleksandar S. Tomic

People's Observatory, 11000 Belgrade, Kalemegdan, Republic of Serbia

Abstract

Following intentions for introducing new forms of astronomical education in secondary schools and, simultaneously, suggestions for adaptation of astronomical methods of investigation for use by amateurs, the author has written two books, as cited in the title, and published by the state textbook publisher *Zavod za udzbenike* in Belgrade. *Solar System – Practical Exercises* contains ten practical exercises for scholars, with emphasis on the fundamental astronomical parameters in the Solar System. *Astronomy – Practical Works* is the result of the author's 25-year-long experience in guiding a competition, "Science for Youth," covering astronomy for secondary and primary school pupils. It contains 82 practical works, all of which were carried out by scholars with some assistance and guidance by mentors at the public observatories or those with personal instruments. A number of high-quality papers have been written by these pupils and published in the astronomical journal *VASIONA*, edited by the astronomical society Rudjer Boscovich in Belgrade.

Astronomy in the training of teachers and the role of practical rationality in sky observation

Paulo S. Bretones and M. Compiani

Instituto de Geociências/Universidade Estadual de Campinas and Instituto Superior de Ciências Aplicadas Campinas, Campinas, Brazil

Abstract

This work analyzes a teacher-training program that departs from courses based on technical rationality. In 2002 the Instituto Superior de Ciências Aplicadas in Limeira, Brazil, offered a 46-hour astronomy course to high school teachers of science and geography. Following the course, a study group held five meetings to obtain data through assessments, interviews, and accounts by the teachers and records from the classes and meetings. The actions and conceptual changes and the role of practical rationality were then investigated. The data verified that for sky observation, the model of practical rationality within the reflective teacher theoretical framework and tutorial actions leads to knowledge acquisition, conceptual changes, and extracurricular activities. It is important to stress that sky observation has specific features that lead to an equally specific school practice, in which the contents and procedures based on observations and their representation point towards a more practical rationality. Even in a training course for teachers based on technical rationality, the introduction of sky observation deepens the practical rationality and the development of principles that guide the acquisition and the teaching of knowledge about sky observation.

Part II

Connecting astronomy with the public

Introduction

The second day of the Special Session began with a status report delivered by Dennis Crabtree on behalf of the Division XII Working Group on Communicating Astronomy to the Public (WGCAP), which was established in late 2003. After giving the URL for the WGCAP's "effective website ... describing its activities" (www.communicatingastronomy.org), he singled out one of those activities, "to promulgate adoption of the Washington Charter by various professional and amateur astronomy societies, funding agencies, and observatories," mentioned that WGCAP held "a very successful 2005 meeting on Communicating Astronomy with the Public" at ESO and had begun planning for another such meeting in 2007 (since held in Athens), and noted that at the current General Assembly the Working Group had been converted to Commission 55 under Division XII on Union-Wide Activities.

The next speaker, Julieta Fierro of Mexico, won the prize for the liveliest presentation of the session, conveying "Outreach Using Media" with the "passion and ... joy" that she hopes all astronomers will bring to their outreach activities. While her presentation involved her leaping onto tables, throwing books out into the audience, and getting normally staid astronomers up on their feet to swing dance according to her instructions, her message was far from frivolous. She recommended that astronomers hire professional fundraisers to help find funding for astronomy outreach activities, which she believes are an ideal way to excite the general public about science in general. The point of Fierro's "dance lesson" was to model some characteristics of the type of informal learning she believes astronomy educators can provide: the teaching is done incrementally to eager participants, who are given a lot of time to practice, are encouraged to learn from their mistakes, and to accept help from their peers in their quest for mastery. The point of her "book toss" was to encourage other astronomers to follow her example in writing illustrated popular astronomy books to capture the public's attention. Book-writing is not Fierro's only outreach activity; for almost a decade she has hosted, together with a chemist co-host, a 45-minute weekly radio program, aired during the evening rush hour. Even if every astronomer does not host a regular show, everyone can be available for interviews. She also recommended television for public outreach. Noting that many media personalities lack a strong science background, she encouraged astronomers to give guest lectures at journalism schools. Fierro reported on a graduate program for public outreach, offering both a master's and a doctorate, offered by Mexico's national university, with students majoring in at least one branch of the media and one scientific field. Since Fierro's presentation was mainly directed at astronomers in developing nations, she concluded by urging public policy makers to fund scientific outreach activities, since only a scientifically literate society can flourish. Noting that "if women are not prejudiced against science, their children will perform better at school," she also called for special programs for women.

In "Integrating Audio and Video Podcasting into Existing E/PO Programs," Aaron Price of the American Association of Variable Star Observers, in Cambridge, Massachusetts, USA, makes a compelling case for "including these new technologies" into efforts to bring astronomy to a broad public. Although passive learning is often looked at askance, Price believes that one advantage of podcasts is the fact that "users can play a passive role in these new technologies." "Time-shifting" enables the listener to absorb the content when and where it is most convenient, making the technology "perfect for those living busy lifestyles." He believes that podcasting should take its place among "blogs, social networking community posts, virtual communities, etc." to make the public aware of cutting-edge astronomy. He urges using "each technology for that which it is uniquely suited."

In "Hands-on Science Communication," Lars Lindberg Christensen named several astronomy-related "fundamental issues with a great popular appeal," including "How was the world created? How did life arise? Are we alone? How does it all end?" He noted the growing importance of communicating science to the public, not only to attract future scientists to the field but also to create support for public funding of science. He also noted that university statutes in some countries are being redrafted "to include communication with the public as the third mandatory function besides research and education." Christensen then identified several interesting lessons learned from the daily work at the Education and Public Outreach (EPO) of the European Space Agency's Hubble Space Telescope. These included the most effective flow of communication from scientist to public, the criteria for a successful press release, the benefits to an EPO office of a "commercial approach," the appropriate skills base in a modern EPO office, and the more efficient use of modern technology for communicating science.

Terry Mahoney, host of a spring 2002 meeting on communicating astronomy in Tenerife, Canary Islands, Spain, argued passionately about the need to remember that content must be given priority over "flashy graphics," despite the temptation to do the opposite. In "Getting a Word in Edgeways: the Survival of Discourse in Audiovisual Astronomy," he urged astronomers to remember that the "wide variety of audiovisual techniques" available to stimulate an audience can sometimes end up overwhelming without truly conveying meaning. His concluding "tips on improving discourse" include such good advice as making "the discourse drive the illustrations," using "only those images and audiovisual aids that further or enhance the discourse," avoiding the use of sound effects that merely distract, and remembering how dramatic silence can be in making a point.

Silvia Torres-Peimbert presented a "Critical Evaluation of the New Hall of Astronomy for the Science Museum" of the University of Mexico, which opened to the public in December 2004. The Science Museum as a whole, which opened in December 1992, covers a wide range of topics, ranging from mathematics to agriculture, and also includes a library, a 3-D theater, auditoriums, and outdoor displays. The original Hall of Astronomy's displays were primarily focused on the Solar System and paid little attention to the Universe beyond. The current renovation "comprises 60% of the space assigned to astronomy"; in a later stage the Solar System displays will be renovated. Divided into three sections (the Sun, stars, and interstellar matter; galaxies, clusters, and the Universe as a whole; and astronomical tools), the primary displays of the new Hall of Astronomy convey the idea that the Universe and all its components "are undergoing continuous evolution." Additional displays include "a representation of the vastness of space [through a powers-of-ten display of pictures], a time line from the Big Bang to the present epoch [laid out in a long strip on the floor], and some video clips from local astronomers [answering frequently asked questions and explaining their own research]," as well as a section on the history of astronomy [in a cartoon display, including the pre-Columbian astronomers]. The exhibit's goals were to "attract young students to science, ... present modern-day astronomical results and show that astronomy is an active science..., show that we can interpret cosmic phenomena by means of the laws of physics..., show modern-day concepts of the structure of the Universe and its constituents, ... show that modern technology has played a major role in increasing our knowledge of the Universe." Torres and her coauthor carried out their evaluation of the new Hall of Astronomy by observing the behavior of a sample of 50 visitors and by interpreting a sample of 100 visitors' responses to a questionnaire they distributed. From their observation of visitor behavior, the authors concluded that "the favorite displays were those that are larger and interactive." They attribute the lack of interest in the time line "to the fact that it is not well advertised." From analyzing the questionnaire, the authors concluded that "many topics are too complicated" for visitors under 15, but that older visitors "found interesting new information." While "not everybody is attracted to the most challenging displays," interested visitors enjoyed the opportunity to delve deeply into these subjects. The authors also concluded that the new hall "can be of assistance to science teachers."

A special lecture on astronomy education research by Timothy Slater of the Conceptual Astronomy and Physics Education Research (CAPER) Team at the University of Arizona, USA, was the final presentation of the second day. The main thrust of "Revitalizing Astronomy Teaching Through Research on Student Understanding" is that the lecture–tutorial model for teaching introductory astronomy is more effective for a majority of students (excluding, interestingly enough, "those most likely to become faculty themselves") than the traditional lecture-only model. Not only do students learn more, but the opportunity to engage in Socratic dialogue during tutorial also leads students "to reason critically about difficult concepts in astronomy and astrobiology." The lecture–tutorial model "does not require any outside equipment or drastic course revision" on the part of the instructor, and is more "learner-centered" than the lecture-only model, transforming passive listeners to active participants in the learning process. Slater made the interesting point that although the course is called "introductory astronomy," for many of the more than 200 000 students who take the course annually "it is their terminal course in astronomy, and in fact marks the end of their formal education in science." For that reason, introductory astronomy "represents an opportunity to engender the excitement of scientific inquiry in students who have chosen to avoid science courses throughout their academic career." Since many future schoolteachers take this course, it is also an opportunity to model "effective instructional strategies" for them. Like other participants in the session, Slater noted that professors tend to believe that their students learn more than they actually do, since students not only lack the basic vocabulary of astronomy to begin with but also come in with misconceptions that impede their grasp of basic concepts. In noting the "small cognitive steps" students can make in the lecture–tutorial model and the method of "having students work collaboratively in pairs in order to capitalize on the benefits of social interactions," Slater echoed points made earlier by Fierro. During the collaborative-learning tutorials, which take place in the regular lecture hall, the professor steps out of the lecturer role and into the role of facilitator, "circulating among the student groups, interacting with students, posing guiding questions when needed, and keeping students on task." In end-of-semester course evaluations, students "frequently commented positively on the lecture–tutorials, even without being prompted," generally noting how much better they understand material after hashing out their difficulties with classmates.

This part of the book ends with a description of the Japanese TENPLA project: *Activities for Popularization of Astronomy* by Kazuhisa Kamegai, M. Hiramatsu, N. Takanashi, and K. Tsukada. They show some examples of unique or unusual activities they use to popularize astronomy for the public through the cooperative efforts of students of astronomy, young astronomers, and social education facilities, including science museums and planetariums. TENPLA = TEN + PLA; TEN comes from "Tenmon-gaku," which means astronomy in Japanese, and PLA means planetarium. The activities include astronomical toilet paper, a set of playing cards to demonstrate astronomical phenomena, astronomy cafés, and visiting lectures to children in hospitals. See also short descriptions of posters, pages 203–205.

22

The IAU Working Group on communicating astronomy with the public: status report

Dennis R. Crabtree, Lars Lindberg Christensen, and Ian Robson

National Research Council Canada, Victoria, Canada; European Southern Observatory, Garching, Germany; UK Astronomy Technology Centre, Edinburgh, UK

Abstract: The IAU Working Group on Communicating Astronomy with the Public (WGCAP), a Division XII Working Group, was created in 2004 following an early October 2003 conference, "Communicating Astronomy to the Public," held at the United States National Academy of Sciences in Washington, DC, The Working Group's Mission Statement is as follows. (1) To encourage and enable a much larger fraction of the astronomical community to take an active role in explaining what we do (and why) to our fellow citizens. (2) To act as an international, impartial coordinating entity that furthers the recognition of outreach and public communication on all levels in astronomy. (3) To encourage international collaborations on outreach and public communication. (4) To endorse standards, best practices, and requirements for public communication. The Status Report reports on the achievements and progress made since the working group's formation and presents the group's plans for the next three years.

22.1 Background

The IAU Working Group on Communicating Astronomy with the Public (WGCAP) originated as one of the outcomes of the "Communicating Astronomy to the Public" meeting held in Washington, DC, in 2003. The other important outcome of that meeting was the formulation of the "Washington Charter," which is discussed later. WGCAP is under Division XII (Union-wide Activities) of the IAU and is separate from Commission 46. Dennis Crabtree and Ian Robson are the Co-chairs for the group, with Lars Lindberg Christensen as the convener and Executive Secretary.

WGCAP's mission statement is as follows.

- To encourage and enable a much larger fraction of the astronomical community to take an active role in explaining what we do (and why) to our fellow citizens.
- To act as an international, impartial coordinating entity that furthers the recognition of outreach and public communication on all levels in astronomy.
- To encourage international collaborations on outreach and public communication.
- To endorse standards, best practices, and requirements for public communication.

WGCAP has an organizing committee consisting of IAU members with a wide geographical representation. The Organizing Committee is:

1. Dennis Crabtree (NRC-HIA, Canada)
2. Ian Robson (UK ATC, UK)
3. Lars Lindberg Christensen (ESO/ESA, Denmark/Germany)
4. Michael West (Gemini Observatory, USA)

Innovation in Astronomy Education, eds. Jay M. Pasachoff, Rosa M. Ros, and Naomi Pasachoff. Published by Cambridge University Press.

5. Claus Madsen (ESO, Germany)
6. Rick Fienberg (Sky and Telescope, USA)
7. Michael Burton (University of New South Wales, Australia)
8. Guillermo Bosch (Facultad de Cs. Astronomicas y Geofisica, Argentina)
9. Kazuhiro Sekiguchi (Subaru, Japan)
10. Tony Fairall (South Africa)
11. Ashok Pati (India)
12. Birgitta Nordström (Sweden/Denmark)
13. Jin Zhu (Director, Beijing Planetarium, China)

The initial goals for WGCAP for the period from early 2004 to the IAU General Assembly in Prague were the following:

- promulgation to and adoption of the Washington Charter by societies, agencies, etc.;
- development of a website to promote WGCAP and advertise activities;
- organization of some form of repository for data;
- organization of a third CAP meeting in 2005.

22.2 The Washington Charter

The Washington Charter arose from the second "Communicating Astronomy to the Public" conference, held in Washington, DC, in October 2003. The original charter outlines Principles of Action for individuals and organizations that conduct astronomical research and that *have a compelling obligation to communicate their results and efforts with the public for the benefit of all.*

WGCAP actively sought and received many endorsements of the original charter. A few organizations were hesitant in their endorsement because of the strong language used in the charter; the American Astronomical Society, for example, declined endorsing the charter in its original form. Rick Fienberg (*Sky and Telescope*) of our Organizing Committee led the effort to edit the charter to meet the concerns of the AAS and other organizations. Rick's deft editing softened the language without altering the strong message contained in the charter.

The revised version of the charter was sent to the original group that drafted the charter at the Washington meeting. They approved of the changes and also passed *ownership* of the charter to the WGCAP.

The preamble to the charter clearly states the rationale for the priority that those of us involved in astronomical research need to give to communicating with the public. *As our world grows ever more complex and the pace of scientific discovery and technological change quickens, the global community of professional astronomers needs to communicate more effectively with the public. Astronomy enriches our culture, nourishes a scientific outlook in society, and addresses important questions about humanity's place in the universe. It contributes to areas of immediate practicality, including industry, medicine, and security, and it introduces young people to quantitative reasoning and attracts them to scientific and technical careers. Sharing what we learn about the universe is an investment in our fellow citizens, our institutions, and our future. Individuals and organizations that conduct astronomical research – especially those receiving public funding for this research – have a responsibility to communicate their results and efforts with the public for the benefit of all.*

The AAS endorsed the revised charter at their June 2006 meeting in Calgary, Canada. At the moment, the charter has been endorsed by seventeen professional and amateur astronomy societies and by eight universities, laboratories, research organizations, and other institutions.

A current list of endorsers and other information about the Washington Charter can be found at www.communicatingastronomy.org/washington_charter/index.html.

22.3 Communicating astronomy to the public – 2005

Over one hundred astronomers, public information officers, planetarium specialists, and image processing gurus descended on ESO Garching in June for CAP2005. This was the third international conference addressing astronomy outreach. (The first such conference was held in Tenerife in spring 2002.) The main aim was to bring together specialists from the various strands of astronomy that undertake outreach in the broadest sense. The four-day conference was a resounding success. Much was achieved, and the work of ESO was better appreciated (especially from the non-European perspective) through a tour of the facility. Some of the highlights of the local environs were much enjoyed, including the conference dinner at the Deutsche Museum's aviation museum, "Flugwerft Schleißheim" (with cockpit tours of an F-4 Phantom), and a splendid (somewhat liquid) evening at the Augustinerkeller, one of the largest Biergartens in Munich.

The plenary sessions covered a number of key themes for the meeting. Each session began with talks by invited speakers. All the presentations were of an extremely high standard, both in terms of content and presentational style. The sessions were: Setting the Scene; The TV Broadcast Media; What Makes a Good News Story?; The Role of the Observatories; Innovations; The Role of Planetaria; Challenges and New Ideas; Keeping our Credibility – Release of News; The Education Arena; Astronomical Images – Beauty is in the Eye of the Beholder; Cutting-edge Audiovisuals; Virtual Repositories.

A most successful discussion on credibility and the general theme of communication ethics took place in the session "Keeping our Credibility," where we were delighted to field a star-studded panel, including the ESO Director General, Dr. Catherine Cesarsky.

Technology and the power of the Web were much to the fore. On the same day as each talk was presented, its PowerPoint presentation was posted online on the conference website: www.communicatingastronomy.org/cap2005. The conference was also broadcast as a live webcast and thus available worldwide. If occasionally some of the speakers clearly forgot this and made controversial statements, there were some hasty interjections of the words "WEBCAST, WEBCAST" from the front row, to much amusement from the audience.

22.4 Virtual repository

WGCAP made great progress in designing the framework for a virtual repository of images, graphics, video, etc. material. This repository will allow outreach resources across projects and country borders to be *cataloged* in a virtual repository and accessed by educators, press, students, and public through specialized visual tools combined with search engines.

The International Virtual Observatory Alliance has made available for public review an IVOA Proposed Recommendation. This document describes a standard for Astronomical Outreach Imagery (AOI) metadata that can span both "photographic" images produced from science data and "artwork" illustrations. This standard will allow individual image files to be cataloged and offered through searchable databases like the Virtual Observatory (VO). The standard includes both the metadata schema for describing outreach images and the method by which the metadata may be embedded within the image file. Embedded metadata are commonly in use in digital photography and the publication industry, and the standard described here easily integrates into those workflows. For data-derived images, full World Coordinate System (WCS) tags can be used to describe fully the position, orientation, and scale of the image while allowing for a variety of applications requiring the full coordinate context.

22.5 International Year of Astronomy

UNESCO and the International Astronomical Union first endorsed the proclamation of 2009 as the International Year of Astronomy (IYA 2009). In 1997, the United Nations General Assembly also endorsed the proclamation.

After the General Assembly in Sydney, the IAU formed a small Working Group to pursue IYA 2009. This WG asked WGCAP for advice relating to IYA 2009, including:

- suggest a draft plan of actions for IAU's involvement with the 2009 IYA;
- consider how to interface with national representatives/groups;
- thoughts about an IAU 2009 Website, since set up at www.astronomy2009.org.

Our main conclusion is that the IAU is the natural organization to undertake the role as a catalyst and coordinator of 2009 IYA on the global scale.

We further suggested the following as a draft mission statement for IYA 2009:

- illustrate the cultural influence astronomy over time, and connect science and culture, by showing the evolution of our place in the Universe from the geocentric world view to the modern view of the world;
- show that astronomy is one of most captivating branches of natural science and excellently suited to show the fascinating aspects of science;
- remind humanity, through the majestic dimensions and violent events on the scale of the Universe, that we are ourselves responsible for the future of our planet;
- portray astronomers as a global family of international collaborations.

We recommended that the IAU is the natural organization to undertake the practical work and function as a catalyst and coordinator of IYA 2009 on the global scale. Individual countries will be undertaking their own initiatives considering their own national needs.

The IAU's role should *not* be to lead national events or to manage end-productions, such as production of TV documentaries.

22.6 CAP 2007

The fourth Communicating Astronomy to the Public Conference (CAP2007) was held in Athens, Greece, October 8–12, 2007. The conference attracted over two hundred participants and had as an overall theme the International Year of Astronomy 2009. The conference was a tremendous success and bodes well for IYA 2009. The next communicating Astronomy to the Public Conference will be held in Cape Town in the spring of 2010.

22.7 Commission 55

At the IAU General Assembly in Prague, the Division XII Working Group on Communicating Astronomy with the Public was converted to a full-fledged Commission. Commission 55 remains under Division XII. The list of officers for Division XII includes:

- President – Ian Robson (UK)
- Vice President – Dennis Crabtree (Canada)
- Secretary – Lars Lindberg Christensen (Denmark/ESO).

The Organizing Committee for Commission 55 is composed of the following:

Richard T. Fienberg	USA
Anne Green	Australia
Ajit K. Kembhavi	India
Birgitta Nordström	Denmark
Augusto Damineli Neto	Brazil
Oscar Alvarez-Pomares	Cuba
Kazuhiro Sekiguchi	Japan
Patricia A. Whitelock	South Africa
Jin Zhu	China

Now that our effort is a Commission, we expect to form Working Groups to lead our different programs. Stay tuned to our website for more information!

Comments

John Percy: How will you ensure that every country is *effectively* represented in the plans for International Year of Astronomy 2009?

Ian Robson: We are going to great lengths to ensure each country has a "single point of contact" (SPoC) and these are required to submit reports of progress to the IYA coordinator on a monthly basis; also the coordinator is now working with all countries that do not have an SPoC (or have an SPoC that is not effective) to get people involved who will actually deliver for IYA 2009.

23

Astronomy outreach: informal education

Julieta Fierro

Instituto de Astronomía, UNAM, Apdo. Postal 70-264, C. P. 04510, Mexico, DF

Abstract: In this paper we shall address ways in which astronomy can be conveyed to the general public in developing nations through the use of radio and television. We believe that the workings of the Cosmos are an effective way to interest the public in science because of their general appeal. This paper is based on the idea that outreach is part of informal education and therefore must be encouraged, since it is the way adults learn throughout their lives. We must take advantage of year 2009 to address astronomy in Galileo's honor. I believe outreach should be carried out in the same way we ourselves enjoy learning about subjects outside our own field of expertise. It must be done with passion and with the joy of giving. The gift that outreach conveys is knowledge.

23.1 Fundraising

Usually outreach activities have scarce funding so one has to obtain extra resources in order to carry it out in an effective way. Fundraising has not been part of Mexico's culture. My suggestion is to hire a professional fundraiser, who will know where the resources are, how to ask efficiently, fill out forms, produce appropriate DVDs, and what to give in exchange, depending on the donor. The fundraiser will require a fixed salary and pertinent information. One must keep in mind that it will take the fundraiser several months to obtain the first resources.

Donors' expectations include wise use of their resources and appropriate prestige-enhancing recognition. They have no interest in a project that is doomed to failure and have to choose among many worthy causes. A proficient fundraiser can talk up the important social benefits of contributing to educational causes.

23.2 Outreach

Several members of IAU Commission 46 have noted that the natural appeal and beauty of astronomy make it a natural for outreach efforts. Because of its multidisciplinary nature, astronomy can be used to explain other fields of science. Because it explores fundamental questions – where did we come from? what will our future be? – it is ideal for popularization. Its appeal also lies in the way the human mind solves mysteries such as stellar distances and evolution.

Outreach is a fundamental part of education. Most adults learn in an informal way throughout their lives, so it is important to have a wide variety of science activities for the public, addressed to a variety of audiences, including other researchers, policy-makers, women, teachers, and children. In other words, the greater the variety of approaches taken in astronomy outreach, the more likely it is to succeed with the general public.

Innovation in Astronomy Education, eds. Jay M. Pasachoff, Rosa M. Ros, and Naomi Pasachoff. Published by Cambridge University Press.

23.3 Informal learning

In order to grasp the way in which informal learning is conveyed, it can be instructive to observe how gifted teachers from a completely different discipline work. Consider, for example, lessons in ballroom dancing. What are some characteristics of such lessons?

(a) They are a voluntary activity.
(b) People are happy to participate, even if they feel tired or clumsy.
(c) Teaching is done in sequences: basic positions, steps, and choreography.
(d) There is always time for lots of practice.
(e) Students are encouraged to learn from their mistakes.
(f) Peers help each other.

These strategies can be applied to other disciplines, such as reading. Reading has to be appealing; one has to undertake reading by following a series of steps, beginning with individual words, which must be pronounced and understood, and advancing on to sentences, paragraphs, and finally whole texts. One reason that reading can be difficult is that, as a generally solitary activity, it requires an internal dialogue. By contrast, ballroom dancing is a social activity; one's partner can provide immediate feedback. When one reads about science on one's own, one has to figure out if one is understanding the text and what it makes us relate to and think about.

All the same, astronomy books are excellent outreach tools. Because they deal with fascinating topics and are often beautifully illustrated, they can encourage adults to read. Remember that literacy is far from complete in developing nations, and society as a whole benefits when adults are stimulated to read.

23.4 Variety

In order to have a successful outreach program, it is important to combine a wide variety of mediums and ways of conveying knowledge: written materials, websites, radio, television, workshops, public lectures, museums, planetaria, plays, etc.

Know your audience and respect their differing needs. For young people, hands-on activities are generally most effective, especially if they can take something home after the activity. Adults trained in science, by contrast, are likely to feel more at ease with written materials.

23.5 Radio

Radio is a great medium for outreach. One-minute programs can be pre-recorded with ease on the phone at home or at the office. Interesting information for program content can be found in magazines and on the web pages of research institutes. Although with practice one can improvise, I recommend preparing with care. I suggest that beginners write down the main ideas they wish to include in their broadcasts, just as they would with a formal presentation. It is advisable to begin with a brief introduction and, after addressing the subject concisely, to wind up with a well-worded conclusion. The attention span of listeners is limited, particularly if they are doing some other activity while they listen.

For eight years I have been broadcasting a 45-minute weekly program during the evening, when many people are driving home from work and retired people are listening to the radio. I share the responsibility for the program with a chemist colleague, Luis Manuel Guerra. Among other advantages to the collaboration, listeners hear an alternation between a male and a female voice, possibly making the commute home through heavy traffic more palatable after a long work day. Three times a month we interview a researcher; once a month the two of us talk about interesting things we have read. I believe the second kind of program, a dialogue between the two hosts, tends to be more effective than the interviews, since we, unlike some of the researchers we interview, are very comfortable with the medium of radio. Nevertheless, having the specialist in the studio can be great, especially if he or she uses common language and does not mind being interrupted in order to clarify ideas. In other words, researchers should be available for interviews and must take it seriously; the media are our allies. The advantage of long programs is that people can phone or send an e-mail, so there is feedback. During the programs I always offer books for free to the audience.

23.6 Television

Television is ideal for public outreach since it reaches millions. The problem is that it takes up lots of time, since the image is so important, so one must learn to be patient and listen to what the experts in the media have to say. One of the common practices on several stations is to interview a scientist. Unfortunately, communication schools usually do not teach science, so the interviewers do not necessarily ask interesting questions. One of the ways in which researchers can improve public understanding of science is by lecturing at schools where professional journalists are trained.

Once a week I have a three-minute feature on the nightly news (which is the third-ranking most-seen kind of program, after soap operas and sports). I believe the success of my participation comes from the fact that I have been on TV hundreds of times, so I feel at ease. I talk about what I want with total freedom, and I do some kind of demonstration. The demonstration is important. Even though it is easier just to talk, television has an image to be watched, and the viewer's attention increases if there is some kind of activity involving changes of color and movement.

23.7 Books

Outreach programs to enhance public understanding of science should make use of the printed word in addition to the audiovisual media. In developing nations, magazines and books for different age groups should be widely available in local languages; otherwise people tend to believe that science is practiced only abroad.

Unfortunately, it is common in some schools to favor rote memory over independent thought. Good outreach programs should encourage creative thinking, and good books make people think. One must remember that the goals of popularization are different from those of training professionals. While formal education takes years of good training and practice, a book can increase a reader's appreciation for a topic without necessarily covering the whole subject in a structured or profound way.

23.8 Graduate programs

Mexico's National University, UNAM, offers a graduate program in public outreach, with degrees at both the master's and the Ph.D. level. It admits students from a wide variety of fields, since we believe that communication among different specialists is good for innovation. Each student learns about at least one of the media (radio, television, journalism, exhibits, etc.) and one field of science. The University of the State of Jalisco also offers a master's degree in popularization.

23.9 Conclusion

In developing nations, public understanding of science is fundamental to reach social benefits for all. As professional scientists, we must make sure that policy-makers who control the purse that funds basic research understand that science is a key to development. We must also make sure that special outreach programs are available to women. If women are not prejudiced against science, their children will perform better at school. Astronomy is so attractive in terms of both the images that it presents and the ideas it explores, making it a natural subject to stimulate public interest in science in general.

Further reading

Tonda, J., Sánchez, A. M., and Chávez, N. (eds.), *Antología de la divulgación de la Ciencia en México*, 2002, (Mexico City: DGDC, UNAM).

Delgado, H. and Fierro, J., 2004, *Volcanes y Temblores en México*. (Mexico City: Editorial SITESA).

Domínguez, H. and Fierro, J., 2005, *Albert Einstein: Un Científico de Nuestro tiempo*. (Mexico City: Editorial Lectorum).

Domínguez, H. and Fierro, J., 2006, La luz de las estrellas. *Correo del Maestro*, (Mexico City: Ediciones La Vasija).

Fierro, J. and Domínguez, H., 2007, *Galileo y el Telescopio, 400 Años de Ciencia*, (México: Uribe y Ferrari Editores).

Fierro, J. and Vital, V. 2006, *Palabras para Conocer el Mundo*, (Mexico City: Editorial Santillana).

Fierro, J. 2005a, *Lo Grandioso del Sonido, Gran Paseo por la Ciencia*, (Mexico City: Editorial Nuevo México).

Fierro, J., 2005b, *Lo Grandioso de la Luz, Gran Paseo por la Ciencia*, (Mexico City: Editorial Nuevo México).

Fierro, J., 2005c, *Lo Grandioso del Tiempo, Gran Paseo por la Ciencia*, (Mexico City: Editorial Nuevo México).

Fierro, J., 2004, *El Sol, la Luna y las Estrellas*, (Mexico City: DGDC, Colección Ciencia para Maestros).

Fierro, J., 2001, *La Astronomía de México*, (Mexico City: Lectorum).

Fierro, J. and Sánchez Valenzuela, A., 2006, *Cartas Astrales, un Romance Científico del Tercer Tipo*, (Mexico City: Editorial Alfaguara).

Tonda, J. and Fierro, J., 2005, *El Libro de las Cochinadas*, Ilustraciones de José Luis Perujo, (Mexico City: ADN Editores).

Comments

Douglas Duncan: I further encourage use of radio! It is much easier to get air-time to discuss something interesting than on TV.

John Percy: There are media and there are media. Intelligent, influential people learn a great deal from high-quality newspapers and TV; they are *very* influential.

24

Integrating audio and video podcasting into existing E/PO programs

Aaron Price

A AVSO, 49 Bay State Road, Cambridge, MA 02138, USA; also at Wright Center for Science Education, Tufts University, Medford, MA 02155, USA

Abstract: Podcasting presents an opportunity to connect with an audience that is traditionally interested in science yet often disconnected from it, owing to limitations on time. Users can play a passive role in these new technologies, which is perfect for those living busy lifestyles. At the same time, the affordability and simplicity of the technologies give astronomers a rare opportunity to share their discoveries directly with the public, bypassing many traditional gatekeepers. We will give a brief summary of the new technologies and offer an integrated approach to including these new technologies into existing education and public outreach (E/PO) programs using new media.

24.1 Introduction

Although podcasting, which was developed in the summer of 2004, is still a relatively new technology, it has experienced phenomenally rapid growth. "Podcast" was the *New Oxford American Dictionary*'s Word of the Year for 2005 (Oxford University Press, 2005). The Pew Internet and American Life Project estimates six million American adults downloaded a podcast in 2005 (Rainie and Madden, 2005), and Arbitron estimates the total to be 27 million in 2006 (Arbitron, 2006).

Podcasts are downloaded on a schedule set by the listener, which can range from minutes to months. If new content is available on a subscribed podcast channel, the client downloads the content to the local computer and automatically loads it into his or her MP3-playing software (such as iTunes) or onto a connected portable MP3 player (such as an iPod).

This process makes the listener's role in content acquisition passive, while giving the listener complete control over how and where he or she listens to content. Once the audio file is on a home computer or MP3 player, it can be listened to at leisure (i.e., on demand). An MP3 can be paused, rewound, or fast-forwarded, similar to a television show on a DVR device. This is often referred to as time-shifting.

Time-shifting allows a listener to fit the show into his or her lifestyle. After initially subscribing, listeners do not need to do anything further but maintain Internet access. New shows appear in a listener's audio player as produced. In addition, if a user has a portable MP3 player, the shows can be heard away from the computer. Subscribers do *not* need a portable MP3 player to listen to a podcast, however; it simply makes it more convenient. Forty percent of Slacker Astronomy listeners do not use portable MP3 players (Gay *et al.*, 2006). Any web-browser or computer produced in the last few years should include built-in support for playing MP3s. Largely because of this convenience, many consumers get their astronomy news exclusively from these technologies. For example, roughly two-thirds of the ~15 000

Innovation in Astronomy Education, eds. Jay M. Pasachoff, Rosa M. Ros, and Naomi Pasachoff. Published by Cambridge University Press.

weekly listeners to the Slacker Astronomy podcast claim that podcast as their sole channel for astronomical news (Price *et al.*, 2006).

24.2 Audio podcasting

Audio podcasting is an area where astronomy has also made inroads, but not at the level of penetration astronomy has seen in the blogosphere. The first sustained astronomically themed podcast was likely Science@NASA (Barry, 2005). However, Science@NASA was an audio recording of the Science@NASA web page. The first original-content astronomical podcast was Slacker Astronomy. Since then, there have been a few additional consistently produced, original-content astronomical podcasts, usually produced by institutions (interestingly, the same ones that traditionally have avoided blogs). But, other than AstronomyCast (Gay and Cain, 2007) and Slacker Astronomy, no other audio astronomical podcast has consistently challenged the other science podcasts for high positions in the iTunes podcast popularity rankings.

24.3 Video podcasting

Astronomy podcasts have had greater success in market penetration through video podcasting (a.k.a. "vlogging" and "vodcasting"). Video podcasts by NASA and PBS's NOVA have emerged as popular shows in the iTunes rankings. At first glance, it is no surprise that such successful podcasts come from established entities with their own high-quality production departments and multimedia expertise. However, video podcasts are just as simple to produce as audio podcasts. Short, professionally produced shows can be created using free software from Apple (iMovie), and these days video can be shot on most consumer-level digital cameras. Video and animations that have already been released into the public domain (such as most NASA materials) can be easily dropped into the video podcast to provide high-quality illustration of the topic.

But even simpler videos can work. The most popular Slacker Astronomy video illustrates the supernova process using Lite Brite diagrams photographed with a digital camera. This video has been seen by over 30 000 people and won a few weekly and monthly site awards when it was first placed online at YouTube.

All video content creators should upload their content to YouTube. Currently, the astronomy offerings on YouTube are dominated by silly jokes by teenagers or dry presentations about telescope-making, content that can be enhanced by a collection of quality astronomical videos. As viewers leave the television they are increasingly getting their video entertainment online through sites like YouTube. YouTube is currently the site du jour for video downloads, with millions of daily visitors and a market value in the billions (Google, 2006). Uploading to YouTube is very simple and takes only a few minutes. YouTube accepts video in a wide variety of formats and in lengths up to 10 minutes (or 100MB in size, whichever comes first).

24.4 Integration into existing E/PO programs and other new media technologies

Like all technologies, podcasting is only a tool to be used as part of a greater strategy for education and public outreach. It should be used along with blogs, social networking community posts, virtual communities, etc. For example, let's take a faux press release about the first light from a powerful new observatory on top of a mountain. A blog could have been maintained through construction of the observatory and used to announce first light. In it, construction workers, technical crew, scientists, etc. can give first-hand, informal accounts of what the day (or night) was like and how it felt to see the fruits of their labors. This blog should then be mentioned and linked in posts on astronomical communities at MySpace, LiveJournal, and other social networking sites. An audio podcast can be released that first captures the applause in the control room and the sound of the motors running, and then presents a dialogue with various staff about challenges they had in construction and how they overcame them (such detailed discussions don't work well in blogs). A video podcast can show the equipment and the first light image along with narration of what was seen. Finally, a replica of the observatory can be created in Second Life so users can see what it looks like, explore what the mountain terrain is like, experience the weather conditions, learn facts about high-altitude observing and why mountaintops are important, etc.

The point behind the integration is to use each technology only for those things to which it is uniquely suited. Do not force a blog entry into an audio podcast or produce a video podcast of two people talking (unless one of the people is famous – then you have star power!). Some announcements will not be suited for all of the technologies. Choose the right tool for the job.

References

Arbitron, Inc., 2006. 27 million American podcast listeners; podcast users young and rich. Press release. www.podcastingnews.com/archives/2006/04/arbitron_27_mil.html.

Barry, 2005. Science @ NASA feature stories podcast. science.nasa.gov/podcast.htm.

Gay, P. L. and Cain, F., 2007, AstronomyCast. www.astronomycast.com.

Gay, P. L., Price, A., and Searle, T., 2006, Astronomy podcasting: a low-cost tool for affecting attitudes in diverse audiences, *Astronomy Education Review*, **5**(1).

Google, Inc., 2006. Google to acquire YouTube for $1.65 billion in stock. Press release. www.google.com/press/pressrel/google_youtube.html.

Oxford University Press, 2005. 'Podcast' is the word of the year. Press release. www.oup.com/us/brochure/NOAD_podcast/?view=usa.

Price, A., Gay, P., Searle, T., and Brissenden, G., 2006, A history and informal assessment of the Slacker Astronomy podcast, *Astronomy Education Review*, **5**(1).

Rainie, L. and Madden, M., 2005, Reports: online activities and pursuits, Pew Internet and American Life Project. www.pewinternet.org/PPF/r/154/report_display.asp.

Comments

John M. Pasachoff: I made chapter-by-chapter podcasts for my text, Pasachoff and Filippenko, *The Cosmos: Astronomy in the New Millennium*, 3rd edn. (2007), www.solarcorona.com, available free at drm.williams.edu/cdm4/pcpod/feeder.php?id=COSMOS or by Power-Searching on iTunes.

25

The IAU's communication strategy, hands-on science communication, and the communication of the planet definition discussion

Lars Lindberg Christensen

ESA Hubble/JWST, ESO/ST-ECF, Karl Schwarzschild Strasse 2, D-85748 Garching bei München, Germany

Abstract: Many of the most important questions studied in astronomy touch on fundamental issues with great popular appeal, e.g.: How was the world created? How did life arise? Are we alone? How does it all end? Communication of astronomy to the public is important and will play an ever-greater role in the coming years. The communication of achieved results is now seen frequently as a natural and mandatory activity to inform the public, attract funding, and recruit science students. In some countries, university statutes are even being rewritten to include communication with the public as a third mandatory function, complementing research and education. The three parts of this essay address the many exciting communication developments that have recently taken place within the International Astronomical Union. Section 25.1 contains a general introduction to the recent science communication initiatives from the IAU. Section 25.2 briefly reviews the essential elements of a communications office and strategy, based on the lessons learned at the European Space Agency's Hubble Space Telescope EPO (Education and Public Outreach) office in Munich. Section 25.3 discusses the particular communication of the planet definition debate that took place at the 2006 IAU General Assembly while our Special Session was in progress.

25.1 The IAU and public communication

25.1.1 Why communicate science?

More than ever we live in an era of unprecedented scientific progress, and the growing impact of technology has brought science ever more into our daily lives. It directly influences the quality of people's lives and makes information about scientific and technical issues crucial for them to make informed decisions. Awareness and knowledge of science is one of the foundations on which we build our modern society. New students have to be recruited to answer the steadily increasing need for science and engineering manpower in our globalized world. In the real world the game – to "convert" young and old minds to see the beauty of the Universe – is naturally a complex interplay between many different "tools," or science communication products, in many different contexts. At the risk of over-simplifying, the process of "netting" new, young minds can be summarized in four crucial steps: inspiration → engagement → education → employment. Inspiration and engagement usually occur in an informal context, e.g., at a science center or a planetarium; education in primary and secondary schools; and employment at university and beyond. We never grow too old to learn! There is no doubt that the IAU has a vital role to play in the coming years – most likely in the context of the lower parts of the triangle in Figure 25.1.

Innovation in Astronomy Education, eds. Jay M. Pasachoff, Rosa M. Ros, and Naomi Pasachoff. Published by Cambridge University Press.

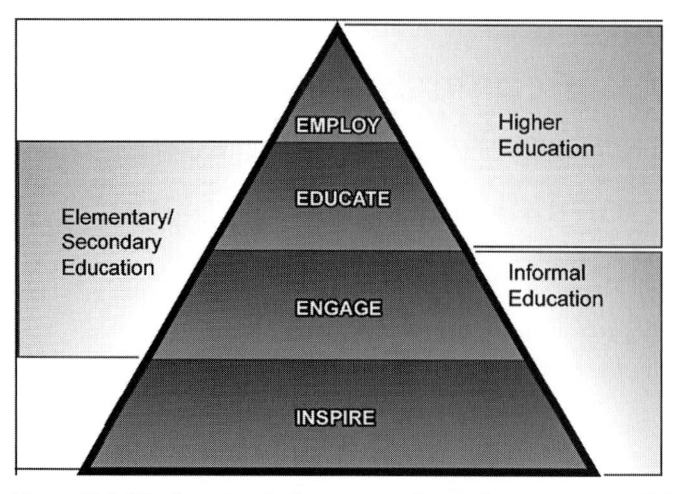

Figure 25.1 The four steps in the process of capturing new young minds for science from inspiration to employment (from ROSES, 2006).

25.1.2 The IAU's communication strategy

The IAU, founded in 1919, is, with almost 10 000 members, the world's largest professional body for astronomers. Its mission is to promote and safeguard the science of astronomy in all its aspects through international cooperation. The IAU's public communication should support its overall mission and activities.

By coordinating and facilitating communication (in contrast to producing communication products) and by connecting IAU executives and scientists with the media, hobbyists, educators, outreach professionals, and laypeople, the IAU should be a catalyst for global collaboration in education and public communication. Another important communication goal is to provide information about the IAU's role as the global astronomical authority and clearing house for science, nomenclature, and standardisation.

The three main communication players in the IAU are Commission 55 (Communicating Astronomy with the Public, www.communicating astronomy.org/), Commission 46 (Education, physics.open.ac.uk/IAU46/), and the IAU Press Office (www.iau.org/PRESS.402.0.html).

25.1.3 The International Year of Astronomy 2009

In the autumn of 2006, the IAU initiated one of its main communication projects by declaring the year 2009 as the International Year of Astronomy (IYA2009): a global celebration of astronomy and its contributions to society and culture. The aim is to stimulate worldwide interest, not only in astronomy but also in science in general, with a particular slant towards young people.

There are indeed very good reasons for celebrating a global year of astronomy. Today we live in what may be the most remarkable age of astronomical discovery in history. One hundred years ago we barely knew of the existence of our own Milky Way. Today we know that many billions of galaxies make up our Universe, and that it originated approximately 13.7 billion years ago. One hundred years ago we had no means of answering the age-old question: are there other solar systems in the Universe? Today we know of over 200 planets around other stars in our Milky Way. One hundred years ago we studied the sky using only optical telescopes, the human eye, and photographic plates. Today we observe the Universe with telescopes, both on Earth and in space, with advanced digital detectors that are sensitive to all wavelengths from high-energy gamma rays through to low-frequency radio emission. Our view of the Universe is now more fully polychromatic than ever before.

The year 2009 will be the 400th anniversary of the remarkable discoveries made by Galileo Galilei that changed astronomy forever. He turned one of his telescopes to the night sky and saw mountains and craters on the Moon, a plethora of stars invisible to the naked eye, and moons around Jupiter. IYA2009 will mark the monumental leap forward that followed Galileo's first use of the telescope for astronomical observations, and portray astronomy as a global endeavor that unites astronomers in an international, multicultural family of scientists working together to find answers to some of the most fundamental questions ever asked. IYA2009 is, first and foremost, a series of activities for people all around the world. It aims to convey the excitement of personal discovery, the pleasure of sharing fundamental knowledge about the Universe and our place in it, and the value of scientific culture.

While the IAU will organize a small number of truly global or international events, such as the Opening and Closing Events, the main activities will take place at the national level and will be coordinated by the IYA2009 National Nodes in close contact with the IAU. The majority of IYA2009 activities will take place on several levels: locally, regionally, and nationally. Several countries have already formed national committees to prepare activities for 2009. These committees are collaborations between professional and amateur astronomers, science centers, and science communicators. At the global level the IAU will play a leading role as a catalyst and coordinator. The IYA2009 was endorsed by the UNESCO General Conference in 2005; it was confirmed by the General Assembly of the United Nations in 2007, so the IYA will be able to benefit fully from an endorsement by the highest international body.

The main IYA2009 website has opened and more information (including a pdf brochure and a sign-up sheet for an e-mail newsletter) is available at www. astronomy2009.org.

25.2 Hands-on science communication

25.2.1 Communication of astronomy

Astronomy, one of the greatest adventures in the history of mankind, has a unique, almost magnetic attraction for young and old. Space is an all-action, violent arena with exotic phenomena that are counter-intuitive, spectacular, mystifying, intriguing, dazzling, and fascinating. There is a large element of discovery in astronomy. The field is extremely fast-moving, delivering new results on a daily basis. It touches some of the largest and most exciting philosophical questions of the human race: Where do we come from? Where will we end? How did life arise? Is there life elsewhere in the Universe? There is no doubt that astronomy has special advantages among the natural sciences and that it can lead the way and be a front-runner in science communication. See Figure 25.2.

25.2.2 The information flow

There are usually four players in the science communication that reaches the public via the media: the scientist, the PIO (Public Information Officer), the journalist, and the end-user. Typically the information flows from the scientist towards the end-user and is simplified along the way. In a fruitful collaboration, the four players in their interaction exchange

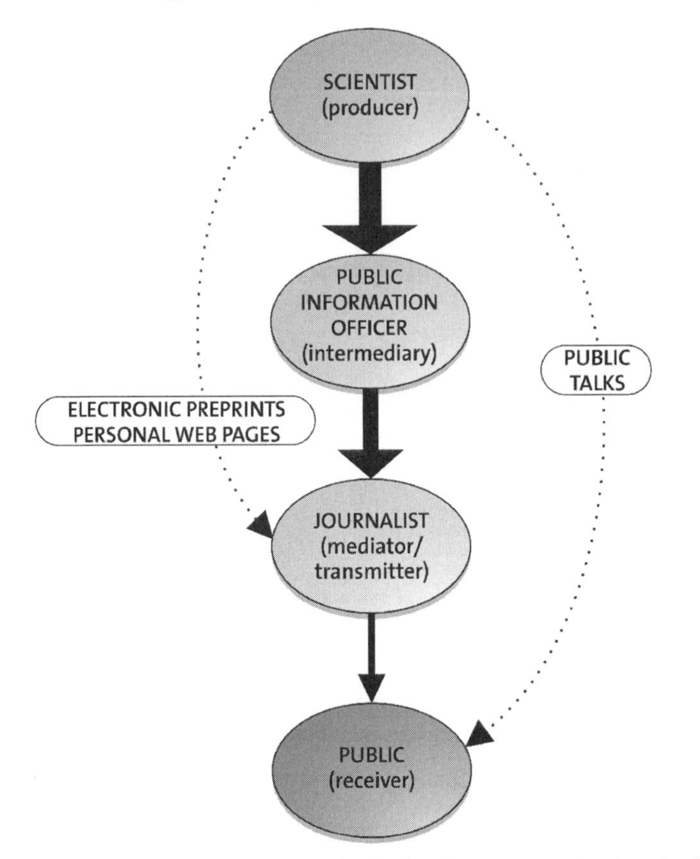

Figure 25.2 The flow of communication in science communication. Credit: L. L. Christensen/ M. Kornmesser.

services – e.g., scientific information, visualization services, science writing, and the like – and create a mutual win–win situation. There is, however, a delicate balance between the players, and there is plenty of room for mistrust to build and problematic issues to arise in the minefield between them. See Figure 25.3.

Some frequently expressed concerns from scientists include:

> "What will my colleagues think?"
> "Will they simplify or distort my results beyond what is reasonable?"
> "I really do not have time for reporters."

There are, however, many good reasons why it makes sense for scientists to participate in public communication:

- to expose the work of his/her specific community;
- to highlight a specific result;
- to highlight the work of an institution;
- to highlight the work of a group;
- to highlight individual efforts (which is perfectly all right!);
- to acknowledge a sponsor;
- to do a favor to the scientific community as a whole (a sense of duty).

Scientist	PIO	Journalist
Values advanced knowledge	Uses the advanced knowledge in a broad context	Values diffuse knowledge
Values technical language	Reshapes technical language into simple language	Values simple language
Values near certain information	Uses facts, but also more speculative indications to give perspective	Values indications
Values quantitative information	Balances facts with emotional and personal accounts	Values qualitative information
Values near complete information	"Cuts through" when the results are trustworthy, but perhaps still not complete	Values incomplete information
Values narrow information	Uses the frontline narrow science to open doors to the broader context	Values comprehensive broad spectrum information
Specialist	Specialist in communicating science to the general public	Generalist
Theorist	Understands theory and applies it in the real world context	Pragmatist
Values knowledge for knowledge's sake	Focuses on the knowledge that is relevant to society	Focuses on what is relevant to society
Is cumulative	Is very picky with which information to accumulate	Is non-cumulative
Is slow	Can develop stories over long time, but always delivers on time	Is fast
Enjoys high professional status	Respects all other actors	Is in the lower ranks of professional status

Figure 25.3 The two main players in science communication, the scientist and the journalist, have very different views on the world. The Public Information Officer sits between them and mediates. Credit: L. L. Christensen. Compiled with inputs from Valenti (1999).

25.2.3 The EPO office

If we look at the education and science communication "space" as a series of activities that can be grouped loosely into categories, different communication products such as promotional videos, news releases, and posters will move along the horizontal axis depending on

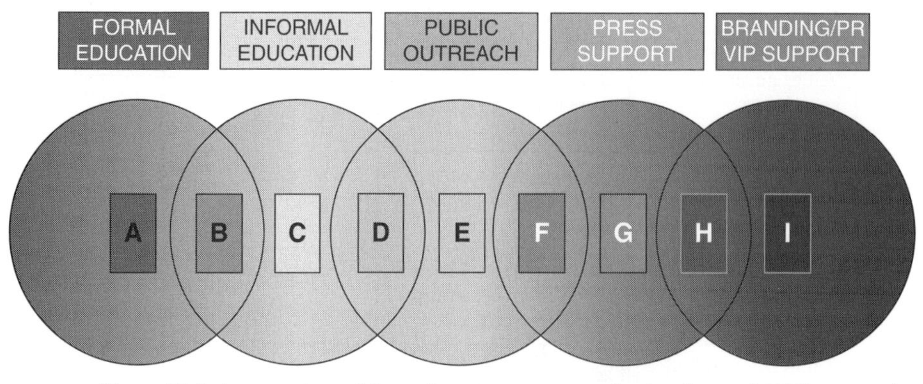

Figure 25.4 An overview of the entire science communication "space." Different products will move along the horizontal axis depending on their target group and content. Curriculum-driven formal education is seen to the left, and the more PR-oriented activities to the right. Inspired by Morrow (2000). Credit: L. L. Christensen/R. Y. Shida.

their target group and content. Curriculum-driven formal education is seen to the left, and the more PR-oriented activities to the right. In Figure 25.4 the letters from A to I designate the different activities or products.

A: Curriculum-driven: textbooks, teacher training, undergraduate courses, . . .
B: Educational programs at planetariums, museums, libraries, parks, . . .
C: Museum exhibits, observing trips (eclipses, comets, . . .), star parties, . . .
D: Planetarium shows, IMAX movies, public talks, hands-on demos, . . .
E: TV/radio documentaries, Podcasts, magazine articles, popular books, Webchats, blogs, cultural/scientific events, CD-ROMs, . . .
F: Photo releases, popular brochures, . . .
G: Press releases, press conferences, press kits, video news releases, media interviews, media courses for scientists, . . .
H: Exhibition booths, technical brochures, newsletters, annual reports, posters, postcards, . . .
I: Merchandise: pins, stickers, caps, t-shirts, bookmarks, mugs, . . .

The production of any science communication product will involve a mix of skills from scientific through graphical to technical skills (see Figure 25.5). The perfect skills base in a professional communication office should, consequently, be a mix of communicators, designers, technical staff, educators, and managers.

25.2.4 Some messages

In the text above I have a provided a framework illustrating the methods and strategies applied in our work in the ESA Hubble office in Munich, Germany. More are described in Christensen (2006). Before moving on to a particular case study of communication in action – the planet definition debate at the 2006 General Assembly – I would like to conclude section 25.2 with three selected messages.

1. Although there is much in common in the worlds of scientists and journalists, there are also many differences, which can lead to conflicts. A good communication collaboration starts with information and an openness to understand the work of the other actors.

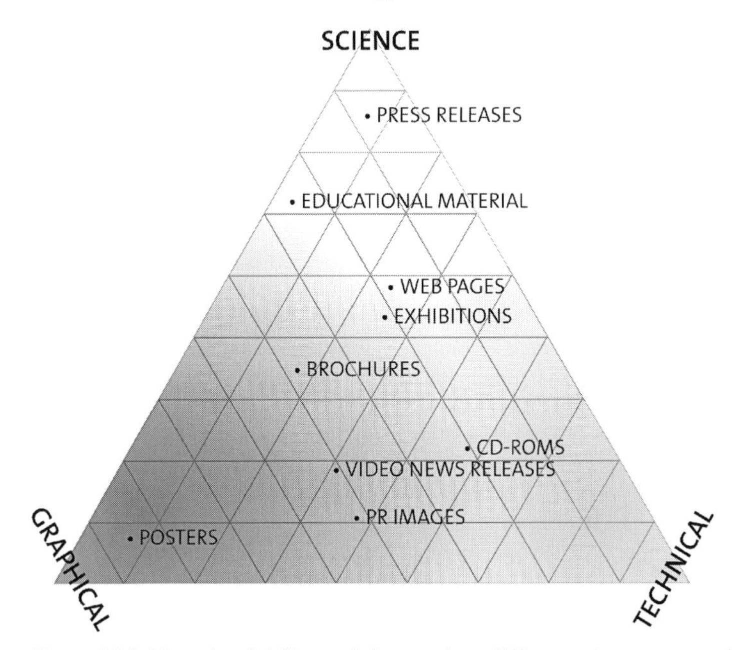

Figure 25.5 The mix of skills needed to produce different science communication products. Credit: L. L. Christensen/M. Kornmesser.

2. Science communicators may benefit greatly from internal collaboration within their community. The advantages of exchanging ideas and sharing resources far outweigh the disadvantages of helping colleagues who, in some senses, may even be in direct competition with you.
3. In the coming years, our knowledge-driven society will demand proper political attention to, and proper funding for, science communication. The future of science depends on our ability to spark scientific interest in the younger generation.

25.3 The communication of the planet definition discussion

As the newly appointed press officer for the International Astronomical Union, I had to build up a press office at the IAU General Assembly in Prague. As a result I witnessed the development of one of the most publicly visible astronomical stories of 2006, perhaps of the decade – the reclassification of Pluto as a dwarf planet. This particularly lively episode can be used as a vivid illustration of the various conflicts and challenges that have to be met and resolved by the Press Office as part of its everyday work.

The 2006 General Assembly of the International Astronomical Union was a General Assembly like no other. Many different interesting scientific topics were discussed – from near-Earth objects to galaxy evolution across the Hubble time. But there was one issue that was by far the most hotly debated in the corridors at the Prague Congress Center, and also took most of the media limelight – how to define a planet. This apparently non-scientific issue obviously has some strong cultural roots and a very high public visibility.

The planet definition debate that took place at the IAU 2006 General Assembly was clearly the hottest topic in many years, if not in the entire history of the IAU. Unfortunately many

Table 25.1 *"Planet definition" timeline*

2001	The American Museum of Natural History decides to leave Pluto out of its scaling walk. The story hits the news and becomes a high-profile item.
2003–2005	Discovery of large trans-Neptunian objects including Sedna and Eris.
2004–2005	The first IAU Planet Definition Committee outlines a few different possible definitions of a planet.
01.07.2006	Second IAU Planet Definition Committee meets in Paris and decides on a definition.
01.08.2006	First Resolution draft approved by the IAU Executive Committee.
14.08.2006	The XXIVth IAU General Assembly starts.
16.08.2006	First Resolution draft is communicated to media and scientists.
18.08.2006	IAU Division III Business Meeting.
22.08.2006	Executive Committee plenary discussion.
24.08.2006 08:00	Press release: "Final Resolution ready for voting."
24.08.2006 ~15:40	Resolution 5A is passed, 5B is not passed. Pluto is defined as a dwarf planet.
24.08.2006 16:21	Press release: "Result of the IAU Resolution votes."
04.09.2006	A petition with almost 400 signatures protesting the decision is delivered to the IAU.

other pieces of interesting science were somewhat overshadowed during the intense planet definition discussion, but it is not possible to control the media.

25.3.1 Rationale for a planet definition

There was no scientific definition of a planet when Pluto was discovered in 1930. The ancients thought of planets as wanderers or moving lights in the sky, and more recently astronomers have considered them simply as bodies orbiting the Sun. There seemed little reason to define more precisely what a planet really was, as it seemed that very little ambiguity could arise. With the advent of modern telescopes, however, it was discovered that Pluto belongs to a vast population of small Solar System objects in the Kuiper belt. Such recent discoveries have prompted astronomers to reconsider the definition of what makes a planet a planet. It was therefore proposed that the term planet should be properly defined, and that the definition should reflect our current understanding of the Solar System.

25.3.2 The events

An approximate timeline of the events around the "planet definition" decision is outlined in Table 25.1.

IAU Resolution 5A implies that a planet in our Solar System (extrasolar planets are specifically not included in this definition) is a celestial body that is in orbit around the Sun and has sufficient mass to become nearly round (due to its self-gravity), and dynamically dominates its orbital zone. The actual interpretation of this definition – especially whether a given body is round enough and dynamically important enough – will have to be discussed by the appropriate IAU body as each new case arises. The resolution also defines a dwarf planet in our Solar System to be a celestial body with sufficient mass to assume a nearly round shape, but not dynamically dominant in its orbital zone. Resolution 5A had the immediate effect that Pluto was reclassified as a dwarf planet along with Ceres and Eris (formerly known as 2003 UB_{313}). See Figure 25.6.

The new definition of a planet has – as predicted – provoked strong reactions from both the public and the astronomical community that still persist years after the General Assembly.

Figure 25.6 A historic moment? The crucial vote that passed IAU GA XXVI Resolution 5A and so reclassified Pluto as a dwarf planet. Image credit: International Astronomical Union/ Robert Hurt (Spitzer Science Center).

Any decision on a topic of this magnitude and importance will inevitably generate a barrage of negative reactions. The current opposition is, in other words, unavoidable. Judging from the ongoing public and internal communication, the main resistance against IAU XXIV Resolution 5 seems to stem from a vocal minority of astronomers.

25.3.3 Crisis communication

Before the General Assembly we had to choose just how open to be with the scientists, press, and public during the process of deciding on the new definition of a planet. An internal working paper written before the General Assembly predicted: "The planet issue has the potential to become an historic event of epic proportions. It may become the hottest astronomy story of the year, or even the decade. It has the potential to change history. Seeing this as a potential historic event, do we fulfill our public duty and inform the world about the process and the decisions openly, or do we keep quiet to protect the slow and thoughtful scientific work process?"

It was already clear that we were dealing with a very special situation. Shortly afterwards, in recognition of the possible negative effects that an improper public communication could have, the situation around the "planet definition" debate was declared a crisis. In crisis communication there are some general rules. (See Christensen, 2006, for more on crisis communication.) The main thing is not to react hastily and let the outside world dominate your decisions. Be proactive rather than reactive. Some guidelines apply:

- communicate internally first to avoid internal confusion and enable all involved to work towards the global goal;
- plan ahead as early as possible;
- react as quickly as possible – the timescale is usually counted in minutes and hours;
- be available via cell phones, e-mail, etc.;

- be credible and fact-based in the external communication;
- apply analytic working methods;
- be transparent, open, and honest;
- be ready to compromise several times along the way in order to achieve the global goal at the end of the process. This point is notoriously difficult to accept as it goes against normal management practice.

25.3.4 Worst–case scenarios

As we were planning for the "planet crisis," we considered a series of worst–case scenarios.

1. Lack of communication
   - A polarized "Them and Us" situation could arise in the media: the press (and public) are largely held outside the process and are not properly informed, leading to public outcry over the secrecy of discussions among senior cigar-smoking astronomers in a "closed club."
   - Leading opinion-makers from cultural, art, and religious backgrounds will speak publicly against "this lab-coat nonsense" and create a global surge of protests. Possible political intervention? Demonstrations? Violence?
2. Communication is too simplistic
   - The issues around the resolution are communicated widely, but its tentative/draft character is omitted in the public communication. In the end a resolution is not passed, and the press and public feels led astray. The IAU comes out looking bad.
3. Broad disagreement
   - The majority of the community disagrees. Resentment? Demonstrations?
   - The majority of the public disagrees. Resistance to the redefinition of the "labeling" of the Solar System, and the modification of geography books. Resentment? Demonstrations?
4. Perception of anti-Americanism
   - Pluto's status will change from "planet" to "dwarf planet," creating a feeling of anti-Americanism on the part of the US (as the IAU is seen by some as a predominantly European organization).
   - Under political pressure, or spontaneously, NASA or the US planetary science community may develop its own categorization for objects in the Solar System – such as developing the "ice dwarf" category using criteria other than those proposed by the IAU.
   - The American Astronomical Society may be asked to develop policies on this and related issues that provide "American" alternatives to the "European" ones of the IAU.
   - US astronomers may be lobbied (for example, by the Planetary Society) to withdraw from the IAU as individual members.
   - An individual member or members of Congress (possibly from Arizona) might be lobbied to move for the US to withdraw from the IAU at a national level.
   - To generate ammunition for political lobbying, the Planetary Society may conduct a poll of the US public on the status of Pluto.
   - The New Horizons team may perceive that a change in Pluto's status will weaken its funding status, and lobby the IAU Executive or members for any change in Pluto's status to be delayed (or, if it is changed, reversed).

- The family of Clyde Tombaugh may protest against Pluto's change in status.
- Flagstaff Observatory is likely to maintain its current displays and materials about Pluto.
- New Mexico State University may continue to refer to Pluto as a planet and Clyde Tombaugh as its discoverer.
- US book publishers, planetariums, and generators of online content may be slow to change their current material on Pluto and its discovery, if they change it at all. They may do this spontaneously: they may also be lobbied to do so.
- Individual schools in the US may be slow to change what they teach about Pluto and its discovery, if they change it at all.

After considering these four hypothetical scenarios the IAU Executive Committee decided to make the process leading to a Resolution as open as possible. Fortunately none of the scenarios played out as visualized above, although scenario 4 came closest, with protests from part of the American astronomy community and the Tombaugh family.

25.3.5 Lessons learned

Many interesting lessons were learned in the press office during the "Pluto affair," especially about the practicalities of setting up a well-functioning pressroom in response to the crisis, but also about the complex ways that information is transmitted from scientists to the press.

Once the IAU Executive Committee – the IAU's highest body – had made the decision to propose the new definition of a planet, the whole issue was somewhat like having two bombs waiting to explode. The first bomb was the public reaction to changes in the worldview – adding or subtracting planets to the Solar System – and the second was the internal tension within the scientific community due to differences of opinion and the appointment of some selected experts to work on the definition. Our job was to try to minimize the negative effects for the IAU and for astronomy, and to maximize the benefits from the two explosions. These explosions themselves were probably unavoidable, but we could at least make sure that the bombs were thrown in a certain direction rather than exploding in our faces. For the first explosion – the public bomb–damage control consisted of keeping the process as open as possible and informing the press about each step of the process as it took place – including the first resolution draft and the ongoing debate. As many as thirty journalists had already signed up weeks before the meeting, and it was well known among science journalists that the definition of a planet was going to be discussed, suggesting a strong outside interest that spoke forcibly for an open communication strategy. It would not have been possible to keep the planet definition debate out of the press. Press releases were issued to deliver all the relevant information; press and public speculation was thus minimized, although not completely eliminated. See Figure 25.7.

It is difficult to speculate how the image of the IAU or the astronomical community might have been affected had a more closed form of public communication been chosen. It is more than likely that the not always constructive messages from many prominent and outspoken astronomers would have reached the press. The open communication did avert most of the potential criticism that the planet definition process took place as closed discussions among senior cigar-smoking astronomers in a "closed club."

With respect to the second bomb, the strong reaction from the scientific community was somewhat underestimated by most of the Executive Committee and the Press Officer. The majority of us also did not anticipate the significant changes to Resolution 5 that took place

Figure 25.7 The GA 2006 pressroom at a time of hectic activity. The episode was later included in Jenny Hogan's (left) blog on nature.com.

during the General Assembly. With 20–20 hindsight, the draft aspect of Resolution 5 could have been stressed more in the initial press release. The "inreach" aspect – sharing the draft resolution earlier with the community (especially Divisions I and III) – could perhaps have been given more emphasis, but this was difficult for two reasons:

(1) the Executive Committee feared that the resolution text would leak to the entire community and to the public, without the Executive Committee and the Planet Definition Committee having a chance to add the necessary scientific context, historical background and interpretation;
(2) the resolution itself was drafted shortly before the General Assembly, and practical considerations made it difficult to initiate discussions with the hundreds of members of Division I and III (collecting e-mailing lists etc.).

25.3.6 Outcome

The planet definition affair has definitely had some negative effects. Astronomers and scientists in general have been publicly portrayed as being in disagreement, arguing, and, at times, even being childish in their discussions. The positive side of this is that astronomers and scientists have appeared as human beings and far from their usual "lab-coat" image. The IAU has also been publicly accused of being a "closed club" that only represents a fraction of astronomers.

In my opinion, however, the positive effects outweigh the negative by far. One of the most important outcomes of the public communication from the General Assembly is that the public today has a much better knowledge of the Union and its mission as the authority on fundamental astronomical issues. The enormous public interest in the planet definition story is perhaps best illustrated by the large number of cartoon jokes/caricatures appearing in the international newspapers. It is the first time in many years to my knowledge that any scientific topic has penetrated so deeply into the public conscience. The effect of this is not to be underestimated. Scientific issues are usually notoriously difficult to get on the front pages (although astronomy usually stands a better chance than most other sciences). The value of this is enormous, despite the unavoidable negative effects described above.

For once, a large fraction of the demographic segment inattentive to science was exposed to science. A small-scale poll among friends and family found that everyone had heard of the Pluto story, and most even offered an opinion about it. This is an important consequence and should not be underestimated. In terms of public communication, it is vital that the current high awareness of the Solar System be used to promote scientific issues. There is great potential to use this debate to teach about the Solar System: about the fact that it is still in formation; about debris; about asteroids; about dwarf planets, Kuiper Belt objects, trans-Neptunian objects, planets, and more. This is a great opportunity to teach that science is not static, and that when new discoveries are made, science must change. In the longer term, the increased awareness of the IAU as a result of the "Pluto Affair" can be used to further the interest in the International Year of Astronomy 2009.

The re-classification of Pluto as a dwarf planet should not be seen as a demotion. Pluto is now the prototype for a whole new class of objects. Pluto is a swan, not an ugly duckling, and we should all celebrate that it has finally been placed in a class of its own. After all Pluto is still Pluto, and what we decide to call it changes nothing about Pluto itself.

Acknowledgments

I would like to thank Martin Kornmesser and Raquel Yumi Shida for helping with the graphics and the pressroom team for their dedication, hard work, and good spirits during the General Assembly! I would also like to thank the Executive Committee and the Definition of a Planet Committee, especially Richard Binzel, for the incredible amount of hard work they put into the planet definition debate and the various topics dealing with communication. ESO and the ESO Public Affairs Department deserve special thanks, as they partly sponsored the loan of a large fraction of the pressroom equipment and partly sponsored and arranged the transport of the IAU exhibition.

References

Christensen, L. L., 2006, *Hands-on Guide for Science Communicators*, New York: Springer.

Morrow, C., 2000, Guidance for E/PO program planning and proposal writing (Boulder, CO: Space Science Institute), www.spacescience.org/education/extra/resources_scientists_cd/source/Venn.pdf and tinyurl.-com/2r68hd.

ROSES (Research Opportunities in Space and Earth Science), 2006, NASA www.nasa.gov/audience/foreducators/postsecondary/features/F_plenty_of_Roses_for_you.html.

Valenti J. M., 1999, Commentary: how well do scientists communicate to media?, *Science Communication*, **21**, 172–178.

Comments

J. V. Narlikar: There is an International Union of Science Communicators (IUSC) set up recently. You will find its website at www.IUSC.in. Its aims are similar to those described in this lecture. I strongly urge those interested in science communication to join it.

Jay M. Pasachoff: See also en.wikipedia.org/wiki/International_Union_for_Science_Communicators.

Magda Stavinschi:

1. Why is the meeting for 2009 International Year of Astronomy in parallel with other ones?
2. What links are there with Commissions 41 and 50?

Lars Lindberg Christensen:

1. It is indeed never optimal to have meetings run in parallel, but given the number of parallel tracks at the General Assembly (up to 14 simultaneous ongoing meetings) and the many people that had to be present, this was the best compromise that could be achieved.
2. The IYA2009 organisers will seek to connect up with as many groups and bodies inside and outside the IAU – especially IAU commissions 41 and 50. The best thing to do for those interested is to propose specific IYA2009 initiatives and try to raise funding for them as early as possible.

26

Getting a word in edgeways: the survival of discourse in audiovisual astronomy

T. J. Mahoney

Instituto de Astrofísica de Canarias, E-38200 La Laguna, Tenerife, Spain

Abstract: Astronomy communicators and educators today enjoy a sophisticated array of audiovisual packages for getting their message across to audiences, but is the ratio of text to illustration always effectively balanced? Images often fail to illustrate the meaning of the text and instead simply adorn it. Images often drive discourse, instead of discourse determining the choice and quantity of illustrations. Astronomy educators involved in selecting audiovisual aids in their teaching, or in the preparation of such products, need to weigh the effectiveness of the media they use. I suggest some criteria for getting the balance right between text and audiovisual content. I further argue that the text of a discourse must always be firmly in the driver's seat. I look at examples in printed and audiovisual matter where the influence of imagery and flashy graphics has led to presentation being given priority over content.

26.1 Introduction: the ultimate "minds-on" activity

Astronomy is a wonderfully practical subject for the lay person and offers many spectacular sights and phenomena within the reach of small telescopes or the naked eye. With the advent of the Internet, which opens up the possibilities of the Virtual Observatory and a plethora of public access astronomical and planetary science databases, astronomy has become even more of a "hands-on" subject.

Nevertheless, there are aspects of the Universe that appeal more to the mind's eye, and that cannot be seen in a physical sense. Such questions as the origin of the Universe, the existence of sentient beings in other planetary systems, etc., have all fallen within the scope of astronomical science, which has made great strides over the past century. Many recent observations and theories deal with objects that either cannot be seen at all (e.g., black holes, dark matter, and dark energy) or can only be dimly discerned (e.g., brown dwarfs). Indeed, much of modern astronomy is carried out at wavelengths that are totally invisible to the human eye. Although there are graphic techniques available to visualize the invisible, it is clear that these introduce an element of unreality into the phenomena concerned (see Henbest, 2005), and there is necessarily a heavy reliance on words to describe what is being represented. It is an audience's *mind* that must in the end be appealed to through the medium of words.

Astronomy is the ultimate "minds-on" activity.

26.2 What is astronomy *really* about?

Our understanding of the Universe is hindered by two factors.

(1) Images of the sky are linear, but our appreciation of length and size on a cosmic scale is logarithmic. (Astronomers use parsecs, kiloparsecs, and megaparsecs when talking about stars, individual galaxies, and clusters of galaxies, respectively.)

Innovation in Astronomy Education, eds. Jay M. Pasachoff, Rosa M. Ros, and Naomi Pasachoff. Published by Cambridge University Press.

(2) The Universe presents itself to our senses as a two-dimensional surface (the sky). Our current four-dimensional appreciation of the Cosmos has necesitated highly refined observing and calculational techniques to break the two-dimensional degeneracy forced on us by our earthbound perspective. We need consciously to acquire knowledge about the three dimensions of space and the added dimension of time; we could never intuit them unaided.

The study of astronomy is required to give a proper sense of scale and to effect a truly meaningful de-projection of our two-dimensional sky perspective. Images of the sky by themselves tell us nothing about the nature of the Universe and are likely to deceive us unless we possess sufficient background knowledge to help us interpret the imagery. Words are the most effective vehicle for acquiring and transmitting this knowledge.

Humans have evolved spoken and written language to enable them to pass their thoughts down the generations by means of discourse. They have developed pictorial and (more recently) animation skills, too, that help in the transmission of ideas. Pictures, artefacts, and animation sequences by themselves, however, are poor media for discourse: how much more vivid and meaningful would the cave paintings of Altamira or the circles of Stonehenge be to us had their creators been able to leave written accounts explaining the motivations behind these legacies! It is likewise with astronomical images. Stunning as are the images taken with the Hubble Space Telescope, without the aid of much explanatory text they are remarkably uninformative except to those with sufficient astronomical background to decipher the information contained in them. Even astronomers themselves have to struggle to comprehend some of the images. (The Hubble Deep Field, for example, is still a topic of frenetic research activity.)

Words are the framework of discourse. Images and audiovisual content may either complement the text (provided they are used illustratively rather than decoratively) or even take discourse beyond the purely verbal, but they can never replace text and often need text for their interpretation. In his comparison of painting and poetry, English essayist William Hazlitt (1778–1830) might well have been talking about imagery and discourse instead: "Painting gives the object itelf; poetry what it implies." Words themselves, of course, are limited in what they can express. And, besides, do not some who work with astronomical images speak of the "language of images" (Rector *et al.*, 2005; Levay, 2005)? So can astronomical images not be made to "speak"? Of course they can and do speak, and they can often say pictorially what would takes reams of text to explain in words. To the often-heard cry from some popularizers, "Never show a spectrum!," Fosbury (2005) replies that much of modern astronomy is plagued with "difficult concepts," such as spectroscopy, polarimetry, and observation at invisible wavelengths, and explains how such ideas can be put across in terms of simple examples. A blue sky is an example of polarized light that would have a direct and emotional appeal to any audience. Once the audience has been regaled with stunning images and video sequences, the presenter must always be prepared with an answer to the question, "But how do you know that?" (Fosbury, 2005).

26.3 Text, sound, and image: three elements of modern discourse

In addition to traditional text and images, popular astronomy communicators nowadays have a wide variety of audiovisual techniques at their command to help get their message across to the public. This very choice, however, can often cause the message to be overwhelmed by images, sounds, or video sequences that do not really illustrate meaning.

These new communicational tools and techniques are clearly of great use in the classroom and lecture hall – provided they are used wisely and always with a view to illustrating the message of the text rather than merely adorning the page (or screen). Such distraction is mere noise drowning out the underlying signal (which, one hopes, contains an intelligent message). As in astronomical observing, the highest possible signal-to-noise ratio should be the communicator's overriding aim at all times. According to the *Concise Oxford Dictionary*, the definition of an illustration is "an example serving to elucidate."

The litmus test of any illustration is, then, "Does it elucidate?"

26.3.1 Text

If it is to be effective, written or oral discourse requires a structured text with specific communicational goals. But even in a well-structured text with a clearly stated aim there are many potential obstacles to clarity and impact of message, including:

- disproportionately large images,
- over-use of images,
- irrelevant images that serve only as a distraction,
- uncaptioned images,
- unlabelled or illegible graph axes,
- indiscriminate use of designer fonts,
- titles and section headers overworked to render them "catchy" (very common in science magazines).

Images as mere adornment are a pointless distraction in any form of discourse. Overly large images or the use of too many illustrations is wasteful of valuable space that might otherwise be dedicated to useful text. In order to sell in sufficient quantities to make a profit, popular magazines and non-fiction for children must look attractive and eye-catching. But professional periodicals are increasingly being given the glossy-magazine look and editorial philosophy, an example being the transformation of the former *Quarterly Journal of the Royal Astronomical Society* into its shiny successor *A&G* (now the official title, shortened from *Astronomy and Geophysics*, "giving the magazine a snappier look,"[1] according to an editorial comment; Bowler, 2005).

26.3.2 Sound

Radio communicators rely exclusively on words and sound effects to get their message across to listeners quickly (time is always limited in radio broadcasting). Two tools they use are the "sound bite" and the "word picture." The sound bite is an incisive sentence or phrase that maintains the listener's attention; the word picture uses vivid pictorial vocabulary to help listeners visualize what they are hearing. (See Rodríguez Hidalgo, 2005 for a clear discussion of radio outreach.) In education at the undergraduate level, the Open University has demonstrated

[1] *A&G* had ceased to be a "journal" except in name, with the mission statement of the first issue (*Astronomy and Geophysics*, 1997), in which special emphasis was given to topical (news) items. About 20% of page space is given over to pictures, as compared to about 5% in *Quarterly Journal*. Another magazine-type feature is the use of explanatory boxes, which would be out of place in a journal aimed at peers. While review articles are still a feature of *A&G*, authors are no longer free, as they were with *QJ*, to take the page-space they need to develop their themes: article length is restricted to a maximum length of 3000–5000 words. Each issue of *QJ* was as long as it needed to be, whereas *A&G* is fairly fixed in its length. In comparison with *QJ*, overuse of eye-catching images in *A&G* has led to a substantial reduction in quantity of discourse, as might be expected when a journal becomes a magazine.

with its unparalleled radio broadcasts and audio study aids that sound, in combination with printed text, is an excellent medium for serious discourse in a learning environment.

While the sound bite and word picture are useful tools, they are not by themselves discourse. The sound bite is for emphasis, and word pictures are illustrations that enhance, but are no substitute for, the underlying narrative. Both tools need to be crafted into the discourse to give the occasional fillip to listener interest. Sound effects might occasionally be helpful to the narrative but must always be relevant and not just background noise.

If used wisely, sound can be used to great effect in discourse. Ballesteros and Luque (in the session on General Strategies for Effective Teaching at this conference; see Chapter 18) give excellent examples of the way in which sound and sonifications can be made to enhance presentations on such topics as space missions and radio signals from various kinds of celestial objects, including pulsars, black holes, and aurorae.

Silence can also be used to dramatic and eloquent effect (e.g., the terrifying absence of sound in a nuclear explosion before the arrival of the shock wave, as portrayed in the film *The War Game*, 1965).

26.3.3 Images and animations

The Space Age reached its fiftieth anniversary in October 2007. The fifties also brought about revolutionary advances in radio astronomy. In a relatively short time, astronomy opened up new windows right across the electromagnetic spectrum. Prior to the fifties, virtually all astronomical observations were made in the visible (i.e., that part of the spectrum to which the human eye is sensitive). The rapid development of astronomy at invisible wavelengths led to an avalanche of discoveries of new kinds of celestial objects, many of them detectable only at invisible wavelengths. How were they to be imaged? Were such "images" truly representative of the objects themselves? Henbest (2005) lists black holes, extrasolar planets, extraterrestrial life, and the Big Bang as themes that set a difficult challenge in terms of visualization for television programs.

These concerns affect all kinds of audiovisual presentations of astronomical topics, whether educational or for entertainment purposes, and are intrinsic to the subject itself. What may be called a structural part of the television industry, however, is the practice of gauging a programme's audience-grabbing potential against high-cost graphical techniques employed in advertisements screened immediately before and after (and possibly during!) the program (see Henbest, 2005). Faced with such well-funded competition, astronomy and other science programs are obliged to follow suit as best they can with spectacular imagery and graphics presented at a breathless pace (to prevent viewers from zapping). Henbest differentiates between the general scientifically-unaware public and the relatively small group of science diehards who actively seek out such programs as the BBC's *Horizon* or *Sky at Night*. *Horizon* itself has now been "livened up" to make its presentation snappier and more appealing to a wider audience, whereas *The Sky at Night*, the world's longest-running television program with the same presenter (Sir Patrick Moore), with its fiftieth anniversary imminent, was rescheduled to the wee hours. It was also available on the web (www.bbc.co.uk/science/space/spaceguide/skyatnight/proginfo.shtml).

As far as popular and semi-popular astronomy programs are concerned, the reasons for such change of tone and rescheduling are purely commercial, and little can probably be done to make programs more serious or to schedule them in prime-time slots (when children might be watching, for example). But educational programs have also suffered. In the UK the Open

University's renowned course-related TV programs are now broadcast well past midnight to make way for more popular programs in the evenings.

Educators who use or refer to popular science programs need to be aware of a number of media malpractices so that they can at least identify them to their students. The list of offenses is a long one and includes:

- over-preoccupation with audience *attention span* (discourse requires audience *interest*);
- overuse of concatenated ultra-short sequences (two seconds or less) that do not give the audience time to keep track of what is going on;
- crude humorous editorial techniques (speeded up sequences, silly sounds, etc.) that are supposedly meant to keep up the audience's attention from flagging but fail to engage its curiosity in the narrative – the documentary-maker's equivalent to the custard pie fight in the early cinema;
- continuous use of music (or even Muzak) to fill a perceived "sound vacuum";
- failure to simplify difficult concepts in a way that is satisfactorily representative (leading to "dumbing down" in the worst instances);
- failure to match image to commentary (very common in news reports on astronomy);
- failure to use images (both still and video) as part of the narrative – the dynamic wallpaper syndrome;
- tongue-in-cheek editing of interviews.

It might make an interesting class exercise to get students to identify examples of hype, misrepresentation, or sheer mistakes in television programs!

26.4 Attention versus interest

The media are concerned with *attention span* and see the public as a dull beast that needs constant prodding and jolting to stop it from falling asleep (or, worse still, changing channels).[2] This they achieve through a battery of camera and editing techniques, as well as a liberal use of all sorts of sounds. Anything serves that will keep the beast's eyes open and its finger off the zapper.

Communicators might consider concentrating instead on maintaining audience *interest*. This goal has been successfully achieved in the past with such television series as Jacob Bronowski's *The Ascent of Man*, Carl Sagan's *Cosmos*, and David Attenborough's *Life on Earth*, all of which had high ratings in their day. Peter Watkins's *The War Game*, a "docu-drama" giving a grim fictional description of the lead-up to and aftermath of thermonuclear war in Britain, was made in 1965. Alarmed at how the public might react to what potentially lay in store for it should the Cold War ever heat up in the way described in the documentary, the BBC waited 20 years before finally broadcasting the program. *The War Game* was a dramatized enactment of possible consequences of thermonuclear war based on evidence from the bombings of Dresden, Hiroshima, and Nagasaki. There were no histrionics, exotic camera effects, or sounds extraneous to the narrative – and it was all shot in black and white. Nevertheless, the documentary made a great impact on audiences and amply demonstrated that a fairly technical theme (albeit one of undeniable human interest) could be handled in such a way as to command audience attention. Each of these documentaries demonstrates that if an audience's interest is aroused, its attention will be assured.

[2] This is no exaggeration. One well-known advertising agent frequently suggested to his colleagues, "Let's get down on all fours and look at it from the client's point of view."

26.5 Some tips on improving discourse

1. Whatever your medium, define your message and shape it into a coherent discourse.
2. Make the discourse drive the illustrations, not vice versa.
3. Use only those images and audiovisual aids that further or enhance the discourse.
4. Exclude all other images and audiovisual products.
5. Always write captions for your illustrations.
6. Label all graphs fully and legibly.
7. Don't use any sound effects (especially silly noises and Muzak) that don't contribute to the discourse; they simply distract.
8. Use of humor should be sparing and always relevant to the discourse.
9. If you do use sound effects, don't forget to use silence; it can be quite dramatic!
10. When selecting or producing video sequences, avoid editorial clowning (speeded-up sequences of people, etc.) and questionable editing of interviews.
11. Bear in mind that radio is much more effective than television in terms of the amount of information it can pack into a time slot, but that television can have great dramatic and visual power (e.g., animations are sometimes essential).

References

Astronomy and Geophysics, 1997, Mission statement, **38**, 5.

Bowler, S., 2005, Editorial: a new look, *A&G*, **46**, 1.4.

Fosbury, R., 2005, Difficult concepts, in I. Robson and L. Lindberg Christensen (eds.), *Communicating Astronomy with the Public* (Munich: ESA/Hubble), p. 130.

Henbest, N., 2005, Science or nonsense? – the role of TV graphics, in T. J. Mahoney (ed.), *Communicating Astronomy* (La Laguna: Instituto de Astrofísica de Canarias), p. 165.

Levay, Z. G., 2005, Hubble and the language of images, in I. Robson and L. Lindberg Christensen (eds.), *Communicating Astronomy with the Public* (Munich: ESA/Hubble), p. 206.

Rector, T. A., Levay, Z. G., Frattare, L. M., English, J. and Pu'uohau-Pummill, K., 2005, Philosophy for the creation of astronomical images, in I. Robson and L. Lindberg Christensen (eds.), *Communicating Astronomy with the Public* (Munich: ESA/Hubble), p. 194.

Rodríguez Hidalgo, I., 2005, Time for space: five minutes a week for astronomy, in T. J. Mahoney (ed.), *Communicating Astronomy*, (La Laguna: Instituto de Astrofísica de Canarias), p. 172.

Comment

Douglas Duncan: There's a practical reason not to use graphics that are too fancy. We cannot compete with computer games, or George Lucas and digital imagery. We should teach using *our* strengths, such as student–student and student–teacher personal communication.

27

A critical evaluation of the new Hall of Astronomy of the University of Mexico Science Museum

Silvia Torres-Peimbert and Consuelo Doddoli

Instituto de Astronomia, UNAM, Apartado Postal 70-264, México 04510, D. F., México; Gral de Divulgación de la Ciencia, UNAM, México 04510, D. F., México

Abstract: On December 2004, the newly designed astronomy hall at Universum, the science museum of the Universidad Nacional Autónoma de México, was opened to the public. It contains displays presented in several sections: (1) Sun, stars, and interstellar matter, (2) clusters, galaxies, and the Universe as a whole, and (3) tools of astronomers. The main concepts developed were the description of each cosmic component and, most importantly, the idea that all components, including the Universe, are undergoing continuous evolution. A set of additional displays rounds off the exhibit: a representation of the vastness of space, a time line from the Big Bang to the present epoch, and some video clips from local astronomers. There is also a section on the history of astronomy. This paper presents the results of an evaluation of the impact of the different elements of this exhibit on visitors.

27.1 Introduction

It is well known that any project undertaken should be evaluated, and if possible any improvements to it should be carried out. In the case of the new Hall of Astronomy for the science museum of the University of México we felt that it was important to assess the impact of the astronomy exhibits on visitors to the museum. We decided, therefore, to observe the behavior of the public attending the hall, and we distributed questionnaires covering different aspects of the exhibit.

27.2 The Hall of Astronomy

27.2.1 Overall description

The science museum of the Universidad Nacional Autónoma de México is located in the cultural section of campus. The museum is comprised of a set of different halls that cover a wide range of topics: from Mathematics, Mechanics, and Chemistry to Energy and Agriculture, among others. It also has a library, a 3-D theater, several auditoriums, and outdoor displays. From the museum's inauguration in December 1992 to August 8, 2006, 7 993 099 people have visited it. Not unexpectedly, the visitors are rather heterogeneous in age, development, and socioeconomic level. Some come to enjoy themselves, others to carry out school assignments; more advanced students also use the museum as a laboratory for their experiments.

Although the Hall of Astronomy was part of the original museum, its displays were mostly centered on the Solar System and were very limited regarding the Universe beyond it. It was thus decided to renew it in two stages, firstly to redesign and construct the section that was not well represented, and in a later stage the Solar System displays will be upgraded. The reconstructed area comprises 60% of the space assigned to astronomy. A more extensive description of this material is given in Torres-Peimbert and Doddoli (2005, 2006).

Innovation in Astronomy Education, eds. Jay M. Pasachoff, Rosa M. Ros, and Naomi Pasachoff. Published by Cambridge University Press.

The themes of the new Hall of Astronomy are: I. Sun, stars and interstellar matter, II. Clusters, galaxies and the Universe, III. Tools of the astronomer. In addition we selected the following topics to complete the hall: (a) a cartoon display to depict important moments of the history of astronomy, (b) a set of pictures that increase in powers of 10, (c) a long strip on the floor to represent the timeline of the Universe, and (d) videorecorded interviews of local astronomers to answer frequently asked questions and to explain their own contributions to science.

27.2.2 Targets of the exhibit

The exhibit was centered on the following ideas.

For the section on stars and the interstellar medium, we prepared (a) descriptions of their most significant characteristics, and (b) the message that all stars undergo constant changes and modify their environment.

For the section on galaxies and the Universe, the purpose was to present (a) a description of clusters, galaxies, and the Universe as a whole, as well as (b) the concept of constant change for all objects involved, and that the Universe itself is undergoing evolution.

For the section on tools of the astronomer, we decided to introduce several displays to describe: (a) the electromagnetic spectrum, and (b) the Doppler effect.

In the history displays, some important astronomers of the past were represented; the pre-Columbian astronomers' achievements were also acknowledged.

Finally, we considered it important to show visitors that Mexican astronomers today are making contributions to modern science.

27.2.3 Goals of the Hall of Astronomy

The goals of the exhibit were:

- to attract young students to science;
- to present to the general public modern-day astronomical results that demonstrate that astronomy is an active science;
- to show that we can interpret the cosmic phenomena by means of the laws of physics;
- to show modern-day concepts of the structure of the Universe and its constituents;
- to show that modern technology has played a major role in increasing our knowledge of the Cosmos.

We knew beforehand that many of the concepts presented are relatively complicated, and that not all visitors – particularly the younger ones – have been exposed to them. We felt, however, that to present a modern vision of astronomy, it was necessary to introduce concepts like multi-frequency observations, etc.

27.3 General evaluation of the Hall of Astronomy

In order to monitor the interest of the public in the different sections, the science museum distributes a daily general survey in the form of a questionnaire for a sample of visitors to fill out. Thus we were able to make use of the global results that apply to the museum prepared by Pérez de Celis (2006).

From January through April 2006 the total number of visitors was 227 354. Of these, 26 657 were surveyed for other questions, while 6011 of them were surveyed for "favorite hall." For this purpose the museum displays have been divided into 22 different topics.

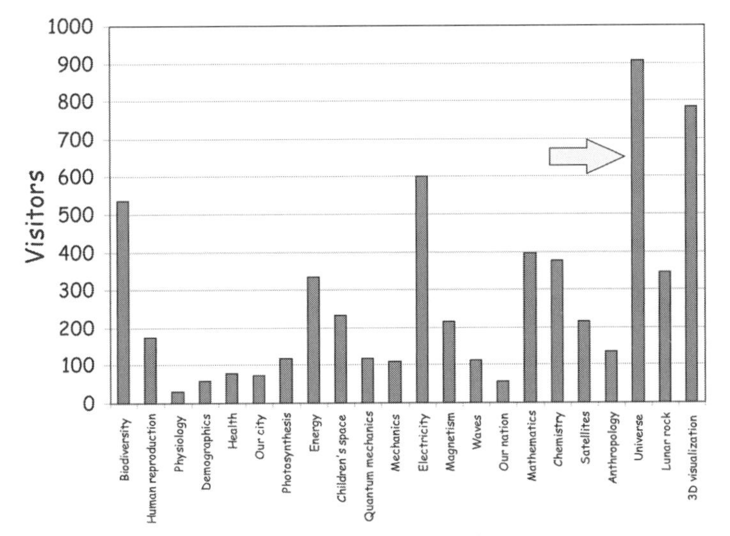

Figure 27.1 Results of the questionnaire about the preferred sections of the museum. This questionnaire was applied to 6011 visitors. The number of visitors to the 3-D visualization hall has been corrected as described in the text. The arrow refers to the Hall of Astronomy. The survey was carried out by Pérez de Celis (2006).

In Figure 27.1 we present a histogram with the results of the survey. The only hall that was opened in mid-survey was the 3-D visualization section. In the histogram presented, we have corrected the preferences of the 3-D visualization section in order to make a fair comparison with the rest of the halls. The result is that 15% of the visitors select the Hall of Astronomy as their favorite one. According to the survey, it is the most popular section of the museum, comparable only to the 3-D visualization hall.

27.4 Detailed evaluation of the Hall of Astronomy

We attempted to measure the interest of the visitors to the Hall of Astronomy in two different modes: (a) we observed the behavior of a sample of 50 visitors, and (b) we asked a sample of 100 visitors to answer a written questionnaire.

27.4.1 Results of the observed behavior of a sample of 50 visitors

The sample of 50 visitors were of different ages, over 12 years old, surveyed at different times of the day and during different days of the week. For each visitor we noted the time spent on each one of the exhibits. We were very careful not to interfere with the visitor; furthermore, the visitor was not aware of our survey. The results of this exercise are presented in Figures 27.2 through 27.4.

In Figure 27.2 we plot a histogram of the number of visitors that interacted with each display vs. the name of the display, presented in decreasing order of interest. The favorite displays were those that are larger and interactive. The most visited one was the infrared camera that provides a live color-coded image of body temperature. This display was approached by more than half of the visitors under observation. The least visited displays are video presentations that are not

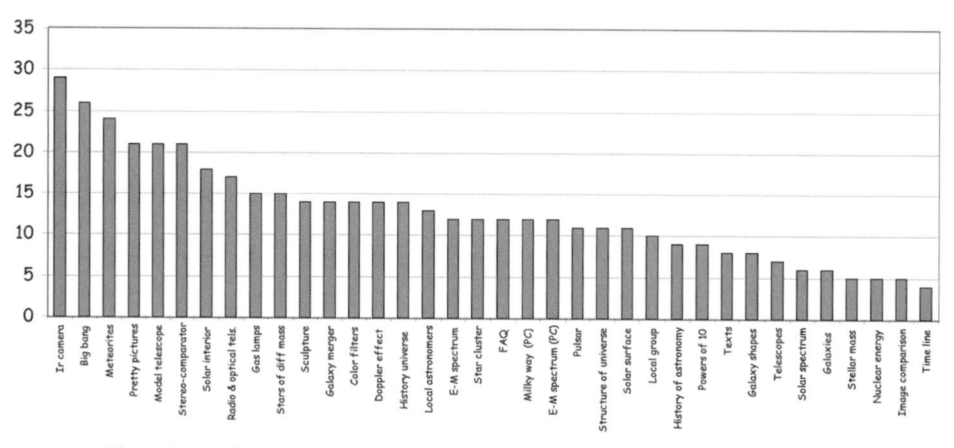

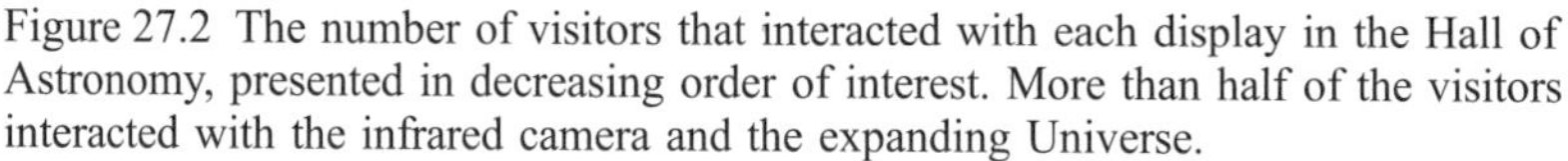
Figure 27.2 The number of visitors that interacted with each display in the Hall of Astronomy, presented in decreasing order of interest. More than half of the visitors interacted with the infrared camera and the expanding Universe.

interactive. Among the least visited ones is the timeline of the Universe; we attribute this result to the fact that it is not well-advertised.

In Figure 27.3 we have plotted the mean time spent on each display, taking into account the full sample of 50 visitors. The plot has been presented in the same order as Figure 27.2, to provide a meaningful comparison. Here the trend has changed substantially. There are two sets of displays that attract the attention of the public: those that were in first and second place in the previous representation (namely, the infrared camera and the expansion of the Universe) and a set of intermediate interest – ones in the previous representation (namely, computer interactive exhibits: electromagnetic spectrum, Milky Way, and attractive videos on the history of the Universe and of real images of solar activity). Although a smaller group of people approached these displays, those that approached interacted with them for a longer period of time.

In Figure 27.4 we have plotted the mean time spent on each display only for the case of those visitors who actually interacted with the display. Here the trend of interest is as expected. Those displays that are more challenging are the ones that keep the visitor longer (electromagnetic spectrum, and the interactive displays with the video interviews of the Mexican astronomers).

27.4.2 Results of the written questionnaire of a sample of 100 visitors

We invited 100 visitors to fill out a questionnaire after visiting the hall. Our form contained a set of multiple choice questions about their opinions on their overall experience, and some open-ended questions to express their concerns about it. Through these interviews we can conclude that indeed we met our goals, at least partially. We made information available. The overall response *for students under 15* was that the astronomical images are very attractive, but that many topics are too complicated. The overall response *for students over 15* was that they found interesting, new information. Furthermore, we have noted that many students are

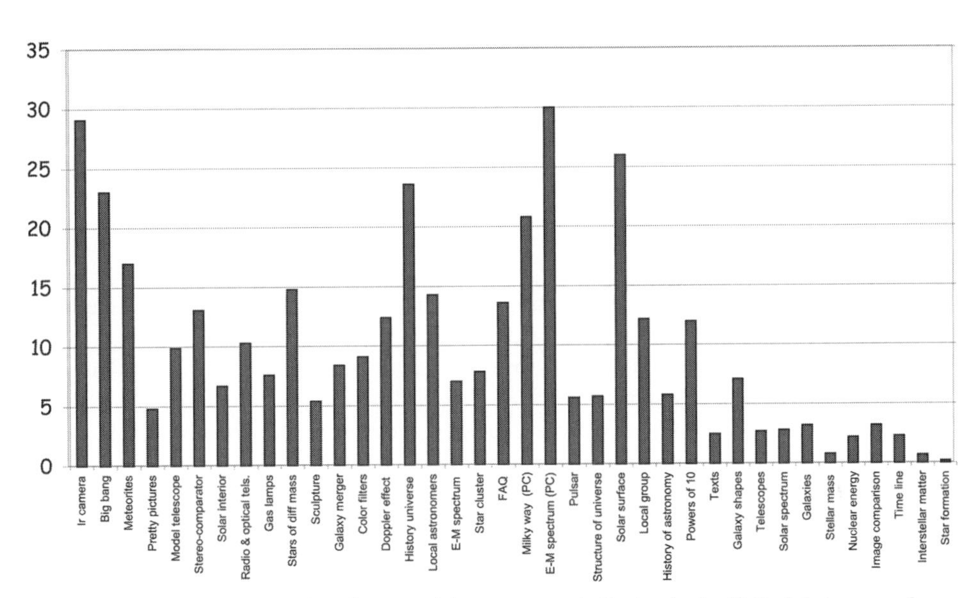

Figure 27.3 Mean time (in seconds) spent on each display in the Hall of Astronomy for a sample of 50 visitors of all ages over 12. Here the values are taking into account all visitors (including those who did not approach the display).

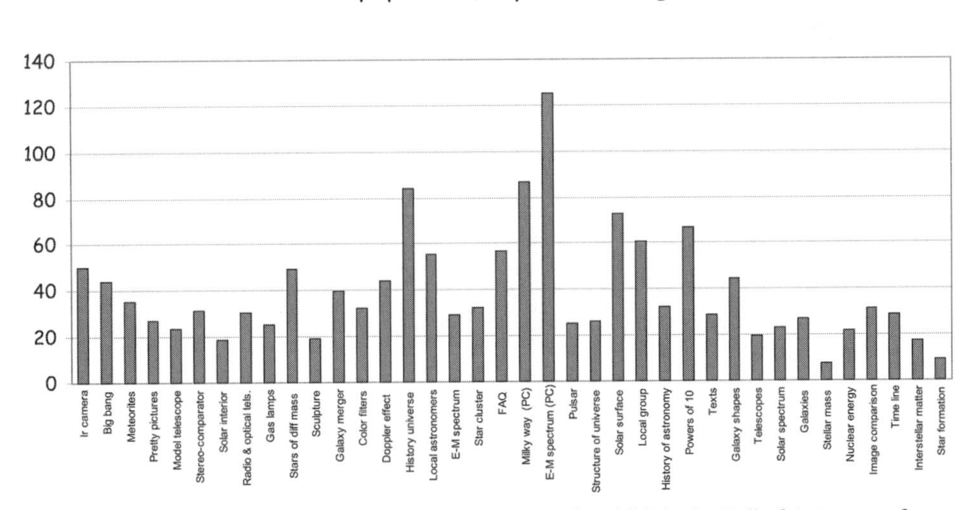

Figure 27.4 Mean time (in seconds) spent on each exhibit in the Hall of Astronomy for a sample of 50 visitors of all ages over 12. For this figure, only those visitors who actually interacted with the displays were included.

being referred to the hall to carry out class assignments. We conclude, therefore, that at least some science teachers find the hall helpful for explaining some concepts.

27.5 Conclusions

We have tried to measure the response of the public by different means available to us. We conclude that:

- in general, the visitors who reach the hall find it very attractive (being the favorite one among the museum visitors);
- the concepts presented in the hall are more appropriate for visitors over 15 years old than for younger visitors;
- from observing their approach to the different displays, we have identified the ones that visually attract the interest of visitors;
- from our measurement of the time spent on each display, we have identified those that catch the interest of visitors;
- we confirmed that not everybody is attracted to the most challenging displays, but these displays allow the interested visitor to delve more deeply into the subject;
- the hall can be of assistance to science teachers.

Acknowledgments

The authors are grateful to M. T. J. Pérez de Celis for making available to us the results of the general survey. STP is grateful to grants Conacyt-46904 and DGAPA-UNAM-IN118405–3.

References

Pérez de Celis, M. T. J., 2006 Informe Opina, (Mexico City: Dir. General de Divulgación de la Ciencia, UNAM).

Torres-Peimbert, S. and Doddoli, C., 2005, *Revista Mexicana de Astronomia y Astrofisica Serie de Conferencias*, **23**, 124.

Torres-Peimbert, S. and Doddoli, C., 2006, La Sala del Universo en Universum (Mexico City: Dir. General de Divulgación de la ciencia, UNAM).

Comments

Jan Vesely: Do you offer also guided tours? Or is it only a "help yourself" exhibition?
Silvia Torres: Yes, on request for school groups.
Rosa M. Ros: Do you have in your museum other sections on astronomy? In particular, do you have a section on archaeoastronomy? On the astronomy of the pre-Spanish period?
Silvia Torres-Peimbert: There are only a few panels on archaeoastronomy. We wanted to emphasize modern results.

28

Revitalizing astronomy teaching through research on student understanding

Timothy F. Slater

Conceptual Astronomy and Physics Education Research (CAPER) Team, Department of Astronomy, University of Arizona, 933 N. Cherry Ave., Tuczon, AZ 85721, USA

Abstract: Which instructional strategies induce the largest conceptual and attitude gains in non-science majoring, undergraduate university students? To determine the effectiveness of lecture-based approaches in astronomy and astrobiology, we found that student scores on a 68-item pre-test/post-test concept inventory showed a statistically significant increase from 30% to 52% correct. In contrast, students evaluated after the use of *Lecture-Tutorials for Introductory Astronomy* increased to 72%. The *Lecture-Tutorials for Introductory Astronomy* are intended for use during lectures by small student groups and to complement existing courses with conventional lectures. Based on extensive research on student understanding, *Lecture-Tutorials for Introductory Astronomy* offer professors an effective, learner-centered, classroom-ready alternative to lectures that does not require any outside equipment or drastic course revision for implementation. Each 15-minute *Lecture-Tutorial for Introductory Astronomy* poses a carefully crafted sequence of conceptually challenging, Socratic-dialogue driven questions, along with graphs and data tables, all designed to encourage students to reason critically about difficult concepts in astronomy and astrobiology.

28.1 Introduction

Because of the perceived ineffectiveness of the lecture-based (instructor-centered) approach to teaching astronomy, the call for a more student-centered approach has been gaining prominence. Our extensive review of the literature, however, suggests that this claim of lecture ineffectiveness has not been validated formally in the specific context of the conventional introductory astronomy course for non-science majors (Bailey and Slater, 2005). As a first response, we in the Conceptual Astronomy and Physics Education Research (CAPER) Team at the University of Arizona have taken it upon ourselves to conduct a case study that documents the degree to which students have conceptual gains on commonly taught astronomy topics as a result of lectures. This can be described as a behaviorist instructional approach in which the student is a passive listener, and the orating instructor acts as the primary source of astronomical knowledge. This research was then used to guide the development of innovative instructional materials.

Every year, more than 200 000 non-science major undergraduate students enroll in an introductory astronomy course (Fraknoi, 2001) in the USA. However, for many of these students, this course is not introductory at all: it is their terminal course in astronomy, and in fact, marks the end of their formal education in science. Introductory astronomy therefore represents an opportunity to engender the excitement of scientific inquiry in students who have chosen to avoid science courses throughout their academic career. For the most part, these students reflect the widest possible cross-section of college students. As such, this course serves

Innovation in Astronomy Education, eds. Jay M. Pasachoff, Rosa M. Ros, and Naomi Pasachoff. Published by Cambridge University Press.

as a unique forum to highlight the intimate relationships among science, technology, and society, while also modeling effective instructional strategies for pre-service teachers who enroll in these courses.

Decades of research on student learning in introductory science courses, including in the area of astronomy, have revealed that faculty routinely overestimate the level of conceptual understanding achieved in these classes. Whereas many faculty believe that students are learning to appreciate the broad landscape of an exciting new field, students are all too often struggling with unfamiliar vocabulary and with the naive ideas they bring to the classroom, which combine to result in a lower level of conceptual understanding of fundamental concepts than faculty hope (Bailey and Slater, 2003, McDermott and Reddish, 1999; Schneps and Sadler, 1987; and Tobias, 1994). This low achievement is often attributed to the over-reliance on lecturing as an instructional mode. The lecturing method is effective only for a minority of students, including those most likely to become faculty themselves. Because lecturing allows students to take passive roles in the classroom, the result is minimal cognitive engagement and low conceptual gains.

A rapidly growing number of faculty are recognizing that the teaching and learning of science must be treated as a complex problem that requires a scholarly approach if they are to be successful. Results from recent surveys of faculty attending college-teaching workshops reveal that while many faculty realize that lecturing alone is insufficient to help all students learn, faculty are largely unaware of what other instructional approaches will benefit their students and still be possible to actually implement in their existing classrooms (Bailey, Jones, Prather, and Slater, 2003). Faculty who study the teaching and learning of science generally recognize that learning in a large-lecture setting can be improved substantially by moving learners from a passive role to a more active role (Bonwell and Eison, 1991). Duncan (1999) eloquently argues that attention-grabbing demonstrations and adept use of multimedia are not enough; truly active learning requires students to do more than passively watch a presentation. Numerous examples of how active learning can help to promote conceptual understanding do exist in the science education research literature. For example, Mazur (1996) developed a series of "ConcepTest" questions for his Harvard physics students to solve in learning teams during lecture. Green (2003) recently extended Mazur's work into the realm of astronomy teaching. Sokoloff and Thornton (1997) demonstrated that physics students showed significant improvement on conceptually challenging questions after using lecture demonstration strategies that are interactive in that they require students to discuss and commit to predictions about what will happen during the demonstration. Mestre (1991) found that to actively engaging students' thought processes, it is most beneficial for students to take the active role in generating their own problems, rather than having the instructor pose problems. At the University of Washington, McDermott (1991) and her colleagues showed that a tutorial-based instructional approach that is informed by research on student misconceptions can produce significant gains in student learning in physics. Francis, Adams, and Noonan (1998) found that the University of Washington tutorials helped students retain conceptual understanding for more than three years after completing a course.

If the documented successes of these alternative instructional strategies clearly point to which sorts of teaching environments are most effective for promoting student learning, then the question of why most faculty are not employing these strategies becomes critical for those of us involved in the scholarship of teaching and learning. One part of the answer is likely limited time and resources. Many of these effective instructional innovations require substantial effort and commitment to implement by the faculty member and the institution. Most faculty at doctoral-granting research institutions necessarily devote significantly more time to their research than to instruction. Further, faculty often lack the pedagogical expertise to implement these teaching strategies successfully. Yet another facet connected to the challenges of implementation stems from a need for institutional commitment in terms of appropriate space, enough equipment for large enrollments, and adequate teaching assistants.

What we identified at the beginning of this project is the need for instructional materials that faculty can easy implement in existing courses that do not necessitate additional institutional support nor require the faculty member to become a pedagogical expert. To this end, we have developed Lecture-Tutorials in an effort to support a learner-centered classroom environment that can improve the effectiveness of the introductory astronomy survey course when implemented by faculty accustomed to using conventional lecture methods. The materials were targeted specifically to serve this sometimes skeptical and certainly busy audience by retaining some sense of the professor-centered instructional style, held dear by many senior faculty, that forms the basis of most teaching environments.

28.2 Design of the lecture-tutorial instructional materials

Earlier research surveying numerous course syllabuses by Slater, Adams, Brissenden, and Duncan (2001) identified the most common topic areas covered by introductory astronomy courses for non-science majors (hereafter referred to as ASTRO 101). These topic areas, in conjunction with previously identified student naive beliefs and reasoning difficulties (see Bailey and Slater, 2005, for an exhaustive review), served to focus and inform the creation of the Lecture-Tutorials.

We endeavored to challenge known student difficulties with carefully designed tasks that focus on having students engage at a cognitively appropriate level to ensure that they confront naive ideas with the goal of achieving a more scientifically accurate and sophisticated understanding. The Lecture-Tutorials are designed around a Socratic-questioning approach that makes use of students' natural language to promote small cognitive steps. Astronomy topics span an enormously wide array of topics in the domain of science. As a result we know that student knowledge of these topics is structured in many different ways, which may include mental models, ontological categories (Chi, 1992), and "knowledge in pieces" (often referred to as phenomenological primitives, after di Sessa, 1993). From this perspective, the questions are written in a sequence intended to guide students' conceptual development. The activities are two to four pages long and are often accompanied by data tables, graphs, or diagrams for students to interpret. Initially, students are asked to examine a novel situation that requires them to reflect on information that they have just heard in lecture. Students provide written responses and explanations to conceptually challenging questions and are repeatedly asked to record answers on diagrams and graphs. One technique we use to help students confront and resolve conceptual and reasoning difficulties is to present the text of a hypothetical "student debate" modeled after work by McDermott *et al.* (1996) The students are directed to critically review the "student debate," which expresses common naive ideas in natural student language. Students are then asked to (a) make explicit whether or not they agree or disagree with each of the hypothetical "student statements," and (b) provide an explanation of their reasoning. From a pedagogical perspective, challenging students to confront their own misconceptions is part of the process of cognitive conflict, which helps mediate meaningful and lasting conceptual change.

Our intent was to create a learner-centered instructional approach that could be implemented seamlessly into existing courses without requiring faculty to give up lecture control wholesale. Lecture-Tutorials were designed to be used in large, fixed-seat, theater-style lecture halls with a single professor and no additional classroom facilitators. This is very different from the 20–30-student class breakout for which many research-based tutorial materials have been developed in introductory physics, and this imposed some strong limitations on our design. Unlike a more typical recitation environment in which student learning is strongly connected to conversations with trained teaching assistants, we needed to ensure that the great majority of student difficulties could be resolved without significant help from the instructor. The cognitive steps are therefore small, and there are a number of built in self-checks to ensure that students have the opportunity to reflect on their developing ideas. Lecture-Tutorials foster the intellectual engagement of students in this challenging instructional setting by having students work collaboratively in pairs in order to capitalize on the benefits of social interactions.

Implementing the Lecture-Tutorials ideally consists of three steps. The first step is to pose a set of conceptually challenging questions – presented to students at the end of an abbreviated lecture on a given topic – to elicit and challenge students' fundamental understanding. If an unsatisfactory percentage of students can correctly answer the questions, the accompanying Lecture-Tutorial should be used. What will be surprising to most faculty is that, even after conventional instruction, the majority of students will still answer seemingly "easy" questions incorrectly (Hewson and Hewson, 1988; Mazur, 1997; Posner *et al.*, 1982). Furthermore, when reading the Lecture-Tutorials themselves, faculty often share an impression about the conceptual difficulty of the materials; however, our research shows that students do not also share this impression.

The second step, and most central to the core of this project, is to use one of the 15-minute, collaborative-learning Lecture-Tutorials in the lecture classroom. During this time, faculty shift from lecturing to facilitating, circulating among the student groups, interacting with students, posing guiding questions when needed, and keeping students on task.

The final step of each Lecture-Tutorial is to debrief the content covered and to bring closure for the students by eliciting student questions and comments. Another common debriefing approach is to make explicit the reasoning needed to understand the concepts fully and provide students with accurate language to describe the phenomenon under investigation. This is an important metacognitive step for both the students and the instructor in that it provides useful insight into how the Lecture-Tutorial experience has impacted student understanding. In addition to extensive implementation notes, an online Instructor's Guide also provides "post-tutorial" questions that can be used to assess the effectiveness of the Lecture-Tutorial before moving on to new material.

28.3 Research design

Two overarching research questions guided the evaluation of the Lecture-Tutorials.

1. What is the effectiveness of a conventional lecture on student understanding?
2. What is the effectiveness of the Lecture-Tutorials on student understanding and the influence on student attitudes?

This study used a mixed-methods (Creswell, 2002), one-group, multiple-measures research design (Freed, Ryan, and Hess, 1991). Students enrolled in ASTRO 101 for non-science majors at a major southwest, Research Level-1, doctoral-granting institution served as the primary data source, with supplementary data coming from five additional field-test sites at varying-size institutions. Data was collected pre-course, post-lecture, and post-Lecture-Tutorial using a 68-item conceptual inventory.

This conceptual inventory uses multiple-choice items over a specific range of topics selected to match those topics commonly taught in an introductory course (Slater, Adams, Brissenden, and Duncan, 2001). The majority of questions on the survey were written by the authors to broadly sample the dominant naive ideas students bring to the ASTRO 101 course. In addition, some items were culled from previously published evaluation instruments, including the Astronomy Diagnostic Test (Hufnagel *et al.*, 2000), the Lunar Phases Concept Inventory (Lindell, 2002), the Project STAR evaluation instruments (Sadler, 1992), and many others (Slater, Carpenter, and Safko, 1996). The important characteristic of the multiple-choice questions we created or selected was that they use attractive distractors based on commonly documented student misconceptions in astronomy (Comins, 2001; Sadler, 1998; Bailey and Slater, 2003) or on phenomenological primitives (Prather, Slater, and Offerdahl, 2002; di Sessa, 1993). The overarching goal was to create a conceptual inventory that probed conceptual understanding rather than elicit only factual recall – in other words, a test the students would rate as conceptually challenging. We endeavored to construct items in the natural language of students, so as to be testing concepts rather than vocabulary. It is not always possible to accomplish this goal, which poses problems when the inventory is to be used prior to instruction. Some of the items do rely on some understanding of vocabulary in order for students to answer them correctly. Normally, questions using technical vocabulary rather than natural student language are inappropriate for an inventory used pre-course. Since the majority of scores on individual questions, however, do not cluster around 25% correct, it seems unlikely that the students are just guessing randomly. In other words, it appears that students do have particular and consistent reasons for answering the questions the way they do, suggesting that the technical language is not completely unfamiliar. Three professional research astronomers who have teaching experience with the target population evaluated each item on a preliminary version of the inventory to establish content validity. Only items which all agreed were "appropriate" were retained.

For the pre-course data, the conceptual inventory was administered on the first day of class using Scantron bubble sheets. To improve the construct validity of the instrument, we split the inventory into two equivalent forms with one-half of the questions on each form. The division of the questions onto two test forms was done to reduce the survey administration time to about 15 minutes. The reason for this decision was to avoid infringing on the inventory's construct validity, which can be threatened if it takes too long for students to complete, in which case they cease to respond thoughtfully to items near the end.

To collect the post-lecture and post-Lecture-Tutorial data, students responded to a subset of two or three closely related multiple-choice items, which came from the original 68-item conceptual inventory. This subset of questions was selected to align with the day's lecture topics and administered immediately following the lecture. This subset of questions was asked again later, after students had completed the corresponding Lecture-Tutorial. These post-Lecture-Tutorial data were often collected on a different day than the post-lecture data were collected. This approach to data collection was relatively straightforward in design, albeit highly complex to carry out in the day-to-day context of a large enrollment course. It was necessary to distribute new computer-scannable bubble sheets to every student at the start of each class meeting. Because of highly varying student attendance as well as requirements for anonymity surrounding human subjects research, we chose to aggregate all student results and use unpaired pre/post data.

It is worth noting that lectures were delivered by the first two authors of the Lecture-Tutorials and characterized by the research team as following the best practices of effective lectures, including demonstrations, animations, and Microsoft® PowerPoint, as described in Adams *et al.* (2004). The lecturers have consistently received evaluation scores well above their departmental average on formal faculty course evaluations by students and were identified as being notably enthusiastic and knowledgeable. What is critical here is that the authors maintained an instructor-centered classroom environment throughout this portion of the study.

A pre-course/post-course Likert-scale style attitude survey was also administered. At the end of the course, a three-component qualitative study was conducted to determine student impressions and beliefs about their learning using the Lecture-Tutorials. One component was a large-group format focus group interview, which was conducted by the University Teaching Effectiveness Center. The second component was an inductive analysis of three clinical interviews of learner-participants conducted by a trained evaluator. Finally, we reviewed the formal end of course evaluations submitted by students.

28.4 Research results

Our first research question examined the effectiveness of a conventional lecture on student understanding of the main topics taught in ASTRO 101. Based on the rhetoric of the ineffectiveness of instructor-centered lectures, we had anticipated that there would be only modest gains in student scores from pre-course to post-lecture. The average score on the 68-item conceptual inventory administered pre-course was 30% correct, whereas the post-lecture average score was 52% correct. Although this 20% gain in average score demonstrates a statistically significant increase in scores ($\alpha < .05$) and could be considered to be a great success for lecture-centered instruction, we are wholly unsatisfied when our students are able to answer only half of these conceptual questions correctly after targeted instruction. Our results serve as evidence that lecture-centered classroom environments are largely ineffective at promoting meaningful and compelling conceptual gains on traditional astronomy topics presented to non-science majors.

The second research question focused on student cognitive and affective gains resulting from the use of Lecture-Tutorials. Although students were able to provide correct responses to about one-half of the items after a lecture, we were not sure how much improvement in scores would result from investing only an additional 15 minutes in learner-centered instruction by doing the Lecture-Tutorials. Surprisingly, we found that students made statistically significant cognitive gains on the 68-item conceptual inventory. The mean percentage correct score increased from 52% immediately following a lecture to 72% after completing the Lecture-Tutorials. This additional 20% gain made by students, above the initial 20% gain achieved after the lecture, is judged to be quite high considering the small amount of class time invested. It is also a noteworthy gain because we believe that student gain scores are non-linear, such that an increase from 50% to 70% correct is much more challenging, for both instructors and students, to achieve than the prior increase from 30% to 50%. Because of the judged conceptual difficulty and attractive nature of the distractors in the questions used, we do not believe that this dramatic increase is simply a result of students' memorizing the answers or because of their having seen the questions as a pre-test (Dokter, Brissenden, Prather, Antonellis, and Richwine, 2003).

The mean scores on the Likert-scale attitude survey were nearly identical pre-course to post-course. In almost every individual item, students showed no statistically significant gain. These results are consistent with other studies using the same instrument (see Zeilik and Bisard, 2000) and lend support to the idea that student attitudes are particularly difficult to impact even when conceptual knowledge increases.

In contrast, the results from the focus group interviews conducted by an external evaluator clearly indicate that many students thoroughly enjoyed the course because it was designed around the Lecture-Tutorials and, to our surprise, several students stated that it was one of the most important parts of the class. Representative comments were:

- "We are able to discuss topics with other students and therefore, we help each other!";
- "Why don't all professors use tutorials during class?";
- "The tutorials were definitely a big part of my learning in the class."

This impression was confirmed by clinical interviews, where three representative comments were:

- "I know the worksheets are real helpful. I found it sometimes hard to talk to as many people as I wanted to talk to and finish the worksheet in time.";
- "And then the tutorials? I don't know who ever thought of that. But it's really how classes should be taught . . .";
- "The tutorials [review concepts] because they break it down. You start with something so simple . . . and then it slowly gets to more."

On the end-of-course evaluations, students frequently commented positively on the Lecture-Tutorials, even without being prompted. Comments were consistent with those from the focus groups and clinical interviews:

- "I really like the way this class is taught. The lectures, which are followed by reinforcing Lecture-Tutorials, make the information so clear to me."
- "I really like the Lecture-Tutorials. They help so much better than the book to understand the material."

From these collective results, we conclude that implementing Lecture-Tutorials in this setting, which somewhat reduced the amount of time available for lecturing, made dramatic and significant positive impacts on both students' understanding and attitudes about learning astronomy.

28.5 Conclusion

Research on how students learn science repeatedly shows that the largest learning gains result when students are active participants in the learning process (e.g., Wandersee *et al.*, 1994). Providing students with opportunities to be active learners, however, is a formidable challenge in the context of the large-enrollment, undergraduate, lecture-based, science survey course for non-science majors. Recognizing that most scientists who teach ASTRO 101 will not completely abandon lecturing as the dominant classroom instructional approach, the philosophy of this project was to develop a suite of activities that would positively impact students' understanding through active engagement while also promoting an instructional intervention that can be easily integrated into a lecture-based course without the need for significant faculty professional development.

These notions on how instructional methods influence student learning are confirmed by our results. From the pre-course, post-lecture responses we find that, even after conventional lectures, student cognitive gains, although statistically significant, are still below our desired expectations, where average post-lecture scores were only 52%. Creating a rich environment for students to engage in learner-centered instruction, however, albeit for only brief periods, provides significant and satisfactory increases in conceptual understanding, as shown by the post-Lecture-Tutorial results, where student scores increased to 72%.

From the overwhelmingly positive responses on our workshop evaluation forms, we infer that the Lecture-Tutorials project is fulfilling an important need within the astronomy teaching community. Our evaluations tell us that although many college faculty feel constrained to use the lecture–based methods that they experienced as students, even though they are aware that innovative approaches exist that would likely further increase student understanding (Bailey, Jones, Prather, and Slater, 2003). Many of these faculty state that the instructional materials developed as part of this project have enabled them to take a major step toward implementing learner-centered instructional strategies in their classrooms.

Acknowledgments

Lecture-Tutorials for Introductory Astronomy were collaboratively developed with generous support from the National Science Foundation (NSF CCLI #9952232 and NSF Geosciences Education #9907755) for introductory astronomy by the Conceptual Astronomy and Physics Education Research (CAPER) Teams at the University of Arizona and Montana State University and with our home institutions. The NSF Chautauqua program, the American Association of Physics Teachers, American Astronomical Society, Astronomical Society of the Pacific, and the American Geophysical Union have graciously given us a forum to share our results and conduct teaching excellence workshops for faculty. Additional funding for workshops has been provided by NASA JPL Navigator, NASA Spitzer, and the National Optical Astronomy Observatory Education and Public Outreach Programs. Numerous individuals contributed to this important project and the national field-testing of the materials.

References

Adams, J. P., Prather, E. E., and Slater, T. F., 2004. *Lecture-Tutorials for Introductory Astronomy*, (Upper Saddle River, NJ: Prentice Hall).

Bailey, J. M. and Slater, T. F., 2003. A review of astronomy education research, *The Astronomy Education Review*, **2**(2).

Bailey, J. M. and Slater, T. F., 2005. Resource letter on astronomy education research. *American Journal of Physics*, **73**(8), 677–685.

Bailey, J. M., Jones, L. V., Prather E. E., and Slater, T. F., 2003. How astronomers view their role as instructors, American Association of Physics Teachers National Conference, Austin, TX, 13 January 2003.

Bonwell, C. and Eison, J., 1991. Active learning: creating excitement in the classroom (ASHE-ERIC Higher Education Report No. 1). (Washington, DC: ASHE).

Chi, M. T. H., 1992. Conceptual change within and across ontological categories: examples from learning and discovery in science. In R. Giere (ed.), *Cognitive Models of Science: Minnesota Studies in the Philosophy of Science*, 129–186. (Minneapolis: University of Minnesota Press).

Comins, N., 2001. *Heavenly Errors*, (New York: Columbia University Press).

Creswell, J. W., 2002. *Research Design: Qualitative, Quantitative, and Mixed Methods Approaches*, (Sage Publications).

di Sessa, A., 1993. Toward an epistemology in physics, *Cognition and Instruction* **10**(2–3), 105–225.

Dokter, E. F. C., Brissenden, G., Prather, E. E., Antonellis, J. C., and Richwine, P., 2003. The use of personal responder devices to assess student understanding, and student beliefs about their effectiveness in Astro 101, *Bulletin of the American Astronomical Society*, **35**(5).
Duncan, D. K., 1999. What to do in a big lecture class besides lecture? *Mercury*, **28**(1), 14–17.
Fraknoi, A., 2001. Enrollments in Astronomy 101 courses. *Astronomy Education Review*, **1**(1), 121–123.
Francis, G., Adams, J. P., and Noonan, E. J., 1998. Do they [FCI Scores] stay fixed? *The Physics Teacher*, **36**(7).
Freed, M. N., Ryan, J. M., and Hess, R. K., 1991. *Handbook of Statistical Procedures and their Computer Applications to Education and the Behavioral Sciences*. (New York: Macmillan Publishing Company.)
Green, P. J., 2003. *Peer Instruction for Astronomy*, (Upper Saddle River, NJ: Prentice Hall Publishing).
Hewson, P. W. and Hewson, G., 1988. An appropriate conception of teaching science: a view from studies of science learning. *Science Education*, **72**(5), 597–614.
Hufnagel, B., Slater, T. F., Deming, G., Adams, J. P., Lindell, R., Brick, C., and Zeilik, M., 2000. Pre-course results from the astronomy diagnostic test, *Pub. of the Astro. Soc. of Australia* **17**(2).
Lindell, R., 2002. Results from the national field-test of the Lunar Phases Concept Inventory, talk presented at the American Association of Physics Teachers meeting in Austin, TX, on January 14, 2003, with abstract listed in *AAPT Announcer* **32**(4), 103.
Mazur, E., 1996. *Peer Instruction: a User's Manual*, (Upper Saddle River, NJ: Prentice Hall Publishing).
McDermott, L. C., 1991, What we teach and what is learned – closing the gap. *American Journal of Physics*, **59**(4), 301–315.
McDermott, L. C. and Reddish, E. F., 1999. Resource Letter PER-1: Physics Education Research", *Am. J. Phys.* **67** (9), 755–767.
McDermott, L. C., and Physics Education Group of the University of Washington 1996. *Physics by Inquiry*. (New York: John Wiley and Sons, Inc.).
Mestre, J. P., 1991. Learning and instruction in precollege physical science. *Physics Today*, **44**(9), 56–62.
Posner, G. J., Strike, K. A., Hewson, P. W., and Gertzog, W. A., 1982, Accommodation of a scientific conception: toward a theory of conceptual change. *Science Education*, **66**(2), 211–227.
Prather, E. E., Slater, T. F., and Offerdahl, E. G., 2002. Hints of a fundamental misconception in cosmology, *Astronomy Education Review* **1**(2).
Sadler, P. M., 1992. The initial knowledge state of high school astronomy students, Ph.D. dissertation, *Harvard School of Education Dissertation Abstracts International*, **53**(05), 1470A (University Microfilms No. AAC-9228416).
Sadler, P. M., 1998, Psychometric models of student conceptions in science: reconciling qualitative studies and distractor-driven assessment instruments, *J. Res. Sci. Teach.* **35**, 265–296.
Schneps, M. H. and Sadler, P. M., 1987. *A Private Universe*. Harvard-Smithsonian Center for Astrophysics, Science Education Department, Science Media Group. Video. (Washington, DC: Annenberg/CPB: Pyramid Film and Video).
Slater, T. F., Adams, J. P., Brissenden, G. and Duncan, D., 2001. What topics are taught in introductory astronomy courses?" *Phys. Teach.* **39**(1), 52–55.
Slater, T. F., Carpenter, J. R. and Safko, J. L., 1996. A constructivist approach to astronomy for elementary school teachers, *J. Geoscience Educ.* **44**(6), 523–528.
Sokoloff, D. R. and Thornton, R. K., 1997. Using interactive lecture demonstrations. *The Physics Teacher*, **35**(6), 340–347.
Tobias, S., 1994. *They're Not Dumb, They're Different: Stalking the Second Tier*, (Tucson, AZ: Research Corp.).
Wandersee, J. H., Mintzes, J. J., and Novak, J. D., 1994, Research on alternative conceptions in Science. In Gabel, D. (ed.), *Handbook of Research on Science Teaching and Learning* (New York: Macmillan).
Zeilik, M. and Bisard, W., 2000. Conceptual change in introductory-level astronomy courses. *Journal of College Science Teaching*, **29** (4), 229–232.

29

The TENPLA project (1): popularization of astronomy under cooperation between students and educators in Japan

M. Hiramatsu, K. Kamegai, N. Takanashi, and K. Tsukada

Department of Astronomy, University of Tokyo, Bunkyo-ku, Tokyo, 113-0033, Japan; Institute of Astronomy, University of Tokyo, Mitaka, Tokyo, 181-0015, Japan; Department of Astronomy and Earth Sciences, Tokyo Gakugei University, Koganei, Tokyo, 184-8501, Japan

Abstract: We present the concepts and products of the TENPLA project, a unique activity in the popularization of astronomy under cooperation between students of astronomy and educators in Japan. The goal of the project is to show the real, latest astronomy, and to let more people become familiar with astronomy. Our mailing list has about 200 participants, including 80 university students. The members share information and exchange views on various educational activities. We have proposed some innovative materials for the popularization of astronomy. One of the representative products is the "Astronomical Toilet Paper (ATP)", on whose paper sheet we print illustrations and detailed description of the life of a star[1]. Now the ATP is available commercially, about 20 000 rolls have been sold in 18 months. See Figure 29.1. Another example of our produst is Astro-KARUTA, a perfect marriage of modern astronomy and Japanese traditional card game. On the cards, various astronomical topics are illustrated. Playing this Karuta, one can become familiar with astronomy naturally. We have many more projects and ideas, still in preliminary stages. All are innovative and novel, and we will continue to develop various ways of popularizing astronomy for the general public, in much more appealing ways.

The prototype of the Astronomical Toilet Paper and the Astro-KARUTA are supported by the Nissan Science Foundation.

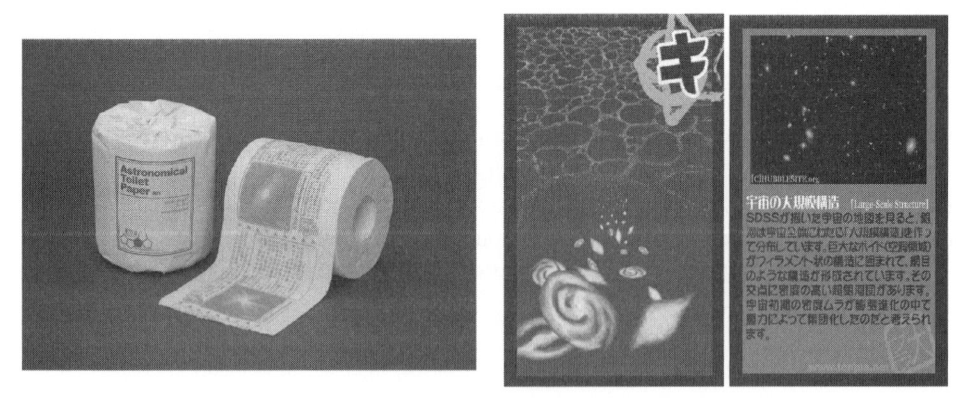

Figure 29.1 Left: The Astronomical Toilet Paper. Right: a playing card of Astro-KARUTA for the Large Scale Structure.

[1] See also a column in the *Science* magazine (10 March 2006 issue, page 1355). The ATP is just in Japanese, but we have an English webpage, www.tenpla.net/atp/indexE.html.

Innovation in Astronomy Education, eds. Jay M. Pasachoff, Rosa M. Ros, and Naomi Pasachoff. Published by Cambridge University Press.

30

The TENPLA project (2): activities for the popularization of astronomy

K. Kamegai, M. Hiramatsu, N. Takanashi, and K. Tsukada

Institute of Astronomy, University of Tokyo, Mitaka, Tokyo 181-0015, Japan; Department of Astronomy, University of Tokyo, Bunkyo-ku, Tokyo, 113-0033, Japan; Department of Astronomy and Earth Sciences, Tokyo Gakugei University, Koganei, Tokyo, 184-8501, Japan

Abstract: We show some examples of unique activities of the TENPLA project in popularizing astronomy for the public under cooperation among students of astronomy, young astronomers, and social education facilities such as science museums and planetariums.

First, we introduce the "Mitaka Starry Sky Club" [1] which is a series of lectures (held twice a month, six times in total), aimed at families with little children, about the latest astronomical discoveries and also how to use an amateur telescope. The purpose of the lectures is to provide an environment in which people can search and enjoy astronomy on their own initiative in their lives. When the participants finish the lectures, TENPLA gives them a license to borrow a 10-cm amateur telescope. It will be used in watching the stars together with their family or friends. TENPLA will also support star parties held independently. This activity is in collaboration with activities for parents with little children.

Second, we have held some science cafés[2] at several public spaces such as bookstores (Figure 30.1), an art museum (Figure 30.2), a local airport (Figure 30.3), and small restaurants. The term "science café" is derived from "Café Scientifique," born in the UK, but we have made some in TENPLA's own style. Participants enjoy the starry sky, lectures by experts of astronomy, and talk freely with other participants and the lecturer. TENPLA provides a friendly atmosphere with some drinks. One of the most striking advantages of our science café is that a number of graduate or undergraduate students can be mobilized to help the participants in discussing astronomy because TENPLA has a lot of university students (as many as 80). For example, TENPLA's first science café was held at a big book store in Sapporo. As many as 200 attended. They first enjoyed lectures by Dr. J. Watanabe and by several students of astronomy. After that, around 20 TENPLA staff, students, and young astronomers went into the audience and discussed interesting topics in astronomy, daily lives in their studies, and so on. This event was a great success and it was a wonderful experience for us.

Third, TENPLA has also provided hospitalized children with traveling lectures on astronomy, because many children are interested in astronomy even though they have no opportunity for learning astronomy or watching stars from their hospital rooms. On March 9th, 2006, nine university students of astronomy and one professor visited the pediatric department in a hospital

[1] This activity is financially supported by Faculty of Science and International Center for Elementary Particle Physics (ICEPP) in the University of Tokyo and the Heisei Foundation for Basic Science.

[2] Jay Pasachoff (editor): See www.sciencecafes.org for information about the science café movement in the United States. The site is jointly sponsored by the Boston education television station's foundation (the WGBH Education Foundation) and Sigma Xi, the honorary scientific society that, among other things, honors B. A. graduates for their research success.

Innovation in Astronomy Education, eds. Jay M. Pasachoff, Rosa M. Ros, and Naomi Pasachoff. Published by Cambridge University Press.

Figure 30.1 Science café at a book store.

Figure 30.2 Science café at Ohara museum.

Figure 30.3 Science café at a local airport.

Figure 30.4 Traveling lecture in a hospital.

in Tokyo (Figure 30.4). We had a lecture on astronomy using many astronomical images and 3-D computer graphics at a playroom in the hospital. About 10 participants joined, including hospitalized children and their parents. For other children who cannot leave their rooms, we visited their individual rooms and talked with them on topics suiting their interests. Most of the children seemed to enjoy the visits and to be interested in astronomy. For example, one of them smiled for the first time since he entered the hospital. This episode shows that this effort has been very successful.

Our TENPLA project will provide more unique activities in the future in order to make bridges between astronomy and the public, especially people who are unfamiliar with astronomy. Our activities are provided directly by students of astronomy and by young astronomers at many scenes of daily life.

Poster highlights

An astronomer in the classroom: Observatoire de Paris's partnership between teachers and astronomers

Alain Doressoundiram and Caroline Barban
Observatoire de Paris, LESIA, CNRS UMR, 8109 Place Jules Janssen, F-92195 Meudon, France

Abstract
The Observatoire de Paris is offering a partnership between teachers and astronomers. The principle is simple: any teacher wishing to undertake a pedagogical project in astronomy, in the classroom or involving the entire school, can request the help of a mentor. An astronomer from the Observatoire de Paris will then follow the progress of the teacher's project and offer advice and scientific support throughout the school year. The projects may take different forms: construction projects (models, instruments), lectures, posters, exhibitions, etc. The type of assistance offered is as varied as the projects: lecture(s) in class, telephone and e-mail

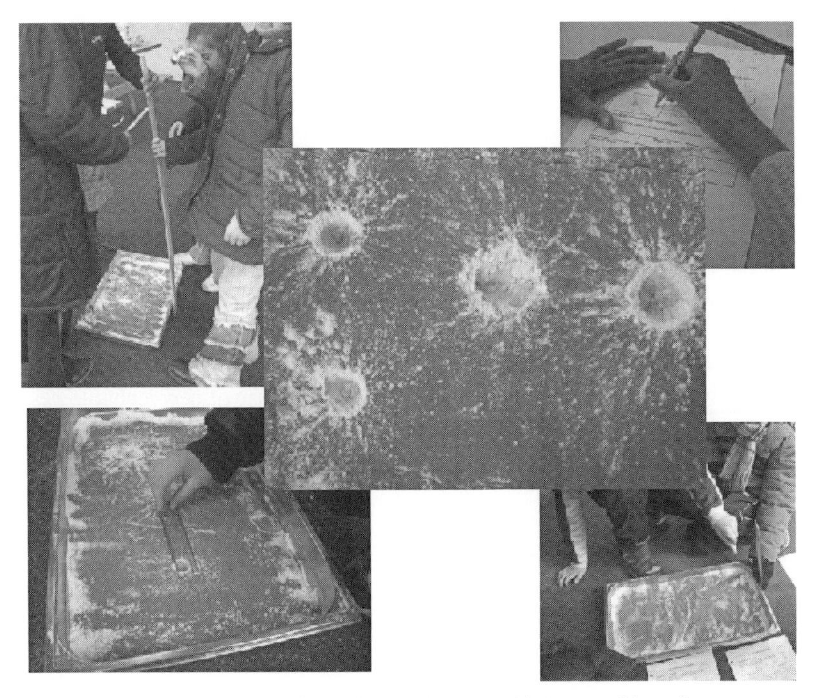

Hands-on activity on crater formation carried out with 8-year old pupils.

Innovation in Astronomy Education, eds. Jay M. Pasachoff, Rosa M. Ros, and Naomi Pasachoff. Published by Cambridge University Press.

exchanges, visits to the Observatoire; an almost made-to-measure approach that delighted the thirty or so groups that benefited from such partnership in the 2005–2006 academic year. And this number is continuously growing. There were a rich variety of projects undertaken, from mounting a show and building a solar clock to visiting a high-altitude observatory, or resolving the mystery of Jupiter's great red spot. The Universe and its mysteries fascinate the young (and the not so young) and provide a multitude of scientific topics that can be exploited in class. Astronomy offers the added advantage of being a multidisciplinary field. Thus, if most projects are generally initiated by a motivated teacher, they are often taken over by teachers in other subjects: Life and Earth Sciences (SVT), history, mathematics, French, and so forth. The project may consist of an astronomy workshop or be part of the school curriculum. Whatever the case, the astronomer's task is not to replace the teacher or the textbooks, but to propose activities or experiments that are easy to implement. Representing the Solar System on a schoolyard scale, for instance, is a perfect way to make youngsters realize that the Universe consists mostly of empty space. There is no shortage of topics, and the students' enthusiasm, seldom absent, is the best reward for the astronomers who willingly devote part of their leisure time to this pedagogical support. See www.obspm.fr/aim/Astro/Parrainage/Parrainage.html.

Astronomy and space sciences in Portugal: communication and education

Pedro Russo and Mariana Barrosa
Navegar Foundation, Portugal

Abstract

Portugal is a country without any tradition in science communication, but nowadays the science community is more conscious of the importance of communication with the public and with decision makers. From the beginning of its establishment, the Navegar Foundation took an important role in bringing the sciences, mainly astronomy, to the public. Our recent past shows different aspects of this achievement and the relations with similar institutions in Europe and the world show us that this is the way. In the year of the public release of the astronomy and astrophysics white paper in Portugal, it is important to think over the present state of astronomy and astrophysics in education and society in Portugal. The Astronomy and Space Sciences in Portugal: Communication and Education Conference has a major role in the construction of a new step in public comprehension of astronomy in particular and science in general. With the development of new ways of communication and education in science, we aim to present the new challenges of the different agencies involved both in Portugal and in Europe. The conference wished to bring together all the national science, education, and popularization agents, from schools to the private sector and the media, and stimulate the use of new technologies and new approaches in the teaching of astronomy and space science by making it more attractive and all-embracing. The conclusions of this conference can be a case study; and there are a large number of lessons to be learned that can be used in new and innovative ways to teach astronomy.

Gemini Observatory outreach

Maria Antonieta Garcia
Gemini Observatory, Southern Operations Center, c/o AURA, Inc., Casilla 603, La Evena, Chile

Abstract

The Gemini Observatory is an international partnership of seven countries, including the United States, United Kingdom, Canada, Chile, Australia, Brazil, and Argentina. Gemini consists of twin 8-meter optical/infrared telescopes located at two of the best locations on our planet for astronomical research: Hawaii and Chile. Together these telescopes can access the entire sky. Gemini puts considerable effort into educational programming for the two host communities (Hawaii and Region IV of Chile), resulting in an increased awareness in those communities of the benefits of having observatories nearby. Once a year Gemini's staff trains about 30 to 40 teachers in basic astronomy, and provides them with resources to enhance their students' educational experience. Selected teachers have participated in Gemini's *StarTeachers* exchange program, in which they make reciprocal visits to the Gemini host communities (in Chile and Hawaii), and experience the execution of Gemini's astronomical research programs while sharing insights on each other's culture and educational systems. The teachers also use Gemini's Internet2 technologies to present live, interactive videoconference classes involving students in both countries. A virtual tour of Gemini facilities is available on CD-ROM in English and Chilean Spanish, with a French version currently under development. A pilot program, *Live from Gemini*, is available to educational groups with access to a high-speed IP address or access to videoconference equipment. Participants in the program will be hosted by a Gemini staff member in one of our control rooms and be able to talk to the astronomers on site. For more information, please visit www.gemini.edu.

Part III

Effective use of instruction and information technology

Introduction

The first of five presentations relating to Part III was made by Douglas Pierce-Price, representing a group of collaborators from Garching, Germany, and Santiago, Chile. In "ESO's Astronomy Education Programme," he described several educational activities admininstered by the Educational Office of the European Organization for Astronomical Research in the Southern Hemisphere (ESO), some of which are run collaboratively with the European Association for Astronomy Education (EAAE). Among the recent activities of the ESO, which is headquartered in Garching with three astronomical observatories in Chile, were the coordination of over 1500 teams of international observers at the time of the 2004 Transit of Venus and the celebration of the World Year of Physics in 2005 by providing equipment for measuring solar radiation levels to students in schools throughout Chile. For several years ESO and EAAE have jointly sponsored "Catch a Star!," an international competition aimed at developing "interest in science and astronomy through investigation and teamwork." Student teams from around the world do research on an astronomical object or theme, "and discuss how large telescopes such as those of ESO can play a part in studying it." Although some of the contributions are judged by an international jury (with awards including travel to ESO's VLT facility in Chile or to observatories in Europe), some prizes are also awarded by lottery in order "to avoid a sense of elitism." In autumn 2005, ESO and its partners in the EIROforum (a partnership of Europe's seven largest intergovernmental research organizations) sponsored the first "Science on Stage," a science education festival. The following such festival took place in spring 2007 in Grenoble. ESO also provides astronomy-related articles for *Science in School* (a new European science education journal for teachers, scientists, and others) and produces "Journey Across the Solar System" (a series of informational sheets) and a series of astronomy exercises "based on real data" from the VLT or HST, the former in collaboration with EAAE and the latter in collaboration with ESA. A new undertaking is ALMA ITP, where ALMA stands for the Atacama Large Millimeter/submillimeter Array – a new astronomical facility under construction in Chile's Atacama desert by ESO "as part of a global collaboration" – and ITP stands for Interdisciplinary Teaching Project. The teaching material "will highlight the links between twenty-first-century astronomy and the topics in engineering, earth sciences, biology, medicine, history, and culture" related to ALMA's location in the Earth's driest spot. The goal of ALMA ITP is "to introduce scientific topics to students as part of other school subjects, and also to put scientific research into a wider context." Like other participants in the session, Pierce-Price spoke about the importance of such educational activities "to ensure that future citizens, whatever their careers, have the scientific literacy they need to make informed decisions about issues related to science."

Mary Ann Kadooka from the Institute for Astronomy at the University of Hawaii spoke about the special challenges and rewards of running an astronomy educational outreach program in Hawaii, which is very culturally diverse. In "US Student Astronomy Research and Remote Observing Projects," Kadooka explained how the NSF's Toward Other Planetary Systems (TOPS) teacher enhancement workshops, held over 18-day periods between 1999 and 2003, were the first major effort to introduce math and science teachers from Hawaii and the Pacific Islands to astronomy. The participants, some of whom returned for several years, became "a master teacher cadre to serve as the backbone of our student project efforts today." Some of these continue to mentor former students, encouraging them to attend star parties, public lectures in astronomy, and other science-oriented events. Among the ongoing partnerships resulting from the TOPS program are those with the Bishop Museum, the Hawaii Astronomical Society, the American Association of Variable Star Observers, the Faulkes Telescope North (now part of the Las Cumbres Observatory), and NASA's Deep Impact Mission. Demonstrating that the PRO-AM relationship advocated by Fienberg and Bennett on the first day of the special session is already in effect in some locations, Kadooka described the contributions to astronomy education in Hawaii made by amateur astronomers. Hawaii's TOPS-trained teachers have mentored many student participants in science fairs in Hawaii and in Oregon (where one of Hawaii's master teachers now teaches in a Portland private school, where she offers an elective science research course). A NASA IDEAS (Initiative to Develop Education through Astronomy and Space Science) grant awarded in May 2006 is making possible outreach to a new inter-island target group of students from grades 7–10, including Native Hawaiians and "at-risk, rural students." To that end, mini-workshops were offered in autumn 2006 to interested and committed teachers on the islands of Maui and Molokai. To maximize the chances of entering "exemplary student astronomy research projects" in the 2008 Hawaii State Science Fair, and with the hopes of having some of these make the cut for the 2008 Intel International Science Fair, students committed to doing research were recruited for a summer 2007 week-long workshop on astrobiology.

Like other special session participants, Richard Gelderman of Western Kentucky University, USA, believes that "our primary job as teachers is to prepare tomorrow's citizens, rather than to prepare tomorrow's scientists" and that "astronomy is perhaps the field of science where the biggest contribution can be made toward the creation of a scientifically literate society." A proponent of "hands-on, minds-on astronomy experiences," he advocates giving students "greater access to astronomical telescopes." In "Global Network of Autonomous Observatories Dedicated to Student Research," he described the Bradford Robotic Telescope and "the beginnings of global networks of autonomous observatories." The Bradford Robotic Telescope, inaugurated in 1993, has had a positive impact on science education by providing for "the general public's requests for queue-scheduled service observations as well as remote, manual operation," thus "offering access to an autonomous observatory coupled with well-designed projects and guided activities." The opportunities that should become available from the proposed networks of autonomous observatories will doubtless enhance "hands-on astronomical education," but "other successful ways of engaging and teaching young people" should not be overlooked. Gelderman described some of these other educational possibilities, such as having students build their own "simple, low-cost" telescopes, exposing them to webcam technology, and – echoing Ferlet's earlier paper – engaging them in the Hands-On Universe program.

David H. McKinnon and coauthor Lena Danaia, from Charles Sturt University, Bathurst, Australia, described in "Remote Telescopes in Education: Report of an Australian Study" how "the use of remote telescopes can be harnessed to impact in positive ways the attitudes of students." Echoing the earlier paper by Lomb *et al.*, McKinnon lamented the decline in science enrollments "during the post-compulsory years of education" in Australia and elsewhere. In the mid-1990s McKinnon built the Charles Sturt University (CSU) Remote Telescope, which enabled students to get their own images of celestial objects. Impressed by "the motivational impact that the control aspect of the CSU Remote Telescope had on primary age students," the Federal Government of Australia commissioned a study that developed educational materials for students and their teachers and assessed some outcomes in those who used them. Students who were given access to remote control of the CSU telescope not only acquired "a significantly greater ability to explain astronomical phenomena," but also "increased their astronomical knowledge significantly" and significantly improved their "attitudes towards science in general and astronomy in particular." McKinnon concluded by expressing the hope that access for students to the Las Cumbres Observatory's global telescope network will confirm "the hypothesis that a love of astronomy can be engendered in more students."

The final paper related to the "effective use of instruction and information technology" was "Visualizing Large Astronomical Data Holdings," by C. A. Christian and collaborators. As "huge quantities of observed or simulated data" become available, it is important both for educators and researchers to optimize the visual display of the astronomical information. Christian described several tools available for scientific visualization, including the Sloan Digital Sky Survey Navigate Tool, which enables users of the SkyServer website to browse, create finding charts, and display portions of the sky in the form of catalog data; World Wind, a NASA Learning Technologies project, which allows exploration of Mars and the sky "as represented by a number of all-sky surveys"; Digital Universe Atlas, from the American Museum of Natural History's Hayden Planetarium, which enables users to browse the sky, find brown dwarfs, and carry out a number of educational activities; Science on a Sphere, developed by NOAA, which already makes it possible to examine data and phenomena on Earth, Mars, and the Moon, and soon will be adapted for all-sky survey data; and Google Earth, which currently enables the display of astronomical images, and should eventually "allow interfaces to data archives, press release archives, and possibly the National Virtual Observatory." McKinnon concluded by expressing the belief that "with the emergence of large multi-wavelength all-sky survey data and collections of data . . . visualization tools can be important for public understanding of science and education both in formal classroom and informal science center settings." See also short descriptions of posters, pages 245–250.

31

ESO's astronomy education program

Douglas Pierce-Price, Claus Madsen, Henri Boffin, and Gonzalo Argandoña

ESO, Karl-Schwarzschild-Strasse 2, D-85748 Garching bei München, Germany; ESO, Alonso de Cordova 3107, Vitacura, Casilla 19001, Santiago 19, Chile

Abstract: ESO, the European Organization for Astronomical Research in the Southern Hemisphere, has operated a program of astronomy education for some years, with a dedicated Educational Office established in 2001. We organize a range of activities, which we will highlight and discuss in this presentation. Many are run in collaboration with the European Association for Astronomy Education (EAAE), such as the "Catch a Star!" competition for schools, which has now been running for many years. A new endeavor is the ALMA Interdisciplinary Teaching Project (ITP). In conjunction with the EAAE, we are creating a set of interdisciplinary teaching materials based around the Atacama Large Millimeter Array project. The unprecedented astronomical observations planned with ALMA, as well as the uniqueness of its site high in the Atacama Desert, offer excellent opportunities for interdisciplinary teaching that also encompass physics, engineering, Earth sciences, life sciences, and culture. Another ongoing project in which ESO takes part is the "Science on Stage" European science education festival, organized by the EIROforum – the group of seven major European Intergovernmental Research Organizations, of which ESO is a member. This is part of the European Science Teaching Initiative, along with *Science in School*, a newly launched European journal for science educators. Overviews of these projects will be given, including results and lessons learnt.

31.1 Introduction

ESO is the European Organization for Astronomical Research in the Southern Hemisphere. Created in 1962, it currently has 11 member states. Its headquarters are in Garching by Munich, Germany, and it operates astronomical observatories at three sites in Chile: La Silla, Paranal, and Chajnantor.

ESO's Educational Office is part of the Public Affairs Department at ESO Headquarters. It was established in July 2001 to build on ESO's previous involvement in educational projects, such as those that had previously been arranged in the context of European Science Weeks sponsored by the European Commission. The aim of the Educational Office is to support astronomy and astrophysics education, in particular at the high-school level.

The European Association for Astronomy Education (EAAE), established following a meeting in Garching, is a partner of ESO. It runs various projects, such as the EAAE Summer Schools for teachers, in which ESO is also involved.

ESO is a member of the EIROforum, a partnership of Europe's seven largest intergovernmental research organisations. In EIROforum, these organisations pursue joint initiatives, combine resources, and share best practices. One of the joint educational initiatives is the European Science Teaching Initiative, whose components "Science on Stage" and *Science in School* are discussed in Sections 31.4 and 31.5 respectively. The seven EIROforum members are CERN, EFDA, EMBL, ESA, ESO, ESRF, and ILL.

Innovation in Astronomy Education, eds. Jay M. Pasachoff, Rosa M. Ros, and Naomi Pasachoff. Published by Cambridge University Press.

31.2 Activities in Chile

As ESO's observational facilities are based in Chile, we also have a series of educational activities there (see for example Figure 31.1), including a program of astronomy training for schoolteachers, which is currently being further developed in an arrangement with the Chilean Ministry of Education. We also work with science museums and planetaria in Santiago and northern Chile, where ESO is active. In addition, we arrange events for the general public such as star parties, and videoconferences to the remote observatory sites. In this way, people who would not be able to travel to the telescopes have the chance to see and speak with people at the observatory, experience a little of what life is like at these sites, and feel a personal connection with the astronomers.

As part of the World Year of Physics 2005, ESO organised a program called *"100 Anos, 100 Colegios"* (100 Years, 100 Schools). Students in schools from the north to the south of Chile were given equipment to measure levels of solar radiation along the entire length of the country.

31.3 Catch a Star!

"Catch a Star!" is an international competition run by ESO and the EAAE, which is in its seventh year. Students from all over the world are welcome to take part. An aim of the program is to encourage students to work together, learning about astronomy and discovering things for themselves by researching information. "Catch a Star!" actually includes more than one competition, so there is something for everyone, no matter what his or her level.

Student teams can develop a project by writing about a chosen topic in astronomy, which might be an astronomical object such as a nebula, star, planet, or moon, or a more general theme such as "black holes" or "star formation." The teams research their topics, and discuss how large telescopes such as those of ESO can play a part in advancing knowledge about it. The most important goal is to develop an interest in science and astronomy through

Figure 31.1 Students attending a Science Summer Camp at the ESO Very Large Telescope on Paranal, Chile.

Figure 31.2 A selection of winning entries for "Catch a Star!" 2005–06. Clockwise from left: Yuriy Baluk; Karolis Markauskas; Rumen Stamatov, Alexandra Georgieva, with Petar Todorov; Gilles Backes; Denitsa Georgieva, Rositsa Zhekova, Tanya Nikolova, with Dimitar Kokotanekov; Edina Budai, Andrea Szabo, Judit Szulagyi, with Akos Kereszturi.

investigation and teamwork. For this reason, to make the program inclusive and to avoid a sense of elitism, one section of the competition has prizes awarded by lottery.

There is also another section in which the prizes are awarded by an international jury. These jury-awarded prizes include major travel prizes, such as a trip for the winning team to visit ESO's Very Large Telescope facility on the Paranal mountaintop in Chile. Additional travel prizes have included trips to Wendelstein Observatory in Germany, Königsleiten Observatory in Austria, and the Hispano-German Observatory at Calar Alto in Spain.

Younger students are invited to take part in an additional "Catch a Star!" drawing and painting competition, with prizes awarded with the help of a public Web-based vote. Some of the winning entries in both the drawing and painting competition, and the project writing competition, are shown in Figure 31.2.

In the 2005–2006 competition, more than 130 teams from 24 countries worldwide submitted written projects about astronomical topics. Given the importance of gender issues in science, and especially physics, it is encouraging to note that girls did particularly well in the competition. For example, 10 out of the 11 students who won travel prizes are girls. There was also, as we have consistently seen in ESO's educational projects, a strong showing from central and eastern European nations. (www.eso.org/catchastar)

31.4 Science on Stage

Science on Stage is a science education festival, organized by ESO and its partners in the EIROforum (a partnership of Europe's seven largest intergovernmental research organizations), and partly supported by the European Commission as part of the NUCLEUS project's European Science Teaching Initiative (ESTI). Science on Stage has developed from the previous extremely successful series of Physics on Stage events, which has now been expanded to encompass all the sciences.

The first Science on Stage festival was held at CERN in Geneva in November 2005, where almost five hundred science educators from more than 25 countries across Europe, and Canada, met to share innovative science teaching techniques. The event included a program of presentations and workshops, and a fair where delegates could demonstrate experiments and teaching projects. An international jury presented the "European Science Teaching Awards" to the best projects at the fair, in a ceremony attended by Jean-Michel Baer, the Director of Science and Society in the EC Directorate General for Research. The prizes, including special awards from the seven EIROforum organisations (ESO among them), had a total value of 17 000 euros.

The second festival, "Science on Stage 2," took place in Grenoble from 2–6 April 2007. The European Synchrotron Research Facility and the Institut Laue-Langevin were the local organizers from the EIROforum.

31.5 *Science in School*

The second part of the European Science Teaching Initiative is *Science in School*, a new European science education journal for teachers, scientists, and others. Issues include teaching materials, articles about cutting-edge science, interviews, and reviews. ESO is also strongly involved with the journal, contributing and arranging for astronomy-related articles. *Science in School* is published quarterly, with a total hardcopy print run of 20 000, and free access to the online version of the journal. Recent ESO related articles include a description of operations at the VLT on Paranal (Pierce-Price, 2006), and a discussion of the search for Earthlike exoplanets (Jørgensen, 2006).

31.6 Other astronomy teaching material

In conjunction with the EAAE, ESO produces "Journey Across the Solar System," a series of information sheets. In collaboration with ESA, ESO also produces a series of astronomy exercises. These provide students with the opportunity to work with real data from the Very Large Telescope (VLT) or the Hubble Space Telescope. For example, a recent exercise invites students to investigate the orbits of stars close to the supermassive black hole Sgr A* in the Galactic Centre, as observed by the VLT.

31.7 ALMA Interdisciplinary Teaching Project

ESO is currently working on the "ALMA ITP," an interdisciplinary teaching project (ITP) based on the Atacama Large Millimeter/submillimeter Array (ALMA). As part of a global collaboration, ESO is currently building ALMA, a major new astronomical facility, at an altitude of 5 000 meters in Chile's Atacama Desert – the driest place on Earth. Building and operating a modern research observatory like this one far exceeds the conception held by many members of the public about what doing astronomy involves.

ESO and the EAAE are working with teachers to produce exciting interdisciplinary teaching material about ALMA. This material will highlight the links between twenty-first-century astronomy and the topics in engineering, Earth sciences, biology, medicine, history, and culture that spring from ALMA's location in the Atacama Desert. The aim is to introduce scientific topics to students as part of other school subjects, and also to put scientific research into a wider context.

Many teaching opportunities are linked with the ALMA project. The site is on one of the highest plateaus in the world, in the Chilean Andes. It has fascinating geology, surrounded by mountains, including an active volcano. The extreme dryness of the Atacama Desert can be studied in the context of its geography and meteorology. The area supports unique flora and fauna, found in well-defined zones at different altitudes, and adapted to the different environments. The local ecosystem can be studied in biology lessons. Despite the forbidding conditions, humans also live in the area. Human presence in the surrounding area can be traced back 11 000 years, with permanent settlements from 4 000 BCE. The indigenous Atacameña culture flourished between 300 and 900 CE, and the nearest town of San Pedro de Atacama has a museum with Atacameña artefacts and mummies – some of the oldest in the world. Students can, therefore, investigate the rich culture and history of the site, from the earliest times to the modern day.

Life for humans at these high altitudes is difficult. At the height of the ALMA site, the air pressure is only 55% of that at sea-level, and the available oxygen is only 52%. The effects of these conditions on human physiology and other high-altitude medicine topics such as mountain sickness can be studied in biology classes.

Following a workshop held at ESO in Garching, which was attended by volunteer teachers, EAAE members, and other experts, we have some preliminary material in the form of educational worksheets. We are currently considering possible ways to develop the ALMA ITP further, and what the most effective and useful format would be for the final products.

31.8 Venus Transit 2004

For the 2004 Venus Transit, ESO organized an international program to involve the public in this astronomical event. The project was launched by ESO, the EAAE, the IMCCE of the Paris Observatory in France (Institut de Mécanique Céléste et de Calcul des Ephémérides), as well as the Astronomical Institute of the Academy of Sciences of the Czech Republic.

An observing campaign coordinated submissions from 1510 teams of observers from about 50 countries worldwide. In addition, the public were invited to contribute to extensive photo and drawing galleries, and to take part in a video contest. The resulting submissions, as well as general information about the Venus Transit, are collected online.

31.9 Conclusions

The ESO telescopes, the scientific results produced by astronomers who use them, and the knowledge of members of the ESO community, are excellent resources for educational projects in the natural sciences in general, and astronomy in particular.

ESO takes part in many astronomy education programs, both independently and as an active partner with the EAAE. These are complemented by joint activities with the EIROforum, such as "Science on Stage" and *Science in School*, through which ESO supports multidisciplinary science educational projects.

The multidisciplinary character of modern astronomical research, and its powerful appeal to the public, mean that organizations like ESO play an important role in science education. This education not only is vital for the training of future scientists but also is needed to ensure that future citizens, whatever their careers, have the necessary scientific literacy to make informed decisions about issues related to science.

References and further reading

European Association for Astronomy Education (EAAE), www.eaae-astro.org/
Catch a Star, www.eso.org/catchastar/

Science on Stage, www.scienceonstage.net/
Science in School, www.scienceinschool.org/
ESO/EAAE Journey Across the Solar System information sheets, www.eso.org/outreach/eduoff/edu-materials/info-solsys/
ESA/ESO Astronomy Exercise series, www.astroex.org/
ALMA Interdisciplinary Teaching Project, www.eso.org/outreach/eduoff/edu-prog/almaitp/
Venus Transit 2004, www.vt-2004.org/
Pierce-Price, D., 2006, Running one of the world's largest telescopes, *Science in School*, **1**, 56.
Jørgensen, U. G., 2006, Are there Earth-like planets around other stars? *Science in School*, **2**, 11.

Comments

Luciana Bianchi: Who's editing the teachers' science magazine?
Douglas Pierce-Price: The *Science in School* editorial board consists of various scientists, science communicators, and educators from across Europe. There are representatives from EIROforum organizations, including ESO, on this board. The editor is Dr. Eleanor Hayes, who is based at the European Molecular Biology Laboratory in Heidelberg, Germany.

Mary Ann Kadooka: Does Europe have science teacher organizations?
Douglas Pierce-Price: There is, of course, the European Association for Astronomy Education (EAAE), mentioned in the talk. For general science teaching, I understand that the largest European organization is the Association for Science Education (ASE), which is based in the United Kingdom.

32

US student astronomy research and remote observing projects

Mary Ann Kadooka, James Bedient, Sophia Hu, Rosa Hemphill, and Karen J. Meech

Institute for Astronomy, 2680 Woodlawn Drive, Honolulu, Hawaii, HI96822, USA; American Association of Variable Star Observers, 1464 Molenu Drive, Honolulu, HI 96818, USA; McKinley High School, 1039 South King St., Honolulu, HI 96815, USA; Oregon Episcopal School, 6300 SW Nicol Rd., Portland, OR 97223, USA; Institute for Astronomy, 2680 Woodlawn Drive, Honolulu, HI 96822, USA

Abstract: We have used astronomy projects to give students authentic research experiences in order to encourage their pursuit of science and technology careers. Initially, we conducted teacher workshops to develop a cadre of teachers who have been instrumental in recruiting students to work on projects. Once identified, these students have been motivated to conduct astronomy research projects with appropriate guidance. Some have worked on these projects during non-school hours and others through a research course. The goal has been for students to meet the objectives of inquiry-based learning, a major US National Science Standard. Case studies are described using event-based learning with the NASA Deep Impact Mission. Hawaii students became active participants investigating comet properties through the NASA Deep Impact Mission. Our students used materials developed by the Deep Impact Education and Public Outreach group. After learning how to use image-processing software, these students obtained Comet 9P/Tempel 1 images in real time from the remote observing Faulkes Telescope North located on Haleakala, Maui, for their projects. Besides conducting time-critical event-based projects, Oregon students have worked on galaxies and sunspots projects. For variable star research, they used images obtained from the remote observing offline mode of Lowell Telescope, located in Flagstaff, Arizona. Essential to these projects has been consistent follow-up required for honing skills in observing, image-processing, analysis, and communication of project results through Science Fair entries. Key to our success has been the network of professional and amateur astronomers and educators collaborating in many ways to mentor our students.

32.1 Introduction

Education should motivate and guide students to become lifelong learners. It should expose students to a variety of areas, including art and music, science and mathematics, computers and technology. This broad exposure allows them to identify their passion, the key motivation to learning. Subsequently, students need guidance about ways to fulfill their passions through education.

Besides having these philosophical goals for individual students, education must also meet the demands of our technological society by producing competent scientists and engineers for our US workforce. The National Science Foundation (NSF) 2006 statistics indicate that the number of students in US colleges earning Ph.D. degrees in astronomy and physics has been increasing over the past 20 years. This percentage increase, however, has been a result of an increase in foreign-born students. In actuality, there has been a decrease in the percentage of total number of students who are US citizens earning doctorates in these two majors.[1]

[1] NSF website with statistics, Table 2–30 – www.nsf.gov/statistics/seind06/pdf_v2.htm.

Innovation in Astronomy Education, eds. Jay M. Pasachoff, Rosa M. Ros, and Naomi Pasachoff. Published by Cambridge University Press.

In Hawaii, the state's cultural diversity compounds the difficulties in increasing the number of scientists and engineers. According to the 2000 US Census, Hawaii's population is 26% Caucasian, 41% Asian, 9% Native Hawaiian and Pacific Islander, and 2% African American. Twenty percent of the people have two or more races.[2] This diversity challenges everyone, yet can add the richness of cultural astronomy to our educational programs.

The astronomy educational outreach program at the University of Hawaii Institute for Astronomy (UH IfA) for grade 7 to 12 students has been based upon the above philosophical, economic, and cultural concerns. We strive to use astronomy research and remote observing projects to spark the passion for astronomy in students through workshops and follow-up. We try to optimize our network of human resources by promoting a team effort.

32.2 Student and teacher workshops

The first step of identifying students with a passion for astronomy will continue to be accomplished through workshops. By focusing on astronomy topics, we expose students and teachers to computer technology, physics and mathematics, photography, and the many other fields relevant to studying the skies. These workshops have been conducted since 1999.

The NSF Toward Other Planetary Systems (TOPS) teacher enhancement workshop, 1999–2003, was the beginning of a major effort to introduce Grade 7–12 science and mathematics teachers from Hawaii and the Pacific Islands to astronomy.[3] Under the direction of astronomy professor Karen J. Meech and with a staff of educators and graduate students, a series of 18-day summer workshops were held on the Big Island of Hawaii. Participants used 6-inch Dobsonian as well as 10-inch Meade telescopes to conduct astronomy projects on lunar photography, Messier objects, variable stars, and other topics suitable for beginners. As the summers progressed, the returning teachers graduated to doing photometry of variable stars and spectral analysis projects. Out of a total of 75 teachers, 33% returned for 2–5 years. These returning teachers formed a master teacher cadre to serve as the backbone of our student project efforts today. Besides this teacher resource, the TOPS program is also responsible for our networking with the following organizations, and these partnerships continue thanks to mutual reciprocity of support services.

1. The Bishop Museum
   - allowed us use of its planetarium and classrooms,[4]
   - has had our astronomers and graduate students volunteer for special events there.
2. The Hawaii Astronomical Society
   - loaned TOPS program telescopes and taught participants how to set up and use them,[5]
   - recruits members at our University of Hawaii Institute for Astronomy Open House.
3. The American Association of Variable Star Observers (AAVSO):
   - its Executive Director conducted variable stars sessions at TOPS workshops, 1999–2003,[6]
   - our UH IfA outreach staff person is a member of the new AAVSO education committee.

[2] Hawaii 2004 population data – quickfacts.census.gov/qfd/states/15000.html.
[3] Toward Other Planetary Systems Program, www.ifa.hawaii.edu/tops/tops.htm.
[4] Bishop Museum – www.bishopmuseum.org/education/education.html.
[5] Hawaii Astronomical Society – www.hawastsoc.org/.
[6] American Association of Variable Star Observers, www.aavso.org/.

In addition to these established collaborations, two major resources emerged to provide us with excellent opportunities for promoting astronomy education and research projects.

1. Faulkes Telescope North (FTN) is a 2-meter professional telescope built on Haleakala on the island of Maui, dedicated to education.[7] Through a memorandum of understanding between the University of Hawaii and the Faulkes organization in Great Britain, this telescope would be available for Hawaii student use 30% of the time. So the 2003 TOPS workshop was modified to train returning teachers with astronomy backgrounds to do image-processing using astrometry and photometry software. This training set the stage for these committed teachers to recruit and mentor students for astronomy projects. Although first light occurred in August, 2003, numerous technical and software problems hampered students' access to FTN. So we collaborated with astronomer Marc Buie to make Lowell Telescope in Flagstaff, Arizona, accessible to our students at certain times.[8]
2. The second additional resource to emerge was the NASA Deep Impact (DI) Mission. After the 760-kilogram probe was launched in January, 2005, it would be struck by Comet 9P/Tempel 1 on July 4, 2005. This was a live experiment in space. Since our TOPS director astronomer, as a comet specialist for the Deep Impact science team, was a co-investigator for a research grant, we were able to obtain an education and public outreach (EPO) rider to this grant with the Space Science Institute. This grant was enhanced by our collaborations with Lucy McFadden, DI EPO manager and University of Maryland astronomer, who led the development and piloting of curriculum activities. Their amateur observing program provided us with excellent introductory materials on how to prepare observing plans and learning terminology for our Deep Impact teacher/student workshop.[9]

Wanting to entice more Hawaii teachers to join our TOPS master teachers cadre, we conducted mini-workshops on Maui, Oahu, and the Big Island of Hawaii during spring, 2005. We focused on informing workshop participants about how FTN could be used for comet research projects. Since we also work in collaboration with the University of Hawaii NASA Astrobiology Institute scientists, the workshops related the study of comet origins to astrobiology. These half-day workshops on a weekend during the school year without follow-up, however, proved to be inadequate for student recruitment. We had to rely upon our TOPS teachers to recruit students interested in astronomy to work on Deep Impact comet projects.

The 2005 summer Deep Impact/Faulkes Telescope workshop held for four days during the DI encounter on July 4 trained Hawaii students on image-processing with Astrometrica[10] and Mira software. Alan Fitzsimmons, University of Belfast astronomer, assisted the students with their real-time observing of Comet Tempel 1 before, during, and after its encounter with the probe. Follow-up sessions with these students were held during the summer and fall, 2005, and spring, 2006, resulting in comet projects entered in 2006 Science Fairs.

32.3 Roles of stakeholders

In order for us to support students with their research projects, we need teachers to be the student contact liaisons. Teachers who guide former students no longer taking courses from

[7] Faulkes Telescope North – www.faulkes-telescope.com/. [8] Lowell Observatory – www.lowell.edu/.
[9] Deep Impact EPO – deepimpact.umd.edu/amateur/. [10] Astrometrica – www.astrometrica.at/.

them support our ability to do our follow-up. One such exemplary teacher, Sophie Hu, at McKinley High School in Honolulu, has been working with us since 1999. The recipient of Hawaii's 2006 Biology Teacher of the Year award, she integrates all sciences into her course. Another teacher, Rosa Hemphill, at Oregon Episcopal School, teaches a research course and credits the TOPS program with giving her the confidence to advise students on astronomy projects since 2002.[11] These teachers are willing to go beyond the classroom by having students attend star parties sponsored by amateur clubs, astronomy public lectures, and other science venues. Such activities can spark the interest needed for lifelong learning. Teachers are the front lines for reaching students and introducing them to astronomy.

Next are the amateur astronomers. Their passion and the expertise developed from years of observing the skies make them excellent resources to assist students with their projects. AAVSO members have research skills and submit photometric data used by professionals, exemplified by James Bedient, an AAVSO member for 20 years. He volunteered to be on our TOPS staff since 2001.[12] Besides helping students learn to use Astrometrica software with images of comets and asteroids, he has been our FTN real-time observing expert, instructing students on its use. He has delighted these students, who are thrilled to see their names listed as observers for FTN images on the Minor Planet Center circulars. Another AAVSO member, Donn Starkey from Indiana, has begun working with us via e-mail and conducted CCD photometry workshops for us in summer, 2007. Having amateurs assist with necessary research skills frees the professionals to focus their time on sharing their astronomy expertise with students.

Since professional astronomers have many educational and research commitments, we have been working on a model to define their role and determining what a reasonable time commitment would be for mentoring students for projects. They are instrumental in giving the student a global overview and helping students maintain a project focus. Our UH IfA astronomers David Jewitt, David Tholen, and Jeffrey Kuhn shared their expertise with students in the fields of comets, orbits, and solar magnetic fields, respectively.[13] Their feedback is helping us to improve the mentoring process.

Instilling the importance of educational outreach with graduate students plays a key role for continued involvement as they become professional astronomers. In 2003, graduate student Catherine Garland worked with the TOPS program and began to mentor Emily Petroff, then an eighth-grader at Oregon Episcopal School. Since becoming a physics/astronomy professor at Castleton College in Vermont in 2004, Garland has continued to e-mail and help this student. Likewise, graduate student James Armstrong helped with the TOPS program. Now on Maui he has worked with Kalama Middle School students. He assisted with our fall, 2006, Maui and Molokai workshops, and he is developing more teaching skills.

Astronomy education outreach requires a coordinator. The coordinator keeps stakeholders informed and involved with education outreach. Providing information, receiving constructive suggestions/concerns, and implementing improvements are key elements the coordinator must provide. The coordinator is the hub that ensures smooth, coordinated implementation and success of outreach efforts.

[11] Oregon Episcopal School – www2.oes.edu/us/hemphilr/. [12] James Bedient – www.bedient.us/.
[13] University of Hawaii Institute for Astronomy (UH IfA) – www.ifa.hawaii.edu.

The roles and responsibilities of all stakeholders will continue to be refined to develop a model program. Assessments of our workshops, follow-up sessions, and mentoring program will be done so improvements can be made.

32.4 Student astronomy research projects

At our 2003 TOPS workshop, 15 returning teachers were trained to do telescope-observing and image-processing. Each was loaned a laptop to support students for astronomy research projects. Seventy-five percent of them complied for the subsequent school year. Out of the total of 50 recruited students, 25 students – or 50% of them – entered their projects in Hawaii school science fairs. The final outcome for the 2004 science fair season was as follows: seven students – 15% of the 50 – had projects recommended for the Hawaii State and Oregon Regional Fairs. When the students freely choose to conduct a research project, the success rate is higher. In Oregon, an elective science research course taught by our TOPS teacher supported a higher percentage of students completing their projects. In Hawaii, a project requirement for a physical science course resulted in a lower percent project completion. Thus, the 2005 Deep Impact/Faulkes Telescope workshop targeted Hawaii students who willingly conducted projects on their own time, since a research course was not available to them.

Table 32.1 is based upon the 2006 Science Fair astronomy projects of two Hawaii TOPS teachers who had six students participate in the 2005 DI/FT Maui workshop and the Oregon TOPS teacher whose students did not participate. Sophia Hu teaches at McKinley High School (MHS), a Honolulu inner city public school, where 50% of the students qualify for the Federal Free Lunch Program. Thomas Chun teaches at Kamehameha School, Hawaii campus (KSH), a private school for Native Hawaiian students. Rosa Hemphill teaches at Oregon Episcopal School (OES) in Portland, Oregon, a private school. Her students were enrolled in the research course so other astronomy projects are listed.

32.4.1 What's next for student astronomy research projects in Hawaii

A NASA IDEAS (Initiative to Develop Education through Astronomy and Space Science) grant awarded in May, 2006, allows us to build on the outreach model we have to develop more astronomy topics for projects as well as a mentor program to support students.

- The target group is grade 7–10 students from the other islands, especially Native Hawaiian and at-risk, rural students.
- Workshops for teachers and students were held during the 2006–2007 school year with eight laptops available for use. A 2007 one-week summer workshop was held.
- Monthly follow-up sessions with students from all islands will use videoconferencing systems. (Students love to see themselves on the TV screen.)
- Amateur and professional astronomers will continue to be recruited, and mentoring workshops will be held.
- Suitable projects will be added to the Faulkes Telescope asteroid and galaxy projects.

Since teenagers identify with youthful professionals, two young physics/astronomy university staff members have been recruited to assist with these workshops. With a passion for education as well as for physics and astronomy, they are working on how best to educate our students and teachers. Michael Nassir, who received the 2005 University of Hawaii Undergraduate Teaching Award for introductory physics courses, teaches astronomy lab.

Table 32.1 *2006 astronomy student projects*

		Astronomy project		Science fair results		
School	Students	Grade	Title	School	State	National
MHS	Iat Ieong	11	Orbit of 9P/Tempel 1	yes	yes	
MHS	Mimi Hang	10	Dust/gas particles	yes	yes	
MHS	S. Voravengseng	10	Photometry of Tempel 1	no		
KSH	Danielle Kuali'i	10	Composition of Tempel 1	yes	yes	
KSH	Clifford Feliciano	10	Tempel 1 Orbit	yes	no	
OES	Elyse Hope	12	Solar magnetic fields	yes	1st	yes
OES	Emily Petroff	10	Dark matter of galaxies	yes	2nd	yes
OES	Patricia Wang	12	Comet orbit models	yes		
OES	Sheng Ng/Minsung Kang	11	Z Ori/DN Ori light curve	yes	yes	
OES	Olivia Haber-Greenwood	9	V380 Ori light curve	yes		

MHS = McKinley High School; OES = Oregon Episcopal School.

The following students received numerous awards for their projects.

- Elyse Hope was one of 40 finalists for the 2006 Intel Science Talent Search, considered to be a junior Nobel competition, for her project using SOHO images to compare the rates of sunspots and the solar magnetic fields movement. Astronomer Jeffrey Kuhn gave her some advice.
- Emily Petroff investigated velocities of different types of galaxies for the first year and is now investigating mass and dark matter of different spiral galaxies from images. She is mentored by Catherine Garland.
- Mimi Hang used BVR filters to do the photometry of comet images to distinguish between the gas and dust particles. She was interested in the composition, but needs to learn spectral analysis research skills. David Jewitt was instrumental in mentoring Mimi for her project.

Catherine Garland, who has doubled the physics enrollment at Castleton College in Vermont in two years, does radio astronomy.

In fall 2006, mini-workshops were held on the islands of Maui and Molokai. We are targeting schools that promote project-based learning and recruiting teachers who have demonstrated a commitment to assist students with projects. Students will also participate, since personal contact with astronomers inspires them. Such was the case in January, 2005, for Mimi Hang, then a ninth-grader at McKinley High School. Fascinated by comets and the Deep Impact Mission as a participant in our mini-workshop, she is now pursuing her passion for astronomy with another project and actively recruiting members of her astronomy/physics club to do projects with us.

On October 6, 2006, the computer technology and physics teacher at Maui High School agreed to assist us with the workshop there. Instead of being limited to our eight laptops, we used a computer lab to accommodate more participants. On October 7, 2006, students and teachers observed and learned the night sky at Haleakala Observatory.[14]

[14] Maui Community College – maui.hawaii.edu/.

Our Molokai workshop in November, 2006, was facilitated by two Molokai teachers who attended our February, 2005, Maui mini-workshop. Our recent collaboration with Na Pua No'eau, a program for gifted and talented Native Hawaiian students, resulted in our use of the Maui Community College Molokai Education Center for our workshop.[15] These contacts enhance our recruiting ability.

In summer 2007, a one-week workshop was held for students who have demonstrated a commitment for doing research. The purpose was to improve their astronomy background and hone their image-processing skills. We wove astrobiology topics into this workshop to demonstrate the integrative nature of astronomy. Students worked with astronomers on their projects, as we all collaborate on how best to mentor students for projects and develop a useful model. We want exemplary student astronomy research projects entered in the 2008 State Science Fair and recommended for the 2008 Intel International Science Fair.

The greatest challenge for our program is keeping all stakeholders motivated and inspired to continue their efforts; overcoming natural attrition by continuing our recruitment of committed mentors, teachers, and students; and stabilizing our funding.

32.5 Summary

The astronomy educational outreach program at the University of Hawaii Institute for Astronomy for grades 7 through 12 has been based upon the expressed philosophical, economic, and cultural concerns. Workshops are being used to identify students with a passion for science and technology and make them lifelong learners. Teachers and amateur and professional astronomers are being recruited and trained in educational strategies with cultural sensitivity to mentor these students for astronomy research projects. Follow-up sessions and guidance for these students will be ongoing to ensure their entry into the science and technology workforce of the US. Most significant has been the growth of our network of human resources and organizations to guide and assist students. We highly value these caring individuals who are instilling their science passion in the hearts and minds of our young teenagers. This rewarding synergism, hopefully, will result in an increased number of excellent student astronomy research projects.

32.6 Federal Grants

NSF ESI-9731083: Toward Other Planetary Systems Teacher Enhancement Grant, 1999–2003.

NASA Ames: NNA04CC08A: Origin, History and Distribution of Water and Its Relation to Life in the Universe, 2003–2008.

STScI HST-EO-10115.09: Inquiry-Based Learning Using Comet Research Projects, 2005–2006.

STScI HST-ED-90297.01-A: Inspiring and Mentoring Students for Faulkes Telescope North Astronomy Projects, 2006–2008.

Comments

John Percy: You have made a strong case for reaching the best and brightest students, and interesting them in STEM careers. But to what extent do your programs teach disadvantaged schools, teachers, and students?

[15] Molokai Education Center – www.hawaii.edu/molokai/.

Mary Kadooka: We have students who attend public schools with over 50% of the population on Federal free lunch programs, are reaching students on rural outer islands of Hawaii, and work with Native Hawaiians at Kamehameha Schools.

Alan Pickwick: The Association for Science Education in the UK has about 30 000 members who are physics, chemistry, and biology teachers. There is a large-scale annual meeting each January (early) with about 5000 attendees. The ASE is by far the largest science teachers' organization in Europe.

33

A global network of autonomous observatories dedicated to student research

Richard Gelderman

Department of Physics and Astronomy, Western Kentucky University, 1906 College Heights Blvd., Bowling Green, KY 42101-1077, USA

Abstract: I discuss progress in providing robotic telescopes, sometimes individual and sometimes in networks, for student use. I also discuss activities for students who have access to automated observatories.

33.1 Introduction

Astronomy is the most approachable of the physical sciences, and thus can be the means to introduce science to a diverse population. The key is to offer an experience with astronomy that maintains student interest and provides a platform for them to develop critical thinking skills. In an attempt to provide the best hands-on, minds-on astronomy experiences, it is desirable to offer students greater access to astronomical telescopes. For multiple decades, astronomers have promised each other the development of global networks of telescopes. In the last decade, without ever fulfilling the initial promise, we have "upped the ante" and promised global networks of robotic telescopes (Maran, 1967). Sometimes the network is to be composed of 20- to 40-cm aperture telescopes; other times the network will include meter-class telescopes. Sometimes the network is exclusive to a select, small group of users; other times the dream is open to any interested parties. We discuss the variety of activities that can be accomplished by students who have access to automated observatories and provide examples of student projects.

33.2 Creating a scientifically literate society

In today's science classes, our primary job as teachers is to prepare tomorrow's citizens, rather than to prepare tomorrow's scientists. Our children are inheriting a world that requires them to master decision-making skills in an increasingly networked and technical culture. They are being asked to communicate effectively, think creatively, and solve technological problems on global scales. To help them keep pace in such a future, efforts such as the US National Science Education Standards (National Research Council, 1995) call for science to be presented through "minds-on" inquiry-based experiences. According to results of current research in science education, students learn science best when they describe objects and events, ask questions, construct explanations, test those explanations against current scientific knowledge, and communicate their ideas to others. Happily, this list is consistent with the skill required for a person to contribute successfully in our modern economy. The skills that citizens must learn in preparation for our networked, high-tech future include many of the same skills that have traditionally been de-emphasized in the formal education system: teamwork, organization, communication, decision making, analysis, and the ability to apply previous knowledge to novel settings.

Innovation in Astronomy Education, eds. Jay M. Pasachoff, Rosa M. Ros, and Naomi Pasachoff. Published by Cambridge University Press.

Astronomy is perhaps the field of science where the biggest contribution can be made toward the creation of a scientifically literate society. With astronomy we have an attractive field of science; a vital scholarly community that researches effective pedagogical approaches; support for finances and resources; and a willing audience of K-12 teachers that are so vital to successful educational reform (see Chapter 11). Most young children are deeply fascinated by astronomy. While few people ever grow up to become astronomers, the beauty and grand scale of the Universe continue to attract people of all ages. Education research in astronomy has advanced greatly in the last decade, building off the foundation of physics education research to establish meaningful research-supported results. Support from NASA has directed a percentage of effort from all its research efforts toward education and public outreach projects. Professional development workshops for teachers tend to be successful, leveraged from the fact that teachers are as interested in and receptive to astronomy as the rest of the population.

33.3 Tiered opportunities: from participation to the challenge of competition

Involving the maximum number of children in excellent science education is one way to create a scientifically literate society. Since there are very few direct examples of excellent science education being experienced by large numbers of students, it is appropriate to examine an analogous situation. All over the world we find examples of youth sports leagues being established for a new sport. It is clear that a successful sports league needs to have a broad base of participation, a scaffolded series of increasing challenges to inspire and motivate the players to build their skills, and a framework of competitions and opportunities for players to test their mastery of the game. Correspondingly, the number and satisfaction of the participants in astronomy-based science education will continue to increase if we provide: (1) open and easy entry level recreational opportunities; (2) a series of challenging activities with steadily increasing challenge and reward; and (3) high-level competitive experiences to allow the best, most motivated students to test their limits and progress toward rewarding careers in our high-tech culture.

The participatory level must attract large numbers of participants, with no special skills or experience required. We must expect that most of the participants will invest very little in the experience, perhaps joining just to be with friends and pass the time. With carefully planned experiences, we can make it so that some of the masses find they really enjoy the experience and/or are better at it than they anticipated.

Steadily increasing the level of challenge is accomplished through carefully designed experiences and play. Skilled coaches must guide the practice and games so that everyone gets the plenty of opportunity to learn. Drills must be both engaging and challenging, with plenty of flexibility to explore various talents while allowing the less precocious players to catch up.

Providing opportunities to excel is the next important step toward a successful science education program. A program that continues to grow must include forums for the most driven, most talented participants to demonstrate and test their ability. Sponsoring agencies need to understand, however, that the cost of hosting local, regional, or even international contests for a small number of the most devoted competitors and their supporters is actually amortized over a much larger population. Such contests establish the most successful competitors as role models for the average participant, trickling down to the challenge and participatory levels in a reinforcing cycle.

33.4 Participatory level experiences – minds-on astronomy experiences

With no special skills or training, groups of people of all ages can be inspired by images of their star sign constellation; of planets, moons and comets; of Messier objects or other "pretty pictures." We can inspire and enthuse people of all ages with pictures of astronomical objects, while simultaneously achieving higher goals. Easy to obtain astronomical images can be used to reinforce other science lessons; such as observing the changing phases of the Moon, locating Apollo landing sites, or witnessing the orbit of the Galilean moons around Jupiter. One of the most basic goals is to get the average citizen to really see the natural world.

33.4.1 Practice sketching to enhance the ability to observe

"We need to use all our faculties to the full – to assimilate with the scientist's brain, the poet's heart, and the painter's eyes." (J. Bronowski, 1974, *The Ascent of Man)* Naturalists of previous generations – biologists, geologists, and astronomers – were all careful, excellent observers who expressed their knowledge through eloquent descriptions and sketches. Those scientists promoted the idea that one could learn to be a better observer by trying to capture on paper some aspects of what is being observed. Good sketching requires accurate rendering of: proportion (shape, relative height versus width, separation, orientation); lightness and darkness; and possibly color. Over the years, various artists and naturalists (e.g., Leslie, 2003; Edwards, 1999; Johnson, 1997) have developed a three-step technique for learning to sketch.

Contour drawing – Observe an image of a galaxy for a few minutes, then put your pencil on your paper and, without looking at your paper and without lifting your pencil, take two minutes to trace the object in great detail.

Gesture drawing – Sketch the essence of the same object in only 5 seconds. Then sketch it in 15 seconds. Sketch it a third time, for 30 seconds.

Memory drawing – Take 3 minutes to draw the same object from memory. Students who follow these steps should begin to make sketches of which they will feel proud. If they still seem intimidated ("I can't draw, this is terrible!"), it often helps to remind them that the goal is to make observations, not masterpieces.

The telescope and microscope are powerful tools for the eyes of experienced users, but learning to use them can frustrate novices and cause them to give up on trying. Galileo realized from the start that it took encouragement and tutoring before his audience was prepared to see the phases of Venus, the satellites orbiting Jupiter, or the "ears" of Saturn. Instructors can demonstrate the power of observation with games. A matching game would involve identifying sketches of the same object through observation of common prominent features. Alternatively, students might be asked to find the differences between two similar images.

33.5 Challenge level experiences

It is easy to evaluate a project where students are simply recreating previous results; e.g., when a student identifies craters on the Moon or measures the mass of Jupiter. Such projects teach important skills and are vital stepping stones to independent research. It is also true that students can learn a great deal and have a wonderful time working on a mentor's research study, but it still is not the same experience as when students are in charge of their own investigation. But, if gently led up the learning curve, students will get the most out of research where they design and execute the project, where they are responsible for asking the question as well as finding the result.

33.5.1 Very simple, very inexpensive telescopes

Telescopes are the symbol of astronomy. But even a small commercial telescope can have a large price tag. While technological advances and mass production make it possible for clubs, schools, and even individuals to own a computerized "go-to" telescope, for most schools the precious resources spent on such a telescope never quite result in a successful astronomy program.

Telescopes are not especially complicated devices, and plans have been created that enable anyone to put together an excellent small refracting telescope without having fully equipped woodshop and metalworking facilities. The mechanical parts and mountings will be made from assorted pieces of hardware and plastic pipe. Only common hand tools are needed (of course power tools are always welcome). The results can rival equipment costing hundreds of dollars.

The Frugal Telescope Maker (astro.wku.edu/frugaltelescope/) has plans for two 61.7-mm diameter refractors, one with a focal length of 700 mm and the other with a focal length of 415 mm. Both of these telescopes can be used on any of three mounts: a portable "bucket-scope" mount, a rugged 4-ft-tall "pier mounting, or a compact version of the "pier" called the "short pier" mount.

It is feasible and arguably important for each student to build her/his own simple, low-cost telescope. The benefits are the joy and motivation of constructing a telescope that is yours; the understanding of optics that comes with commissioning the scope; and the thrill of observing and collecting digital images of the Sun, Moon, and planets.

33.5.2 Webcam astronomy

Webcam technology provides low-cost solutions that enable most science teachers to use digital imaging with their classes. Polish physicists, astronomers, and teachers provide software and detailed, easy-to-follow instructions for how to adapt a Phillips CCD webcam into a powerful astronomical detector. The CCD webcams can be used with a camera lens for wide-field imaging of the sky or terrestrial events, or inserted into the eyepiece of a telescope for high-resolution digital imaging. Full details on the webcams can be located at the TOOLS link on the European Hands-On Universe website (www.euhou.net).

33.5.3 Autonomous astronomical observatories

The Bradford Robotic Telescope (Baruch, 2000; Baruch, 2003) has impacted hundreds of thousands of people with its easy access to high-quality images of the heavens. Inagurated in 1993, the Bradford Robotic Telescope was the first fully autonomous astronomical observatory. The telescope has been available for the general public's requests for queue-scheduled service observations as well as remote, manual operation. Its operation is designed to make the observatory's resources available to the largest number of people by efficiently operating for every moment that environmental conditions allow. The project reports that it positively affects science education through access to an autonomous observatory coupled with well-designed projects and guided activities.

The Bradford Robotic Telescope allows for realistically complex observing programs. Supporting documentation reduces the preparation work required for astronomers to obtain quality data. This example demonstrates that it is possible to provide a global network of telescopes for the general population and to reduce the costs for long-term and large-scale projects.

33.6 Competitions and other opportunities to excel

The most motivated and skilled students need to be provided with opportunities to test themselves. Opportunities to excel should build upon earlier lessons and be visibly structured so that beginning and intermediate learners can plot a path to success for themselves.

33.6.1 Hands-on Universe

Hands-on Universe (HOU) is an educational program that continues to provide the structure and resources for mid- to high-level astronomy experiences. This globalized program is now strongly established in many countries, including the USA, Japan (J-HOU), Europe (EU-HOU; see Chapter 3); and has recently been introduced in Australia, China, and countries in Africa. The HOU program relies on experienced Teacher Resource Agents to lead training workshops for interested teachers. The resulting network of teachers share newly developed activities and join in multi-school collaborations for research investigations.

There are multiple ways students in the HOU program can acquire data. A network of remotely operated telescopes has been established, allowing students in the classroom to make astronomical observations during their day, using a telescope located part way around the world. Using the Internet, HOU participants around the world request observations from an automated optical or robotic telescope, download images from a large image archive, and analyze them with the aid of user-friendly image-processing software. In addition, an archive of images taken on large telescopes is also accessible. Pupils manipulate and measure images in the classroom environment, using the specifically designed software, within pedagogical trans-disciplinary resources constructed in close collaboration between researchers and teachers. Students are encouraged to take part in original, meaningful research projects. For instance, in recent years, students in the HOU program have discovered a Kuiper Belt object and supernovae, and have contributed to multiwavelength campaigns to monitor the brightness of quasars.

33.7 Making use of global networks of telescopes

A number of projects are coming online as the beginnings of global networks of autonomous observatories. The long awaited development of modest scale networks (e.g., Gelderman *et al.*, 2004; Newsam, Carter, and Roche, 2003; Hessman, 2001) has been recently upstaged by the purchase of Telescope Technologies, Ltd. and the Faulkes Telescope Project by the Las Cumbres Observatory Foundation (www.lcogt.net/). The vision of Las Cumbres is a global network of up to eight 2-meter and over thirty 0.5-meter astronomical observatories (Rees *et al.*, 2006). This vision is supported by firm private funding with a 20-year business plan.

For all of these proposed networks, there is a need for an educational strategy to match the proposed resources. Returning to the sports analogy, a strong youth league is not built solely on backyard play or a junior recreational league. Most kids want to be able to see examples of older, more capable kids who have excelled at the sport. Some kids want to compete and have other venues to demonstrate their abilities. A strong sports program, however, requires the recreational league in order to feed the competitive teams and regional tournaments. We can use the power of networked robotic telescopes to create a strong hands-on astronomical education program, but it should consciously follow the structure of other successful ways of engaging and teaching young people.

Acknowledgments

Development of the STARBASE Network has been supported by NASA under grant number NAG-58762.

References

Baruch, J. E. F., 2000, *PASA*, **17**, 119.
Baruch, J., 2003, in T. D. Oswalt (ed.), *The Future of Small Telescopes in the New Millennium*, (Dordrecht: Kluwer), vol. II, p. 75.
Bronowski, J., 1974, *The Ascent of Man*, (Boston, MA: Little, Brown, Inc.)
Edwards, B., 1999, *The New Drawing on the Right Side of the Brain*, (New York: Tarcher/Putnam Press).
Gelderman, R., Carini, M. T., Davis, D. R., Everett, M. E., Guinan, E. F., Howell, S. B., Marchenko, S. V., Mattox, J. R., McGruder III, C. H., and Walter, D. K., 2004, *AN*, **325**, 559.
Hessman, F. V., 2001, in W. P. Chen, C. Lemme, and B. Paczynski (eds.), *Small-Telescope Astronomy on Global Scales*, ASP Conference Series, vol. 246, p. 13.
Johnson, C., 1997, *The Sierra Club Guide to Sketching in Nature*, revised edition, Sierra Club Book (Berkeley, CA: University of California Press).
Leslie, C. W., 2003, *Nature Journal: Guided Journal for Illustrating and Recording Your Observations of the Natural World*, (North Adams, MA: Storey Publishing).
Maran, S. P., 1967, Telescopes and automation, *Science*, **158**, 867.
National Research Council, 1995, *National Science Education Standards*, National Academy Press, p. 2.
Newsam, A., Carter, D., and Roche, P. 2003, in T. D. Oswalt (ed.), *The Future of Small Telescopes in the New Millennium*, (Dordrecht: Kluwer, vol. II, p. 91).
Rees, P. C. T., Conway, P. B., Mansfield, A. G., Mucke-Herzberg, D., Rosing, W., Surrey, P. J., and Taylor, S., 2006, in L. M. Stepp (ed.), *Ground-based and Airborne Telescopes, Proceedings of the SPIE*, **6267** (spie.org).

Comments

John Baruch: There is no competition between the Bradford Robotic Telescope and the Faulkes Telescopes. They are complementary. We expect students to start with the Bradford Robotic telescope and graduate to Faulkes.
Jay Pasachoff: I have often suggested a clear–cloudy switch on a small telescope that students use. Now John Baruch's comment gives me the idea of a reverse direction – a feedback from the pointing of a Bradford or Las Cumbres telescope to a small telescope or telescope model in the classroom, giving the students a 3-D experience. (We saw earlier how students have great difficulty with 3-D ideas like the phases of the Moon or the origin of the seasons; see also my article in *Astronomy Education Review*: aer.noao.edu/cgi-bin/article.pl?id=13.)

Bruce McAdam: This discussion has compared the effectiveness of many years with student access to actual telescopes (eye contact) with only few robotic telescopes accessed by computer links. Valid comparison should wait until robotic real time control by students has had more educational operations – full Faulkes net in use. [The subsequent paper did report on effective use of the Bathurst remote telescope].

David McKinnon: Do you have evidence for the assertion just made that children are just as happy getting an image from a robotic telescope as they are from a remote one?
Richard Gelderman: My definition of a remote telescope is that it is operated from a distance but in real time. My minimum definition for a robotic telescope is that it is one that operates at a distance and has its functions handled as queue jobs instead of as real-time interactions. Thus the question deals with whether children get equal satisfaction if they are sending requests to a telescope in the daytime and getting data returned in the morning but not actually

getting to command the telescope to do its observing. My short answer would have to be "yes." We find that getting data returned is every bit as satisfying as the likely alternative of spending hours waiting for weather to clear or mechanical issues to be resolved during the typical real-time operation of a remotely controlled telescope.

John Percy: I would be interested in seeing *research* data on the relative advantages of (i) looking directly at e.g., Saturn, through a small telescope and (ii) various types of remote telescopes.

Alan Pickwick: The live remote control of a telescope makes for difficult management of the classroom as the teacher needs to be concentrating on the telescope. The use of queue-scheduled telescopes is a much better match as all the students can be busy at their computer screen making their selection of object and also processing their results.

34

Remote telescopes in education: report of an Australian study

David H. McKinnon and Lena Danaia
Charles Sturt University, Bathurst NSW 2795, Australia

Abstract: In 2004, the Australian Federal Department of Education, Science and Training funded a study of the impact of using remote telescopes in education in four educational jurisdictions: The Australian Capital Territory, New South Wales, Queensland, and Victoria. A total of 101 science teachers and 2033 students in grades 7–9 provided pre-intervention data. Students were assessed on their astronomical knowledge, alternative conceptions held, and ability to explain astronomical phenomena. They also provided information about the ways in which science is taught and their attitudes towards the subject. Teachers provided information about the ways in which they teach science. Both students ($N = 1463$)[1] and teachers ($N = 35$), provided the same data after the intervention was completed. The return rate for students and teachers was 71% and 34% respectively. This represents the largest study undertaken involving the use of remote telescopes in education. The intervention comprised a set of educational materials developed at Charles Sturt University (CSU) and access to the CSU Remote Telescope housed at the Bathurst Campus, NSW. Outcomes showed that students had increased their astronomical knowledge significantly ($F(1, 1173) = 201.78$, $p < 0.001$).[2] There was a significant reduction in the students' alternative conceptions ($F(1, 1173) = 27.9$, $p < 0.001$) and the students had acquired a significantly greater ability to explain astronomical phenomena ($F(1, 1173) = 25.66$, $p < 0.001$). There was a significant concomitant increase in students' attitudes towards science in general and astronomy in particular. Discussion centers on the ways in which the use of remote telescopes can be harnessed to impact in positive ways the attitudes of students.

34.1 Introduction

The current study has its origins in the mid-1990s when the first author developed a remote telescope system using control techniques that were different from those then available. The system involved no special software acquisition or installation by the end user. The system was to be usable by students and their teachers in the primary (elementary) school and was built as a partial response to the research literature that had been appearing claiming that astronomy was badly taught in primary schools and that there were many alternative conceptions evident even after formal teaching had taken place (e.g., Dunlop, 2000; Rider,

[1] N indicates the member of student and teacher participants providing both pre- and post-intervention data.

[2] F is based on the variance of the means obtained before and after the intervention, divided by the variance within the sample. If there is no difference between the pre- and post-intervention results then $F = 1$. P denotes the probability of obtaining the value of the F ratio shown given the number of degrees of freedom shown in the brackets. If the F ratio is large and the p value small, then there is a significant difference between the two results. Here, $p < 0.001$, which means that there is less than one chance in 1000 that there is no difference between the pre- and post-intervention results. They are *significantly different*.

Innovation in Astronomy Education, eds. Jay M. Pasachoff, Rosa M. Ros, and Naomi Pasachoff. Published by Cambridge University Press.

2002). A range of educational projects centered on astronomy and addressing the educational outcomes of the primary science curriculum were written and evaluated during 2000. It was clearly evident at this time that teachers were afraid of the technology involved. During 2001, a second study was conducted with modified educational materials and an embedded professional development program on the impact of using these and the telescope on the educational outcomes achieved by the students (Danaia, 2001). It was apparent that the impact on children was highly positive and that they were motivated by the prospect of taking control of the Charles Sturt University (CSU) Remote Telescope to get images of celestial objects (Danaia, 2001; McKinnon, Geissinger, and Danaia, 2002; McKinnon, 2005a).

Various reports commissioned by the Federal Department of Education, Science and Training (DEST) have revealed a disturbing trend in both primary and secondary science education in Australia. Further, enrolments in science during the post-compulsory years of education have been steadily dropping and fewer students are going on to undertake science at the tertiary level. In short, the trend in Australia is similar to the trends in other countries. More specifically, the researchers report that the science taught in secondary schools is largely transmissive rather than investigative and that many students spend their science time copying notes that their teacher dictates or writes on the blackboard (Goodrum, Hackling, and Rennie, 2000). The lack of practical work alienates students: they want much more of that. This depressing picture is described by the authors of the Australian DEST report as the *actual* picture of science education in the country. More importantly, the report paints a picture of what the *ideal* picture of science should look like. This picture consists of nine elements, five of which relate to the science that should be occurring in the classroom.

Specifically, *ideal* science is relevant and inquiry-based and where assessment is complementary to good teaching. The environment is characterized by enjoyment, fulfillment, ownership of, and engagement in, learning, and there is mutual respect between the teacher and students. Three of the remaining four elements relate to teachers and their lifelong professional development requirements, career paths, class sizes, and the employment of appropriate pedagogies. The final element relates to the status that science and science education should be accorded within the society, within the school, and within the curriculum (Goodrum *et al.*, 2000; p. vii).

During 2003, DEST commissioned a research and development study to examine the impact of using remotely controlled telescopes in junior secondary science classes. They were interested by the motivational impact that the control aspect the CSU Remote Telescope had had on primary-age students. The study developed educational materials for both students and their teachers. The learning package comprised a Student Workbook, a Teacher Guide supplemented by a CD-ROM, an interactive website, a communication forum, and a professional development program for teachers. A concurrent research project was undertaken to investigate a number of features of science education five years after the Goodrum *et al.* (2000) study to see what had changed since its publication.

34.2 Pedagogical approach

The learning package included access to, and control of, the CSU remote telescope and two digital cameras that students control online to take pictures of celestial objects they have decided to image. One of the cameras produces wide-angle images of the sky; the other one takes highly magnified images of celestial objects. The focus is on the students' learning how to control the telescope and its cameras to take the astronomical pictures they want. In the process, they have to learn a reasonable amount of astronomy that is tied directly to their science curriculum, though the program does allow extensive cross-curriculum integration, e.g., in English, mathematics, health, and the other sciences. Accompanying this system are two additional cameras: an infrared camera that allows students to see that they are indeed controlling the telescope and an all-sky camera showing the night sky above the observatory.

The second component of the package is a CD-ROM containing software, images, and PowerPoint presentations that support students and teachers in their quest to locate and capture images of celestial objects. In addition, a printed teacher's guide (McKinnon, 2004a) and accompanying student workbook (McKinnon, 2004b), provide an approach to teaching astronomy based on investigation (McKinnon and Danaia, 2005). The student book includes a series of projects that promote student-based inquiry in science. The learning materials are underpinned by a social constructivist approach to teaching and learning science. They provide students with the opportunity to undertake a range of practical investigations such as building scale models of the Solar System, measuring the diameter of the Sun, and investigating craters on the Moon, as well as investigating the causes of the seasons, day and night, and phases of the Moon.

The third component of the package is a website that contains links to resources and a gallery of images that displays the work of students. Also provided are links to the Bureau of Meteorology and special events that are broadcast through video streaming, e.g., comet impacts and the transits of Venus and Mercury.

34.3 Method

A concurrent nested mixed-method approach (Creswell, 2003; Tashakkori and Teddlie, 2003) involving a quasi-experimental non-randomized pre-test/post-test design (Shadish, Cook, and Campbell; 2002) complemented by qualitative data was employed to investigate students' perceptions of junior secondary school science, teachers' perceptions of what happens in the science they teach to students, and students' knowledge of astronomy. In addition, semi-structured interviews were conducted with a small number of students and teachers and documentation collected from participating teachers related to the teaching and learning experiences they employed during the intervention.

34.4 Instruments

The perceptions of students were measured both before and after the intervention using the questionnaire developed by Goodrum *et al.* (2000) to provide a direct comparison with the data they had collected five years previously. This questionnaire was reverse-engineered to produce a parallel version to measure the perceptions of teachers (Danaia, 2006). Four open-ended response items allowed both groups to respond to stimulus questions, e.g., "What do you like/dislike about science in your class?" The same instruments were employed on the post-intervention occasion with suitable changes made to the tenses of verbs to direct the attention of students and teachers to what had been happening during the intervention period.

The Astronomy Diagnostic Test (ADT) was employed to assess students' knowledge of various phenomena before the intervention and again at its conclusion (CAER, 1999, 2001, 2004). The ADT was modified to make it suitable for Southern Hemisphere administration and to elicit more data than the original. Specifically, four additional questions required students to draw and label a diagram that explained day and night, the phases of the Moon, the orbits of the Earth and Moon about the Sun, and the seasons. Students were asked to explain, in writing, what their diagram meant (Dunlop, 2000). The remaining 21 multiple-choice questions of the ADT asked students to provide reasons for their choice of response. Analysis of the reasons given by students allowed an in-depth analysis of their alternative scientific conceptions as well as analyses of the complexity of their reasoning and quality of response.

Six schools were visited during the intervention period. During these visits, groups of students and individual teachers were interviewed using a number of focus questions to ensure that some common data were collected from each teacher or group of students.

34.5 Participants

The participants in this study were 101 science teachers and 2033 students in grades 7, 8, and 9, drawn from 31 public and private high schools in four educational jurisdictions on the east coast of Australia. Of the 2033 participants, 2016 provided usable responses on both pre-intervention instruments. Of these, 1463 students provided responses on the post-intervention occasion. This represents a return rate of 70.9% for the students. For the teachers, the return rate was lower. Of the 101 who agreed to participate in the study, 35 returned the post-intervention questionnaires and supplied data on what they had actually chosen to do from the compendium of projects supplied. Despite numerous e-mails, faxes, and telephone calls requesting the return of completed instruments, the return rate for teachers was a disappointing 34.6%.

34.6 Teacher professional development

Fully funded professional development days were held in each jurisdiction for the participating teachers, at which time the project was explained to them, consent forms were completed, and they provided their first set of questionnaire perception data about the science they taught. At the conclusion of the session, the teachers were provided with enough copies of the ADT and student perception questionnaires to administer to their participating pupils. Teachers were instructed on how to administer these and asked to deliver the completed forms promptly, by post, to the researchers.

During the day, the various components of the project were covered: using the software in image-processing; using the planetarium software; making judgments about the projects to do in consultation with their pupils; making contact with the telescope; using the instructional resources; documenting what they did during the intervention period; continuing to communicate after the day had ended; and the administrative aspects to claim for travel, etc. Systems had been set in place to allow the teachers to communicate with each other and with the researchers by e-mail, telephone, fax, and a forum during the intervention period. The professional development was ongoing throughout the intervention period for the majority of teachers who chose to employ one or more of the various communication methods.

34.7 Data reduction

A research team processed the completed questionnaires and readied them for data entry by an experienced operator at the host institution. Procedures were implemented to ensure that high inter-judge concordance was achieved on the analysis of students' alternative conceptions, the responses to the open-ended questions in both of the perception questionnaires. In all cases, the level of agreement was better than 95%. Thus readied for data entry, the experienced data-entry operator entered the coded data into a form suitable for analysis by the Statistical Package for the Social Sciences (SPSS v12.02).

34.8 Data analysis

Thematic analyses of the written responses provided by students and teachers were undertaken to produce categories of response that could be analyzed statistically. The same technique was employed with the interview data, though these data were not analyzed statistically.

Quantitative data were analyzed using the statistical package SPSS v12.02. Multivariate analysis of variance (MANOVA) procedures with repeated measures on the occasion of testing were employed to compare the pre-/post-intervention data. For the questionnaire designed to elicit students' perceptions about science, non-parametric procedures (cross-tabulation and chi-square) were employed to compare the pattern of responses in the 1999 data collected by Goodrum *et al.* (2000) with the 2004 data set. To reduce the likelihood of a Type I error, given that 42 items were being compared in both sets of analyses, a full Bonferroni correction was employed. That is to say, the generally accepted p-value of 0.05 was substituted by the more rigorous p-value of 0.0012 (0.05/42). Repeated measures procedures were employed to compare the pre-/post-intervention perception data. Correlation analysis (Pearson) was employed to examine the commonality of perception between teachers and their pupils.

Students' responses to the questions in the ADT were analyzed at five levels: whether the answer was correct or incorrect; the alternative scientific conception(s) evident in the written response; the complexity of the written response using the Structure of the Observed Learning Outcome (SOLO) Taxonomy (Biggs and Collis, 1982; Biggs, 1995); the frequency of non-response; and the quality of the written response using a hierarchical scale derived from the SOLO Taxonomy and whether the response was correct or not.

34.9 Results

It is not intended here to present an exhaustive analysis of all of the data that were collected in this large project. Rather, the aim is to present an overview of results as they relate to the aims of Commission 46. An exhaustive analysis is available in McKinnon (2005b) and Danaia (2006). Following this, the discussion will seek to extract what the authors feel is important for further action by the members of Commission 46.

34.9.1 Students' and teachers' perceptions of science

The pattern of responses to the 42 items of the Secondary School Science Questionnaire in 2004 were compared with those reported in the Goodrum *et al.* (2000) study to determine the differences, if any, in students' perceptions of science. Only nine of the 42 items produced significant differences in the pattern of student responses. Students in the 2004 study reported an increased use of computers and the Internet in science compared with the 1999 sample. Despite the increase, only a small proportion of students reported that they used computers and accessed the Internet in science on a weekly basis. That is, the majority of students reported using the technology less frequently than once per week, and most used it less than once per month.

The results also suggest that in the intervening five years, teacher-directed experiments have become significantly more common in secondary school science classes, teachers more frequently use language that is easier to understand and, more often than not, take notice of students' ideas in science. There still appears to be limited opportunity, however, in linking science in secondary school to outside the classroom, with a high percentage of students reporting that they never experienced excursions or listened to guest speakers. A large proportion of students still report that they are rarely excited or curious about the science they experience at secondary school and feel that it lacks relevance. Despite the fact that there were nine items showing significant differences in the pattern of responses, only four of these could be considered to be in the "positive" direction, one that is desired by curriculum developers and science education commentators.

The same univariate analysis procedures were also used to compare the pre- and post-intervention student responses to the 42 rating scale items. The results show that now, for 36 of the 42 items, statistically significant differences were observed in the pattern of student responses from the pre- to the post-intervention occasion. The comparison revealed that the incidence of teacher-directed experiments had significantly reduced and that computer use had significantly increased.

The covariance of the change in response pattern with the introduction of the intervention is difficult to explain other than to attribute it to the components of the intervention. For example, in comparing the 1999 data with the pre-intervention 2004 data, it was found that there had been an approximate doubling in the use of both computers and the Internet. In comparing the pre- with the post-intervention data, there was a further factor of eight increase in both the use of computers and the Internet. This is not surprising given that the students had to use the Internet to execute many of the research projects, access the telescope system, and use computers to process the images that they acquired during their observation sessions.

34.9.2 The astronomy diagnostic test

Analysis of the pre-intervention ADT results revealed an appalling and depressing picture. Of the 25 items in the instrument, 14 can be considered to be knowledge outcomes of the curricula taught in primary school science, which is also covered again in Year 7 science classes in most educational systems. The mean score for all students was four correct out of 14. Mostly, they knew what caused day and night, and many could describe the orbit of the Moon and the Earth about the Sun. Very few knew what caused the phases of the Moon or the seasons and many (97%) thought that the Sun was overhead every day at noon (none lived within the tropics). One conclusion that could be drawn from these results is that the astronomy students are taught in primary school has made no impact. On average, the students possessed seven alternative conceptions that related to day and night, the seasons, the Moon phases, and the orbits of the Earth and Moon.

The number of alternative conceptions did marginally reduce as students moved through high school, but even by Year 9, the mean number of alternative conceptions was 6.5. One has to conclude that even the secondary school astronomy education has little impact on students' alternative conceptions related to such phenomena as the phases of the Moon and the seasons.

Table 34.1 presents the overall results as far as the five dimensions of analysis of the data from the ADT are concerned. The table shows that students' astronomical content knowledge increases significantly for all of the year groups, while the number of alternative conceptions about the astronomical phenomena decreases significantly. Of interest are the remaining three measures related to students' non-responses, the complexity of their responses, and the overall quality of response when one is made. The patterns of the non-responses, complexity, and quality of students' responses for Years 7 and 8 are similar, while for Year 9 they are different.

While the Year 9 students' astronomical content knowledge showed a significant improvement, they were less inclined to explain the reasons for their answers. One could entertain the

Table 34.1 *Results of the ADT data analysis*

	Knowledge of astronomy	Alternative conceptions	Non-responses	Complexity of written explanations	Quality of responses
Year 7	** occasions ↑ ns sex	ns occasions ↓ ** sex f > m	** occasions ↓ ** sex f < m	** occasions ↑ ** sex f > m	** occasions ↑ ns sex
Year 8	** occasions ↑ ** sex f < m	**occasions ↓ * sex f > m	ns occasions ↓ ns sex	ns occasions ↑ ns sex	** occasions ↑ ns sex
Year 9	** occasions ↑ ns sex	** occasions ↓ ns sex	** occasions ↑ ns sex	** occasions ↓ ns sex	ns occasions ↓ ns sex

**indicates a highly significant difference $^{**}p < 0.00067$.
*indicates a significant difference $^{*}p < 0.003$.
ns indicates that the difference is not significant.
↑ indicates an increase from the pre- to the post-intervention occasion of testing.
↓ indicates a decrease from the pre- to the post-intervention occasion of testing.

idea that this may be a maturation effect for the Year 9 students. It is a well known fact, at least in Australia, that Year 9 students are "difficult" to teach. An alternative explanation for these results may be that by Year 9, the majority of students have already learned that science is boring and not to be entertained.

34.9.3 Teachers' and students' perceptions of science

A correlation analysis of student and teacher perceptions of science proved revealing. Seven reliable and valid scales were extracted from the 42-item perception questionnaire computed using exploratory factor analysis on the student data and confirmed on the teacher data using confirmatory factor analysis. See Table 34.2. The scales allowed meaningful comparisons to be undertaken between the student data amalgamated for each class and the data supplied by that teacher. The scales are: perceived relevance; perceived difficulty; teacher-directed experiments; computer use; thoughts about what students need to be able to do; teacher's role; and outside experiences. It is beyond the scope of this paper to describe these and the statistical techniques employed to check their reliability and validity. Nonetheless, it is sufficient here to state that all scales possess high validity and reliability (the authors may be contacted if such data are required). The factors have been normalized on a common scale where 1 = never and 5 = almost every lesson.

The most noticeable feature of the correlation analysis (Pearson) is the lack of agreement between teachers and their students. Only two factors correlate on the pre-intervention occasion at a significant level: "computer use" and the "teacher's role." As far as "computer use" is concerned, both the students and their teachers agree that they are used infrequently. There was a significant increase in the use of computers from the pre- to the post-intervention occasion. There is disagreement, however, as to "how much," with teachers perceiving that there has been much more of an increase in the frequency of use compared with their pupils. Both students and teachers agree on the frequency with which the teacher discharges his or her

Table 34.2 *Teacher and student perceptions of science*

		Occasion 1			Occasion 2		
Scale	Group	M	SD	r	M	SD	r
1. Relevance of science	Teacher	2.953	0.545	–0.079	2.746	0.751	–0.091
	Student	3.249	0.321		3.173	0.458	
2. Difficulty of science	Teacher	2.518	0.447	0.029	2.608	0.622	0.05
	Student	2.923	0.296		3.004	0.32	
3. Teacher-directed experiments	Teacher	3.934	0.485	0.108	3.775	0.65	0.201
	Student	3.902	0.46		3.411	0.637	
4. Computer use in science	Teacher	3.000	0.868	0.680**	4.163	0.624	0.215
	Student	2.446	0.825		3.342	0.625	
5. What students need to be able to do	Teacher	4.141	0.433	0.206	4.106	0.427	0.21
	Student	3.678	0.351		3.555	0.457	
6. Teacher's role in science	Teacher	3.783	0.517	0.298**	3.484	0.447	0.104
	Student	3.368	0.468		3.025	0.538	
7. Outside experiences in science	Teacher	1.684	0.354	–0.096	1.45	0.421	0.421**
	Student	2.495	0.998		2.939	1.062	

**Correlation is significant at the 0.01 level (two-tailed).
*Correlation is significant at the 0.05 level (two-tailed).

role in science classes. On the post-intervention occasion, however, the pattern of agreement has changed, with the perceptions being no longer significantly correlated. With respect to frequency of the outside experiences, both teachers and students agree that there has been an increase, but students perceive that this has happened more often than their teachers. What is clear from the analysis is that the perceptions of students and teachers differ quite markedly on what appears to be happening inside the same classroom.

34.10 Discussion

The brevity of the presentation of results above illustrates the diversity of what appeared to be happening in the science classrooms. Nonetheless, it illustrates that significant changes in attitude towards science happened and were accompanied by significant increases in students' astronomical content knowledge. For the younger participants, there were significant increases in content knowledge, complexity of reasoning, and quality of response, accompanied by significant reductions in non-responses and alternative conceptions about certain astronomical phenomena. The impact of the intervention, however, seems to weaken with increasing grade level. The weakening effect may in part be due to the fact that students in the later grades have already experienced science and have already become alienated from it and the associated transmissive pedagogies employed to "teach" it. Perhaps the Year 7 students saw the intervention as a welcome relief in their short experience with the subject and engaged with it.

In these classes, the students enjoyed what they did; they were enthralled by the prospect of taking images and having them delivered to them immediately over the Internet. They seemed to be engaged and enthusiastic about what they were learning. It was also evident from a deeper analysis of learning outcomes that those teachers who had a physics background or who were already interested in astronomy achieved the best learning outcomes. They commented positively on the learning materials and were enthused by the remote control of the telescope and the image-processing that took place afterwards. The stark contrast in pedagogical style is best summed up by a comment one teacher made on the post-intervention questionnaire, related to the fact that she "did not feel as if she was teaching the students and yet they seemed to be learning" and that it was really interesting to "see another way to teach science." This begs the question of how she had been teaching science before, though it may be inferred from the very high proportion of students who claimed on the pre-intervention questionnaire that they had to "copy notes from the blackboard."

It was clear from this study that there was a range of pedagogies employed, from transmissive ones, where the teacher was in control all of the time, to ones where investigation was the dominant approach. This was most clearly evident in the interview and open-ended question data collected from students and their teachers. The way in which science was taught led many students to express their disappointment and frustration. For example, at one school where three classes were involved, students complained about a number of aspects of the pedagogy. At the interview, one student's comments summed up the disappointment at not being able to take their own images when he said, "We got there and *he* did it all and we were like . . . we chose what *he* took photos of, but *we* didn't do anything. We just *watched* him do it. So it was a bit disappointing – *really* annoying."

It would appear that if Commission 46 wishes to have an impact on astronomy education then steps must be taken to address the pedagogical issues. The teaching of science needs to move beyond the transmissive approaches required by science curricula that "are a mile wide but an inch deep" (Rutherford and Ahlgren, AAAS, 1990).

It was also clear that remote control of the CSU telescope was a motivating feature of the intervention. Pupils seemed to get a real "buzz" from driving the telescope, in controlling its cameras to get "their" images and processing these to show their parents and/or have them posted on the project website. This motivational aspect should not be ignored. Rather, it should be exploited to generate an even greater interest in the subject.

It is clear to us that using a remotely controlled telescope constitutes the *first* step in a developmental program that can end up with more senior students using robotic or completely autonomous telescopes to take images on a repeated basis where they wish to concentrate on the data extraction and the science contained in the picture. We say this for the following reasons. The students involved in this project had never done anything like this before. They had not had any prior experience with remote-control observing, and this aspect motivated them. The students and teachers all got a "kick" out of doing the observing. The specialist human support for such a system is expensive but a *necessary* first step at this time. More economical robotic systems can be used later, where projects that require repeated observations are undertaken.

The developmental approach alluded to above has not yet been investigated, but, with the advent of the Las Cumbres Observatory global telescope network, research can and will be undertaken to test the hypothesis that a love of astronomy can be engendered in more students. Perhaps they may even develop a lifelong interest in science in general rather than the current situation where the majority have had it quashed by Year 9.

References

Rutherford, F. J., and Ahlgren, A., AAAS (American Association for the Advancement of Science) Project 2061, 1990, *Science for All Americans*. New York: Oxford University Press.

Biggs, J. 1995, Assessing for learning: some dimensions underlying new approaches to educational assessment. *The Alberta Journal of Educational Research*, **41** (1), 1–17.
Biggs, J. B. and Collis, K. F., 1982, *Evaluating the Quality of Learning – the SOLO Taxonomy*. New York: Academic Press.
Collaboration for Astronomy Education Research (CAER), 1999, *Astronomy Diagnostic Test*, solar.physics.montana.edu/aae/adt/, modified for Southern Hemisphere 2001 by the University of Sydney, further modified 2004 by Charles Sturt University.
Creswell, J. W., 2003, *Research Design Qualitative, Quantitative, and Mixed Methods Approaches* (second edn.). Thousand Oaks: SAGE Publications.
Danaia, L., 2001, An investigation of primary age students' alternative conceptions of astronomical phenomena: an intervention study. Bachelor of Education Honours Thesis. Bathurst: Charles Sturt University.
Danaia, L., 2006, Students' experiences, perceptions and performance in junior secondary science: an intervention study involving astronomy and a remote telescope. Ph.D. Thesis. Bathurst: Charles Sturt University.
Dunlop, J., 2000, How children observe the universe. *Publications of the Astronomical Society of Australia*, **17** (2), 194–206.
Goodrum, D., Hackling, M., and Rennie, L., 2000, *The Status and Quality of Teaching and Learning of Science in Australian Schools*. A research report prepared for the Department of Education, Training and Youth Affairs. Canberra: Government Printer.
McKinnon, D. H., 2004a, *Practical Astronomy for Years 7, 8 and 9: Using Online Telescopes in Australian Science Classrooms – Teachers' Guide*. Bathurst: DEST and Charles Sturt University.
McKinnon, D. H., 2004b, *Practical Astronomy for Years 7, 8 and 9: Using Online Telescopes in Australian Science Classrooms – Student Workbook*. Bathurst: DEST and Charles Sturt University.
McKinnon, D. H., 2005a, Distance/ Internet Astronomy. In Jay Pasachoff and John Percy (eds.), *Effective Teaching and Learning of Astronomy*. Cambridge, UK: Cambridge University Press.
McKinnon, D. H., 2005b, *The Eye Observatory Remote Telescope Project: Practical Astronomy for Years 7, 8 and 9*. A research and development report prepared for the Department of Education, Science and Training (DEST), Canberra.
McKinnon, D. H. and Danaia, L., 2005, PD and ICTs: we have a ways to go. *Australian Educational Computing*, **20**(1), 24–29.
McKinnon, D. H., Geissinger, H., and Danaia, L., 2002, Helping them understand: astronomy for Grades 5 and 6. *Information Technology in Childhood Education Annual*, **14**, 263–275.
Rider, S., 2002, Perceptions about Moon phases. *Science Scope*, **26**(3), 48–52.
Shadish, W. R., Cook, T. D., and Campbell, D. T., 2002, *Experimental and Quasi-Experimental Designs for Generalized Causal Inference*. Boston: Houghton Mifflin Company.
Tashakkori, A. and Teddlie, C., 2003, *Handbook of Mixed Methods in Social and Behavioral Research*. Thousand Oaks: SAGE Publications.

Comments

Julieta Fierro: How did you decide on what to ask teachers and pupils for your evaluation?
David McKinnon: Research team meetings, reference panel of teachers, collaboration.
Jay Pasachoff: What is the difference between "remote telescopes" and "robotic telescopes"?
David McKinnon: Definitions that I use are as follows:

Remote telescopes: The user controls them in real time, tells them where to point, waits for them to move, tells the cameras the exposure time, takes the picture and sees it in real time as it is downloaded. The user then has to save the image either locally at the telescope or download it.

Robotic telescopes: The user does not get to control the telescope or the cameras. Software executes the observation request at some stage according to decision rules set in software. When the image is taken, the observer is notified and usually retrieves the image from the archive.

Depending on the software decision rules in a robotic system, the user may have to wait some time for an image to be taken.

Alan Pickwick: How many children have interacted with your remote control telescopes?
David McKinnon: There are 1000 hours of use per year, which contacts 20 000 plus children.

35

Visualizing large astronomical data holdings

C. Christian, A. Conti, and N. Gaffney

Space Telescope Science Institute, 3700 San Martin Drive, Baltimore, MD 21218, USA

Abstract: Scientific visualization involves the presentation of interactive or animated digital images for the interpretation of potentially huge quantities of observed or simulated data. Astronomy visualization has been a tool used to convey astrophysical concepts and the data obtained to probe the Cosmos. With the emergence of large astrophysical data archives, improvements in computational power and new technologies for the desktop, visualization of astronomical data is being considered as a new tool for the exploration of data archives with a goal to enhancing education and public understanding of science. We will present results of some exploratory work in the use of visualization technologies from the perspective of education and outreach. These tools are being developed to facilitate scientists' use of such large data repositories as well.

35.1 Introduction

A number of visualization tools have become available to scientists for browsing and locating data. Some of these tools are also in use for educational purposes. We briefly introduce these tools so that educators become aware of the potential for using visualization in education. This work is conducted in collaboration with A. Connolly, S. Krughoff, and R. Scranton of the University of Pittsburgh, and with B. McClendon and A. Moore of Google.

35.2 Tools available

(1) The Sloan Digital Sky Survey Navigate Tool: visual tools for browsing, creating finding charts, and displaying portions of the sky with associated measurements, in the form of catalog data, are available through the SkyServer website (cas.sdss.org/dr5/en/). A suite of educational projects suggested by teachers is available there.

(2) World Wind: NASA's Learning Technologies project, originally created to handle Earth science data, has been adapted to allow exploration of Mars and now the sky, as represented by a number of all-sky surveys. The astronomical application of this technology is in development; samples can be seen at www.worldwindcentral.com/wiki/Add-on:SDSS, where a download is available.

(3) Digital Universe Atlas from American Museum of Natural History (AMNH): a tool related to exploration of the sky in a planetarium setting was developed by AMNH. This tool, called the Digital Universe, is available for download over the Web from the following website: haydenplanetarium.org/universe/. A number of packages are available for browsing the sky, finding brown dwarfs, and accessing a variety of educational activities.

(4) Informal Science – Science on a Sphere at the Maryland Science Center: this technology was developed by NOAA, sos.noaa.gov/, for examining various Earth data and phenomena

Innovation in Astronomy Education, eds. Jay M. Pasachoff, Rosa M. Ros, and Naomi Pasachoff. Published by Cambridge University Press.

such as sea surface temperature and other data available from Earth-looking satellites. This technology has been adapted for viewing Mars and the Moon and soon will be available for all-sky survey data. Maryland Science Center offers a number of presentations and automated programs as described at www.mdsci.org/shows/scienceonasphere.html.

(5) Other experiments: the authors of this paper are interested in pursuing other technologies that are very familiar to computer users in order to allow users to visualize all kinds of astronomical data. In addition to the tools mentioned above, we have begun to experiment with various Google technologies such as those used for Google Maps and Google Earth. We have successfully demonstrated that we can display astronomical images in a Google Earth browser and are developing further templates and scripts to allow interfaces to data archives, press release archives and possibly the National Virtual Observatory.

35.3 Summary

We believe that, with the emergence of large multi-wavelength all-sky survey data and collections of data ("best of the best"), such as press release collections, visualization tools can be important for public understanding of science and education both in formal classroom and informal science center settings.

Acknowledgment

This work is supported by a contract, NAS5-26555, to the Association of Universities for Research in Astronomy, Inc., for the operation of the Hubble Space Telescope at Space Telescope Science Institute and an NSF ITR grant funding the development of the NVO.

Poster highlights

An educational CD-ROM based on the making of the Second Guide Star Catalogue

R. L. Smart, G. Bernardi, and A. Vecchiato

Instituto Nazionale di Astrofisica (INAF), Osservatorio Astronomico di Torino, Strada Osservatorio 20, I-10025 Pino Torinese, Italy

Abstract

In Viaggio fra le Stelle (*Voyage in the Stars*) will be an educational application using multimedia and interactive tools developed using Director and distributed via CD-ROM. The overlying theme will be the making of GSC II (the Second Guide Star Catalogue), the largest collection of celestial objects in existence, whose realization is the product of a ten-year collaboration between the Astronomical Observatory of Turin and the Space Telescope Science Institute of Baltimore, with the support of several other international astronomical institutions. The idea is to take a project such as the construction of an all-sky object catalog, a highly technical and specialized subject that would seem ill-suited for popularization at first sight, and make it accessible to an audience with little if any expertise in the field. See the figure on the next page.

Astronomia.pl portal as a partner for projects aimed at students or the public

Krzysztof Czart and Jan Pomierny

Nicolaus Copernicus University, Torun Center for Astronomy, ul. Gagarina II, 87-100 Torun, Poland; Polish Astronomy Portal, ul. Podgorna 58, 05-822 Milanowek, Poland

Abstract

There are many educational astronomical projects. To reach their target audience they need a proper way of communication, including suitable media. As the role of Internet media continually increases, one of the elements of communication should be an astronomical portal visited by teachers and students. Astronomia.pl portal supports (more or less formally) different projects and actions undertaken to educate or communicate astronomy. Several examples of such projects are Become a Young Copernicus, Kids on the Moon, Faulkes Telescope, Hands-On Universe, and events like Globe at Night and Catch a Star! The portal also organizes its own activities, like quizzes for students. The address of our main website is www.astronomia.pl, with an English version, www.astro nomia.pl/english.

Innovation in Astronomy Education, eds. Jay M. Pasachoff, Rosa M. Ros, and Naomi Pasachoff. Published by Cambridge University Press.

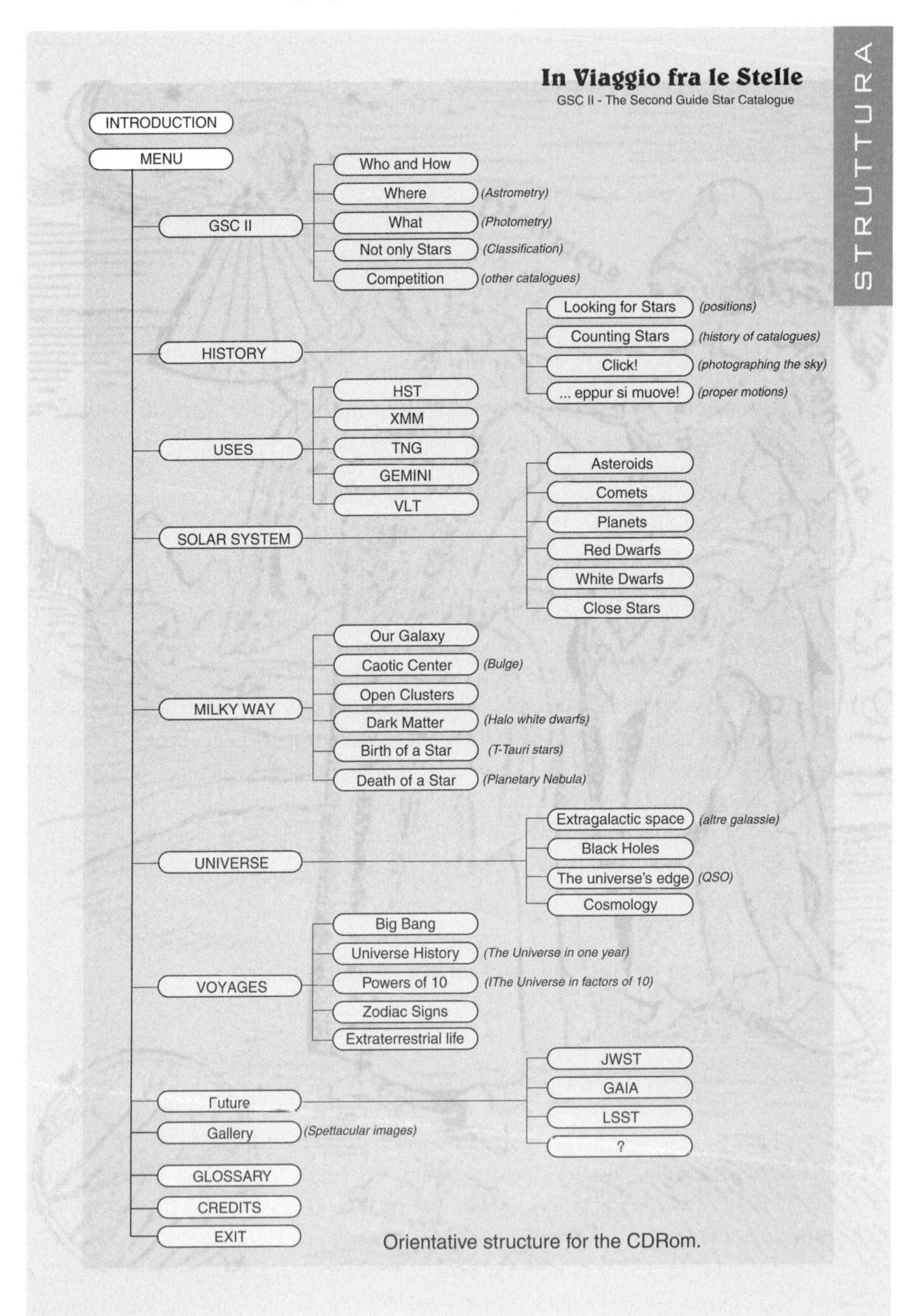

In Viaggo fra le Stelle (*Voyage in the Stars*), schematic.

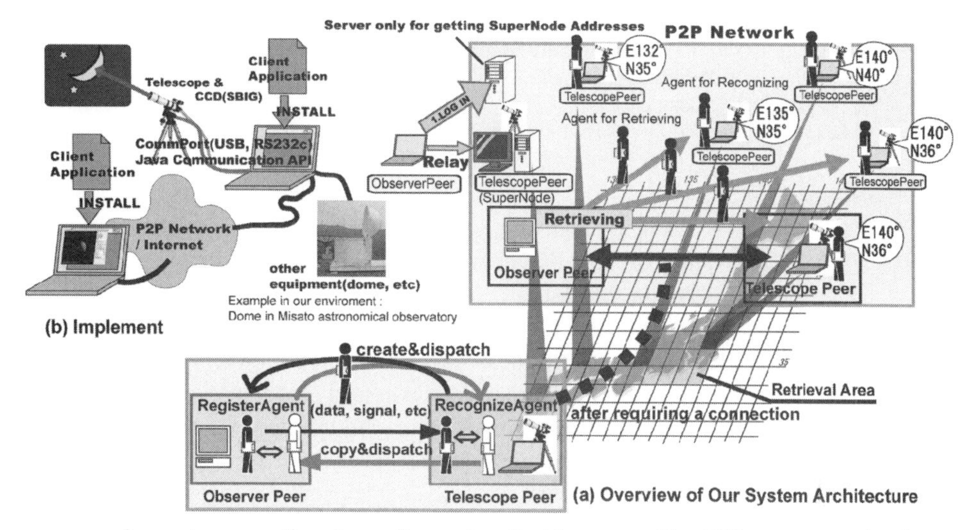

A remote cooperative observation system for telescopes with a P2P agent network using location information.

Development of a remote cooperative observation system for telescopes with P2P agent network by using location information

Takuya Okamoto, Seiichi X. Kato, Yuji Konishi, and Masato Soga

Department of Design and Information Sciences, Wakayama University, 930 Sakaedani, Wakayama, Japan; Department of Medicine, Hyogo College of Medicine, Division of Electronic Information and Engineering, Osaka University, 1-1 Machikaneyama, Toyonaka 560-0043, Osaka, Japan

Abstract

We propose a new remote telescope system with P2P agent network. We named this system e-SpaceCam. We are developing an environment where users can cooperate and share telescopes and the observational data for both education and observation. In the past, remote telescope systems were server-and-client-type systems. If a problem develops with such a system, however, the entire system may break down. Our system network architecture adopts a Hybrid P2P model that has neither client nor servers; all peers on the network have the function of both “client” and “server.” All user nodes, either Observer Peer or Telescope Peer, join the P2P network as Relay Peer. All new user nodes can participate in the P2P network by relaying to Relay Peer. Our system mounts LL-Net (Location-based Logical Network), which can manage the user nodes on the P2P network by a key based on longitude and latitude information. Observers can retrieve the telescope nodes faster by using the key. Java objects are the agents in our system. The various agents autonomously come and go in the P2P network by ATP (Agent Transport Protocol) until they fulfill their missions. We built this new remote telescope system as a Java client application with PIAX (P2P Interactive Agent eXtensions, www.piax.org/). Users can install our client application on their PC. Our system has GUIs like a planetarium to control the remote telescopes.

Image processing for educators in Global Hands-On Universe

J. P. Miller, C. R. Pennypacker, and G. L. White

Hardin Simmons University, Abilene, TX 79698, USA; Lawrence Berkeley Laboratory, University of California, Berkeley, CA 94720, USA; James Cook University, Townsville, QL 4811, Australia

Abstract

A method of image processing to find time-varying objects is being developed for the National Virtual Observatory as part of Global Hands-On Universe (Lawrence Hall of Science; University of California, Berkeley).

Objects that vary in space or time are of prime importance in modern astronomy and astrophysics. Such objects include active galactic nuclei, variable stars, supernovae, or moving objects across a field of view such as an asteroid or comet, or an extrasolar planet transiting its parent star.

The search for these objects is undertaken by acquiring an image of the region of the sky where they occur followed by a second image taken at a later time. Ideally, both images are taken with the same telescope using the same filter and charge-coupled device. The two images are aligned and subtracted with the subtracted image revealing any changes in light during the time period between the two images. We have used a method of Christophe Alard using the image processing software IDL Version 6.2 (Research Systems, Inc.) with the exception of the background correction, which is done on the two images prior to the subtraction.

Testing has been extensive, using images provided by a number of National Virtual Observatory and collaborating projects. They include the Supernovae Trace Cosmic Expansion (Cerro Tololo Inter-American Observatory), Supernovae/Acceleration Program (Lawrence Berkeley National Laboratory), Lowell Observatory Near-Earth Object Search (Lowell Observatory), and the Centre National de la Recherche Scientifique (Paris, France).

Further testing has been done with students, including a May 2006 two-week program at the Lawrence Berkeley National Laboratory. Students from Hardin-Simmons University (Abilene, TX) and Jackson State University (Jackson, MS) used the subtraction method to analyze images from the Cerro Tololo Inter-American Observatory (CTIO) searching for new asteroids and Kuiper Belt objects. In October 2006 students from five US high schools will

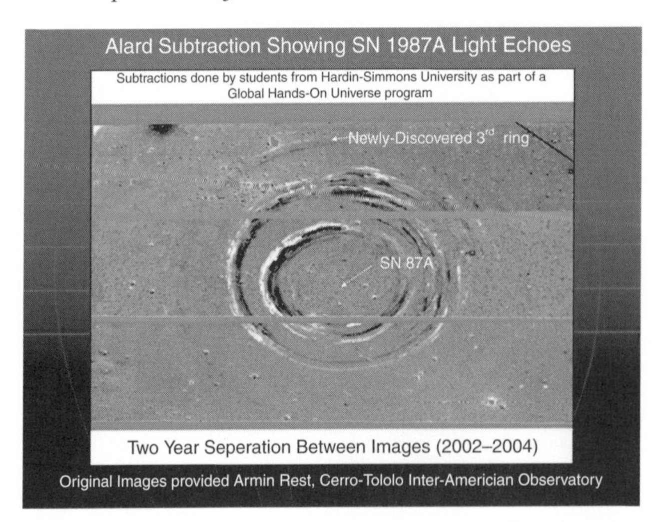

Alard subtraction showing SN 1987A light echoes. The subtractions were done by students from Hardin–Simmons University as part of a Global Hands-On Universe program. There was a two-year separation between the images (2002–4). (Original images provided by Armin Reat, Cerro Tololo Inter-American Observatory.)

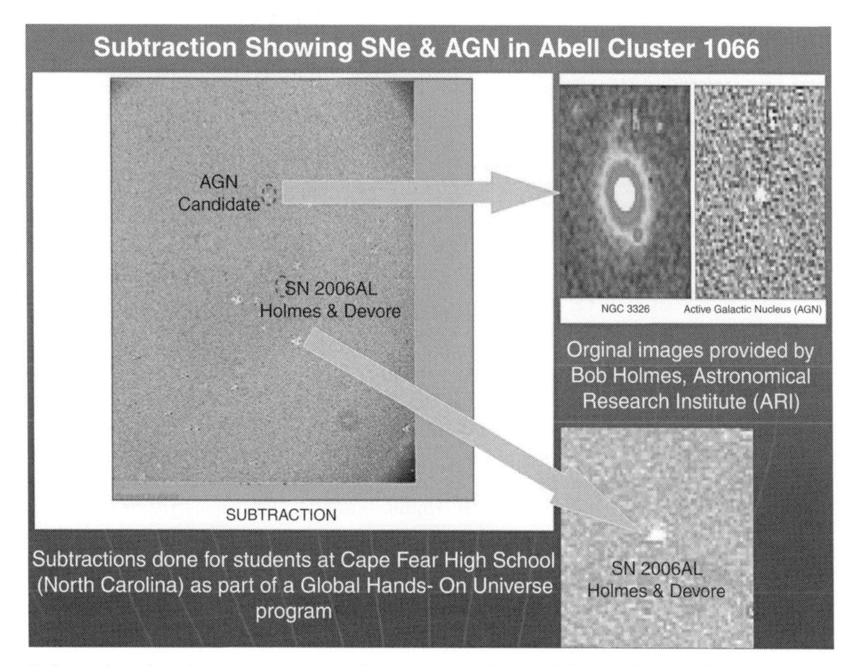

Subtraction showing supernovae and active galactic nuclei in the cluster of galaxies known as Abell 1066. (Subtractions done by students at Cape Fear High School, North Carolina, as part of a Global Hands-On Universe program. Original images provided by Bob Holmes, Astronomical Research Institute, ARI.)

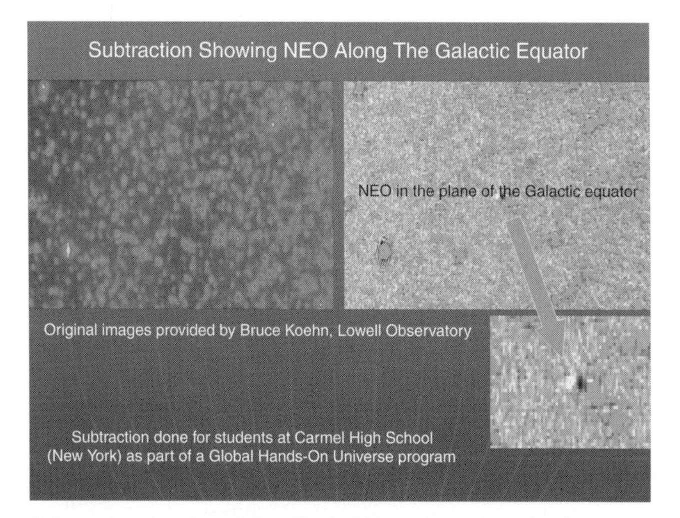

Subtraction showing a Near-Earth Object along the galactic equator. (Subtractions done for students at Carmel High School, New York, as part of a Global Hands-On Universe program. Original images provided by Bruce Koehn, Lowell Observatory.)

use the subtraction method in an asteroid search campaign using CTIO images with seven-day follow-up images to be provided by the Las Cumbres Observatory (Santa Barbara, CA).

During the Spring 2006 semester, students from Cape Fear High School used the method to search for near-Earth objects and supernovae. Using images from the Astronomical Research Institute (Charleston, IL) the method contributed to the original discovery of two supernovae, SN 2006al and SN 2006bi.

Two Global Hands-On Universe programs using Alard image subtractions to find supernovae. (Left) Science teacher Harlan Devore at Cape Fear High School, North Carolina, with his students in the background, analyzing images from the Astronomical Research Institute, Illinois. (Image from Andrew Craft, Fayetteville online.) (Right) Students from Hardin–Simmons University (Texas) and Jackson State University (Mississippi) analyzing Alard image subtraction images at the Lawrence Berkeley National Laboratory (California). (Image courtesy of Holly Ann Fidler, Hardin–Simmons University.)

The Pomona College undergraduate 1-meter telescope, astronomy laboratory, and remote observing program

B. E. Penprase
Department of Physics and Astronomy, Pomona College, Claremont, CA, USA

Abstract

We describe some of the innovations we have been making in undergraduate astronomy education, with the use of our 1-meter telescope, and other laboratories at Pomona College. Pomona – a small liberal arts, undergraduate college in Southern Calfiornia, with approximately 1450 students – is part of the Claremont Colleges, a group of affiliated small colleges in close proximity. The consortial arrangement has enabled Pomona, together with Harvey Mudd and the Joint Science Division of Scripps, Pitzer, and Claremont McKenna College, to own and operate a 1-meter telescope at Wrightwood, CA, in the San Gabriel Mountains, at a high-altitude (2400-meter) site. Our curriculum includes introductory and advanced observing classes, which make use of the 1-meter telescope and an on-campus observatory featuring two 14-inch telescopes. In an attempt to provide training that prepares students either for summer and graduate work in astronomy and astrophysics, or for careers in which enhanced appreciation of science will be an asset, our laboratory program provides students with research-grade equipment and simulates many aspects of astrophysics research, Each semester students in the introductory course perform a sequence of exercises of increasing complexity, primarily with the on-campus Brackett Observatory. Initial naked-eye observations are followed by wide-field imaging with digital 35-mm cameras, and CCD cameras with 35-mm lenses. A new lab using digital SLR cameras at both prime and Cassegrain focus of our 14-inch telescopes teaches students about plate-scale, magnification, and also colors and brightnesses of astrophysical sources. In the advanced classes students typically observe a sequence of stellar spectra illustrating the MK spectral types, and also some emission line nebulae and galaxies. Students also use IRAF and IDL (two professional image-processing programs) during advanced course work. Recently students using our remotely operable 1-meter telescope have observed Pluto's transit of a background star, the Deep Impact collision with comet 9P/Tempel 1, and T-Tauri stars in the infrared, and have measured the parallax of 2004XP14 by simultaneous observations from telescopes in California, Wisconsin, and New York.

Part IV

Practical issues connected with the implementation of the 2003 IAU resolution on the Value of Astronomy Education, passed by the IAU General Assembly, 2003

Introduction

The International Astronomical Union,
considering

- that scientific and mathematical literacy and a workforce trained in science and technology are essential to maintain a healthy population, a sustainable environment, and a prosperous economy in any country;
- that astronomy, when properly taught, nurtures rational, quantitative thinking and an understanding of the history and nature of science, as distinct from reproductive learning and pseudoscience;
- that astronomy has a proven record of attracting young people to an education in science and technology and, on that basis, to careers in space-related and other sciences as well as industry;
- that the cultural, historical, philosophical, and aesthetic values of astronomy help to establish a better understanding between natural science and the arts and humanities;
- that, nevertheless, in many countries, astronomy is not present in the school curriculum and astronomy teachers are often not adequately trained or supported; but
- that many scientific and educational societies and government agencies have produced a variety of well-tested, freely available educational resource material in astronomy at all levels of education;

recommends

- that national educational systems include astronomy as an integral part of the school curriculum at both the elementary (primary) and secondary level, either on its own or as part of another science course;
- that national educational systems and national teachers' unions assist elementary and secondary school teachers to obtain better access to existing and future training resources in astronomy in order to enhance effective teaching and learning in the natural sciences;
- that the National Representatives in the IAU and in the Commission call the attention of their national educational systems to the resources provided by and in astronomy; and
- that members of the Union and all other astronomers contribute to the training of the new, scientifically literate generation by assisting local educators at all levels in conveying the excitement of astronomy and of science in general.

The responses of a number of different countries to the IAU Resolution were discussed in this session, beginning with E. V. Kononovich's "Stellar Evolution for Students of Moscow University." Advanced students specializing in astrophysics and with a strong background in theoretical astrophysics are assigned "a special practicum work" that requires them to solve five problems using a PC program based on Paczynski code and supported by the Web interface, enabling them to use the Internet. Problem 1, which deals with zero-age main sequence (ZAMS) stellar models, requires students "to calculate ZAMS models for three different star masses and two variants of the chemical composition"; problem 2, dealing with main sequence stellar models, requires students "to calculate the evolutionary tracks for three stars with different masses during the time of hydrogen burning in the star core"; problem 3, dealing with "the evolutional peculiarity of stars with different masses and different chemical composition," requires them "to calculate the evolutional tracks for three stars with the same composition and different masses," and in one case "to change the chemical composition"; problem 4, dealing with "structure of red giants and supergiants," asks students "to calculate an evolutionary star track up to the supergiant branch"; and problem 5, relating to the evolutionary model of the present Sun, asks students "to compute the standard solar model . . . using as free parameters that of the convection zone and of the chemical composition."

Moving from the particular to the universal, M. C. Pineda de Carias presented "Astronomy for Everybody: Approach from the CASAO," the acronym for Central America Suyapa Astronomical Observatory, part of the National Autonomous University of Honduras, where currently all professional astronomers in Honduras are employed, along with regional and foreign colleagues. After noting that "Honduras is a country of very young people" and that "most students are at elementary school level" with fewest students in university, she explained that the astronomical observatory of Honduras was inaugurated in 1997. In the ensuing decade the observatory not only received "regional accreditation" but also initiated "a regional program in astronomy and astrophysics at both undergraduate and graduate levels" for Central Americans. She described three different outreach programs, targeting different audiences. One program, aimed at "elementary and secondary school students, teachers, college students, parents, and media communicators," involves two-hour visits by groups of 20 to 100 to CASAO, where an astronomer delivers a lecture "organized and adapted to the interest of the participants," followed by practical activities aimed at familiarizing the visitors with the use of "small telescopes to observe the Sun and planets" and at demonstrating "how the Maya of Central America used stelae" to measure solar time and design solar calendars. The goals of this program include motivating "the study of science, mathematics, and space exploration," arousing "curiosity for learning about what exists" beyond "our own environment," and introducing new resources for teaching and learning science. A second program involves Friday-night visits by the general public to the observatory. The goals of the "Astronomical Nights Program" are to familiarize visitors with the night sky and the visible Universe and give them the opportunity to make naked-eye observations under the guidance of a professional astronomer. Recognizing that most of the beneficiaries of these two programs come from the capital city, Tegucigalpa, the observatory is making an effort to reach children elsewhere in the country. A third program, "Introduction to Astronomy @ Internet," is an online course written in Spanish, offered as an elective to all full-time students in Honduras as part of their education but particularly recommended for elementary and secondary school teachers, as well as to students and teachers elsewhere in Central and Latin America. The course "represents an opportunity . . . to be part of a scientifically literate generation, trained by professional astronomers."

Turkey's response to the 2003 IAU Resolution, "Towards a New Program in Astronomy Education in Secondary Schools in Turkey," was presented by Z. Aslan. Although astronomy was taught on its own at the secondary level in Turkey before 1974, since then it has been incorporated into secondary-school physics courses as well as into elementary science and geography courses on the primary level. Most teachers on those levels, however, have had little formal preparation in astronomy, so that the teaching has not been "very effective." In addition, topics in astronomy "are generally scheduled at the end of a particular course," when there is little time left to devote to them. To remedy the situation, in 2005 the TUBITAK National Observatory (TUG) proposed to the Ministry of Education that a national meeting on "teaching of astronomy and using astronomy to teach physics" be held for teachers of physics and astronomy teachers to coincide with the March 29, 2006, total solar eclipse. The approximately 120 hand-picked schoolteachers and ten schoolchildren "had a beautiful sky to see the eclipse and to carry out . . . experiments."

According to Aslan, the meeting, which involved "astronomers and physicists from Turkish universities and educators from the Ministry of Education (ME) plus three educators from abroad," was "very successful." TUG subsequently submitted to the Ministry of Education a number of proposals, suggesting, among other things, that the Ministry of Education and TUG should collaborate in providing educational materials and in holding summer schools and in-service training for primary and secondary school teachers. Independently, in 2005 ME published a draft for a new primary-school science and technology course, in which astronomy topics, previously taught on the primary level as parts of other subjects and now presented under the rubric "The Earth and the Universe," will make up nearly 10% of the entire course. A similar program for secondary schools is being prepared. Turkey clearly has done much to respond to the IAU 2003 resolution's suggestion that "the National Representatives in the IAU and in the Commission call the attention of their national educational systems to the resources provided by and in astronomy."

Cecilia Scorza, on behalf of colleagues at ESO and the universities of Heidelberg and Leiden, presented "Universe Awareness for Young Children" (UNAWE). This international program, "motivated by the premises that access to the simple knowledge about the Universe is a birthright and that the formative ages of 4 to 10 years play an important role in the development of a human value system," targets "economically disadvantaged young children" in this age group and exposes them to "the inspirational aspects of modern astronomy." The program should be operational by 2009, the International Year of Astronomy, with goals including production of "entertaining material in several languages and cultures," organization of training courses for those who will present the program, and provision of a network for exchange of ideas and experiences. In 2006 pilot projects were carried out in Venezuela and Tunisia to examine UNAWE's feasibility. Even in this early stage, UNAWE's efforts are helping to facilitate the 2003 IAU resolution's recommendation that "members of the Union and all other astronomers contribute to the training of the new, scientifically literate generation by assisting local educators at all levels in conveying the excitement of astronomy and of science in general." More information relating to UNAWE is available at www.UniverseAwareness.org/.

Ahmed A. Hady spoke about "Education in Egypt and the Egyptian Response to Eclipses." He began by summarizing the history of modern astronomy education at the university in Egypt, which began in 1936. The University of Cairo offers the bachelor of science degree in astronomy and physics, in astronomy, and in space science; a master's degree in astrophysics, in theoretical physics, and in space science; and a doctorate. Professional astronomical research is conducted at Cairo University and at Helwan Observatory in a variety of fields. Egyptian scientists participated in international observations of total solar eclipses in Egypt on February 25, 1952 and March 29, 2006. The more recent observations are being coordinated with the ESA/NASA Solar and Heliospheric Observatory (SOHO) and NASA's Transition Region and Coronal Explorer (TRACE) observations to show the magnetic structure of the corona, and since 2006 with Hinode and STEREO.

Paul Baki of Kenya described "Spreading Astronomy Education Through Africa." After explaining how different African societies "practice astronomy largely for understanding and predicting the weather and climatic changes for seasons," he described "some traditional tools that are used by some ethnic communities in East Africa to interpret astronomical phenomena for solving their local problems." He noted that these communities "combine the knowledge of plant and animal behavior changes together with sky knowledge to predict the weather and climate for the coming season." Baki argued that from the African perspective, "it seems that the best way to spread knowledge in astronomy is to begin by appreciating its cultural value." He suggested incorporating the traditional practices into the standard astronomy curriculum, and predicted that by leading to environmental conservation and increased crop yields, as well as to an increase of tourism, such an approach would "get recognition and possible funding from the various African governments."

Located on the Pampa Amarilla in western Argentina, the Pierre Auger Cosmic Ray Observatory not only studies the highest energy particles in the Universe but also participates, according to Beatriz Garcia, in "a wide range of outreach efforts that link schools and the public with the Auger scientists and the science of cosmic rays, particle rays, particle physics, astrophysics, technologies." In "Education at the Pierre Auger Observatory: movies as a Tool in Science Education," Garcia described the use of educational videos for children between 6 and 11 and for general audiences, as well as the use of animation in science teaching and learning. She identified scientific outreach as a means of encouraging "scientific vocations," particularly in countries where they are not accorded a high social status. Garcia asserted that "if we want to help the student-public to think and be able to solve problems, the audio-visual language must be characterized by its originality and the search of new forms of expression that stimulate the imagination." Garcia's essay underlines the importance of implementing the section of the 2003 IAU resolution calling for "members of the Union and all other astronomers" to "contribute to the training of the new, scientifically literate generation by assisting local educators at all levels in conveying the excitement of astronomy and of science in general."

In "Freshman Seminars: Interdisciplinary Engagements in Astronomy," Mary Kay Hemenway explained the role such courses can play in implementing the 2003 IAU resolution. To facilitate the transition of the diverse population of students entering the University of Texas at Austin to college academic and social life, the university offers freshman seminars limited to 15. Instructors invited to participate in the program may design the course of their choosing. The only stipulations are that students must complete a certain number of certain types of writing assignments, and must also attend "sessions on time-management and using the library"; students whose seminars involve only two hours of class time each week must also attend an additional hour-long weekly event. Hemenway reported on two seminars rooted in astronomy. For a seminar focused on the life of Galileo, students modeled "rotation and revolution of Solar System objects with their own bodies"; experimented with lenses to master the concepts of focal length, field of view, and refracting telescopes; compared positions of Jovian satellites as Galileo drew them in *Sidereus Nuncius* with those calculated by the *Starry Night* computer program for the dates and location corresponding to Galileo's depictions; and a dramatic reading of an English translation of Bertolt Brecht's play "Galileo." These classroom activities were enriched by visits to the Blanton Museum of Art to view Italian art of the period, as a springboard to discussing the influence of religious struggles on contemporaneous artists; and to the Harry Ransom Center for Humanities Research, to expose students to original seminal works in the history of astronomy, including, among others, Ptolemy's *Almagest*, Copernicus's *De Revolutionibus*, and several works by Galileo himself. A highlight of this seminar for many students is the mock trial "in which Galileo has the benefit of something he lacked in reality – a defense team." Students not assigned to either the prosecution or defense team participate in the judgment phase. A second seminar, "Astronomy and the Humanities," exposes students to science fiction, which they are asked to contrast with science facts known now and at the time of the writing; to poetry and literature with some astronomical connection; to a range of music with an astronomical theme; and to art work relating to astronomy. Hemenway concluded by noting that while the seminars' goal differs from that of introductory astronomy courses, Astronomy 101 instructors might do well to pick and choose from among the broad humanistic connections the seminars highlight to enrich the teaching of those standard academic offerings.

In "Astronomy for Teachers," Julieta Fierro demonstrated how far Mexico has come in fulfilling the recommendation of the 2003 IAU resolution "that national educational systems and national teachers' unions assist elementary and secondary school teachers to obtain better access to existing and future training resources in astronomy in order to enhance effective teaching and learning in the natural sciences." She described the preparation of books for different audiences: to answer schoolteachers' frequently asked science questions; to build the scientific vocabulary of mothers and small children; to pique the curiosity of middle-school students in science. She explained how Mexico has promoted literacy by setting up classroom libraries to which teachers can add ten books a year from a list of 700 titles previously chosen by a selection committee; by holding book fairs; and by erecting a major library in Mexico City, which offers workships and lectures. She also detailed the offerings made available to high school physics teachers, including a 180-hour-long general astronomy course, offered for five hours each Saturday over the course of the year, and a biannual conference for about 1500 teachers, where astronomy books, magazines, and teaching materials are made available, in addition to lectures, workshops, and demonstrations. She concluded by asserting not only that astronomy materials "must be available for teachers and students" and that "if one wants to improve the quality of education one must take time to work with teachers in order to understand their needs and difficulties," but also that developing nations have to go beyond curriculum reforms to more basic considerations: "children must be well nourished and healthy in order to learn."

John Mattox shows how to use daylight hours for astronomical observations, times when students are more likely to be in class than at night. He discusses daytime observations of venus and of sunspots. See also short descriptions of posters, pages 315–323.

36

Stellar evolution for students of Moscow University

E. V. Kononovich and I. V. Mironova

Sternberg Astronomical Institute, Universitetskij Prospect 13, Moscow 119992, Russia

Abstract: The theory of stellar interiors is a very stimulating topic in the physics and astrophysics curricula. To support the corresponding lecture courses, some practical work was proposed and elaborated in 1991 for advanced students of the physics department at Moscow State University specializing in astrophysics. The work, which requires significant knowledge in theoretical astrophysics, is recommended for fifth-year students. The main purpose of the work is to calculate the evolutionary set of stellar models, including those for the Sun. The PC program, based on the B. Paczynski code and supported by the web interface, allows working via the Internet. The results of the work may be presented in both tabular and graphic forms.

36.1 Introduction

Stellar evolution is one of the most important items on the university astrophysics curriculum. To support the corresponding course of lectures, a special practical work was started at Moscow State University. For this purpose Paczynski's PC code (Paczynski, 1970) was used. The program computes the evolutionary tracks and inner structure of a star for given mass and chemical composition as the initial conditions.

During the semester the students are supposed to solve five problems. The star parameters are usually given by the teacher to individual students. In Table 36.1 the five problems are summarized.

36.2 Problem 1 Zero-age main sequence stellar models (ZAMS)

Students have to calculate ZAMS models for three different star masses and two variants of the chemical composition. The initial data given by the teacher are supposed to lead to star models with convective core, convective envelope, or both.

36.2.1 Exercises for students

1. To compute the chemically homogeneous star models, for the given masses and chemical compositions.
2. To investigate the obtained star models' structure depending on the mass and chemical compositions (radii, effective temperatures, convection and radiative zones).
3. To plot the star parameters on the Hertzsprung–Russell diagram. How does the star location change with mass, and with composition?

36.2.2 Example questions

- In which case does the star have both a convective envelope and a convective core (for what mass and chemical composition)?
- What is a red dwarf?

Innovation in Astronomy Education, eds. Jay M. Pasachoff, Rosa M. Ros, and Naomi Pasachoff. Published by Cambridge University Press.

Table 36.1 *Five problems*

Problem 1	Zero-age main sequence stellar models
Problem 2	Main sequence stellar models
Problem 3	Evolutionary properties for stars of different masses and different chemical compositions
Problem 4	Structure of red giants and supergiants
Problem 5	The standard solar model (the evolutionary model of the present Sun)

36.3 Problem 2 Main sequence stellar models

Students have to calculate the evolutionary tracks for three stars with different masses during the stage in which hydrogen burning takes place in the star's core.

36.3.1 Exercises for students

1. To compute the evolutionary tracks for the given stars and plot the time dependencies for the central density, the surface temperature, the luminosity, the radii, and the central hydrogen and helium abundances. To explain these values' variation during the main sequence stage.
2. To find out the main sequence evolution time for the given stars.

36.3.2 Example question

- Which stars evolve faster and which slower?

36.4 Problem 3 The evolutionary peculiarity of stars with different masses and different chemical composition

The students have to calculate the evolutionary tracks for three stars with the same composition and different masses; in one case they have to change the chemical composition.

36.4.1 Exercises for students

1. To compute the evolutionary star tracks for the different mass values up to the moment when helium burning begins in the star's core, and to time-plot the dependencies of the surface temperature, luminosity, radius, and the central abundances of hydrogen and helium.
2. To compute the evolutionary tracks of a given star for different chemical composition.
3. To specify the basic phases of a star's life and to determine the time intervals of evolutionary phases; all tracks must be plotted in the same figure.

Figure 36.1 presents an example of student results.

36.4.2 Example questions

- What are the subgiants?
- What are the red giants?
- What is the helium flash?

36.5 Problem 4 Structure of red giants and supergiants

Students have to calculate an evolutionary star track up to the supergiant branch.

Table 36.2 *The basic phases of stellar evolution (Student results)*

Masses	Main sequence	Subgiant	Red giant
2 M_{sun}	6.12×10^8	7.43×10^8	8.25×10^8
5 M_{sun}	5.17×10^7	5.57×10^7	5.66×10^7
15 M_{sun}	8.56×10^6	8.82×10^6	8.87×10^6

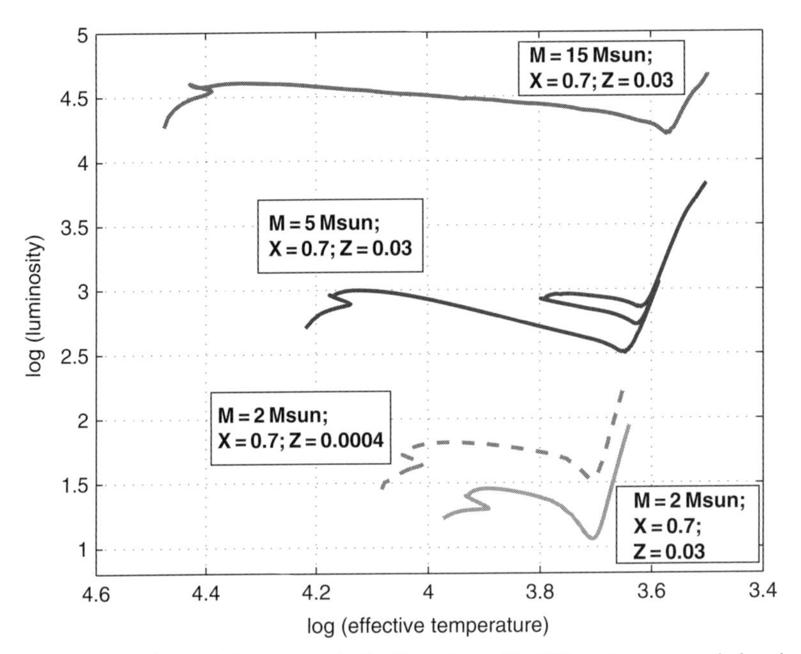

Figure 36.1 The evolutionary tracks for four stars with different masses and chemical compositions. Luminosity vs. effective surface temperature is graphed on a log–log scale. X is the fractional hydrogen abundance, Y (not shown) is the fractional helium abundance, and Z is the fraction of all other elements.

36.5.1 *Exercises for students*

1. To compute the star track up to the supergiant branch.
2. To investigate the inner structure of a star on the red giant branch and on the supergiant branch. To plot upon the radius the density, temperature, luminosity, and chemical abundances of hydrogen and helium.
3. To compare the inner structures of a red giant, a supergiant, and ZAMS star models.

36.5.2 *Example questions*

- What happens after the core helium burning ends?
- What is the horizontal branch?
- What is the asymptotic giant branch?

36.6 Problem 5 The standard solar model (the evolutionary model of the present Sun)

Students have to change two values: the parameter of the convection zone and that of the chemical composition. The standard solar model is required to fit the present Sun luminosity and its radius.

36.6.1 Exercises for students

1. To compute the standard solar model (the evolutionary model of the present Sun) using as free parameters that of the convection zone and of the chemical composition.

36.6.2 Example questions

1. Compare zero-age and present Sun models. How much have the radius and the luminosity of the Sun changed?
2. What was the luminosity of the Sun in the Jurassic period, during the age of the dinosaur? How much does it differ from the present Sun (i.e., 146 up to 200 million years ago)?
3. At the end of the Sun's life on the main sequence, what will be the solar luminosity and radius? What will be the Sun's age?
4. Assuming that an increase of solar luminosity of 3% will be fatal for life on Earth, estimate the time remaining for the existence of human civilization.
5. Estimate the depth and the mass of the convective zone of the Sun at present and at the very beginning of its life (ZAMS model).

Reference

Paczynski, B., 1970, *Acta Astronomica*, **20**(2).

37

Astronomy for everybody: an approach from the CASAO/NAUH view

María Cristina Pineda de Carías

Central America Suyapa Astronomical Observatory, National Autonomous University of Honduras (CASAO/NAUH), Ciudad Universitaria, Tegucigalpa M. D. C., Honduras

Abstract: Astronomy is a science that attracts the attention of people of all ages and from a variety of points of view and interests. At the Central America Suyapa Astronomical Observatory of the National Autonomous University of Honduras (CASAO/NAUH), in addition to the general course of Introduction to Astronomy (AN-111) and the regular courses for a master's degree in Astronomy and Astrophysics, three different academic outreach programs have become important after less than a decade of experience. "Visiting CASAO/NAUH," a program for elementary and secondary schools, involves thrice-weekly astronomers' presentations to groups of from 15 to 100 students and teachers; conferences on selected topics of astronomy, illustrated with real sky and astronomical images; opportunities to observe the Sun, Moon, and planets using a small telescope; and explanations of how contemporary astronomers do their observations, with comparisons drawn to the methods of observing used by the Maya who once inhabited Central America. On Friday nights, the "Astronomical Nights Program," intended for a general public of children, youth, and adults, involves visits to the astronomical observatory, where the visitors learn about the properties of astronomical bodies, the sky during the week, and the differences between making observations using telescopes and with the naked eye alone. "Introduction to Astronomy @ Internet Program" is an online course designed not only for school teachers but also for Central American college and university students who are willing to learn more systematically on their own, using new technologies for studying the sky, the Solar System, the stars, galaxies, and the Universe. In this paper I present a complete description of these programs at CASAO/NAUH, and a discussion of how they contribute to the implementation of the IAU Resolution on the Value of Astronomy Education.

37.1 Introduction

Astronomy is both an old science and a new science that strongly attracts the attention of people of all ages and from a variety of points of views, interests, and fields. Astronomers who attended the XXV General Assembly of the IAU held in Australia in 2003 raised several pertinent issues, including the facts that scientific and mathematical literacy and a workforce trained in science and technology are essential to maintain a healthy population, a sustainable environment, and a prosperous economy in any country; that astronomy, when properly taught, nurtures rational, quantitative thinking and an understanding of the history and nature of science, as distinct from rote learning and pseudoscience; that, nevertheless, in many countries, astronomy is not present in the school curriculum and astronomy teachers are often not adequately trained or supported, but that many scientific and educational societies and government agencies have produced a variety of well-tested, freely available educational resources in astronomy at all levels of education. In recognition of these facts, the 2003 IAU

Innovation in Astronomy Education, eds. Jay M. Pasachoff, Rosa M. Ros, and Naomi Pasachoff. Published by Cambridge University Press.

passed a resolution that recommended the following measures: (1) that astronomy be included as an integral part of the school curriculum at both the primary and secondary level, (2) that elementary and secondary school teachers be assisted to obtain better access to existing and future training resources in astronomy, (3) that the national representatives in the IAU call the attention of their national educational systems to the resources provided by and in astronomy, and (4) that members of the IAU and all other astronomers contribute to the training of a new, scientifically literate generation by assisting local educators at all levels in conveying the excitement of astronomy and of science in general.

In this paper I present a complete description of the three most outstanding astronomy outreach programs offered at CASAO/NAUH for the benefit of everybody and a discussion of how these programs contribute to the implementation of the IAU Resolution on the Value of Astronomy Education from the CASAO/NAUH point of view and experience.

As a strategy, first I will identify the people who are the target for each of the four recommendations above: for the first one, elementary and secondary school students; for the second one, elementary and secondary school teachers; for the third one, the national representative in the IAU; and for the fourth one, all the astronomers contributing towards the training and development of astronomy at different levels of the national educational systems. I will then show how all these sectors of the population of Honduras are involved with the CASAO/NAUH outreach programs.

Looking at the most recent census data of 2001, distributed by age groups as shown in Table 37.1, we can see that almost 50% of the population is under 17 and 77% is under 35. This means that Honduras is a country of very young people.

What is the Honduran population studying? The levels of the Honduran National Educational System are: pre-school (from 5 to 6 years), basic elementary (the traditional six grades from 7 to 12 years), basic secondary (three years from 13 to 15 years), secondary (from 16 to 17 years), undergraduate university studies (average from 18 to 25 years), postgraduate studies (average from 26 to 35 years). In Figure 37.1 we can find the distribution of population studying by age and levels. In the same figure we can see the high number of people out of the school system, and we can also see that most students are at elementary school level. It is also

Table 37.1 *Honduran population by age groups (INE, 2001)*

Ages	Population	As %	Accumulated %
0 to 4	874 288	14.39	14.39
5 to 6	353 681	5.82	20.21
7 to 12	1024 830	16.86	37.07
13 to 15	448 670	7.38	44.45
16 to 17	282 154	4.64	49.09
18 to 25	967 195	15.92	65.01
26 to 35	751 086	12.36	77.37
36 to 45	563 516	9.27	86.64
46 to 60	494 782	8.14	94.78
60 and more	316 683	5.21	100.00
Total	6076 885	100.00	

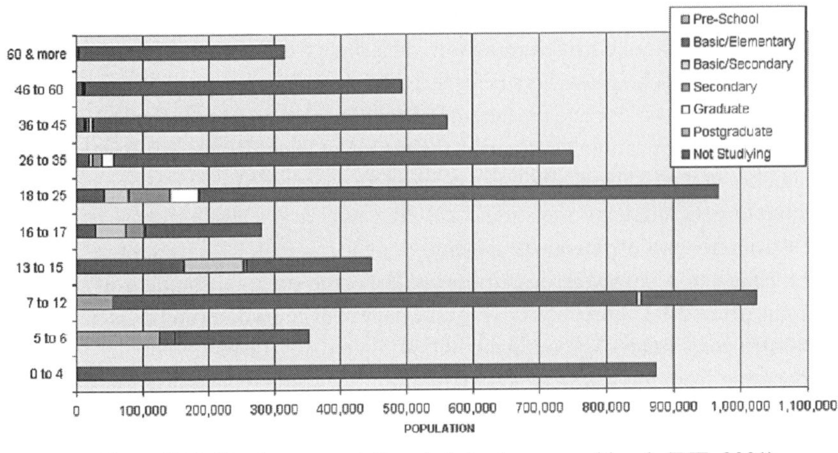

Figure 37.1 Honduran population studying by age and level. (INE, 2001)

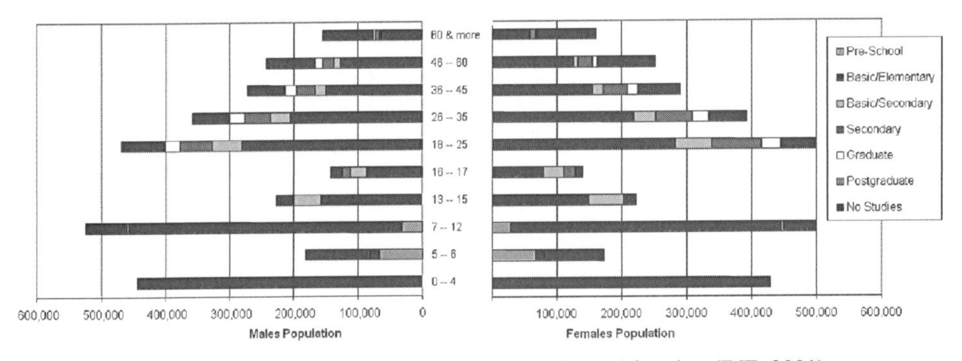

Figure 37.2 Educational levels for Honduran males and females. (INE, 2001)

apparent that the number of people in school decreases as the level rises, there being fewer people at university levels.

By comparing the educational levels of male and females we can find, as is shown in Figure 37.2, that they are almost equally distributed. In both graphs we can see that most people have gone through elementary school, with the male population somewhat higher in the lower group ages. But in the higher age groups the situation is the reverse, with the female group being a little higher. At the secondary school levels, the trend is towards the females being higher. For both sexes, the university population is the smallest.

To describe education in astronomy in Honduras as part of the national educational system, I will present the CASAO/NAUH contribution. This center began to function in 1997 when the astronomical observatory was formally inaugurated. At that time we began to develop a master's program in astronomy and astrophysics (Pineda de Carías, 2001) that now has graduated a small core of astronomers for the region. At the same time, a general course, an introduction to astronomy, open to all university students regardless of their chosen fields, began to be part of all curricula at the National Autonomous University of Honduras. Then, recognizing the high percentage of the population at elementary and secondary school levels, we began with a "Visiting Program to CASAO/NAUH" for this audience. We also organized the "Astronomical Nights Program" for people of all ages, whether in school or not.

In less than ten years, but especially now that we have a regional accreditation, we have begun to work on a Regional Program in Astronomy and Astrophysics at both undergraduate and graduate levels for the benefit of the Central American inhabitants. Most recently, and due to the need for using new technologies and to reach a broader regional group, for the introductory course in astronomy we have incorporated the use of the Internet. In this respect, we are now working on the online course "Introduction to Astronomy @ Internet Program."

I shall now describe each of these three outreach programs, beginning with the schoolchildren's visiting program, moving on to the astronomical nights program, and concluding with the "Introduction to Astronomy @ Internet Program."

37.2 Description of the CASAO/NAUH outreach programs

37.2.1 The Visiting Program to CASAO/NAUH

For about ten years, CASAO/NAUH has received visits from elementary and secondary school students, teachers, college students, parents, and media communicators, who attend in groups from 20 up to more than 100 persons, twice or three times per week, for two hours in the mornings or early afternoons. Each visit, organized and adapted to the interest of the participants, begins with a lecture on different topics such as the Solar System, the Sun, stars and nebulae, or galaxies, illustrated with astronomical images obtained with different telescopes on Earth and spaceborne facilities, in all ranges of the electromagnetic spectrum.

Practical activities are included, so that visitors can learn how to use small telescopes for daytime observations of the Sun and planets. A small telescope equipped with a solar filter and a small Web camera allows students to obtain images of sunspots they can observe on the computer. By using Earth dials, visitors make measurements of the solar time and solar calendars. Of special interest is the use of a replica of a Maya stela located at CASAO/NAUH, in order to learn how the Maya of Central America used stelae for the same purpose.

The goals of the "Visiting Program to CASAO/NAUH" are to present specialized lectures and updated information about space science; to motivate the study of science, mathematics, and space exploration; to stimulate curiosity for learning about what exists beyond our own environment; and to introduce new techniques and methods for teaching and learning science in general.

The quality and level of this well-organized program, which has drawn thousands of visitors each semester to the astronomical observatory, has earned for the "Visiting Program to CASAO/NAUH" the status of a permanent outreach program of the National Autonomous University of Honduras.

37.2.2 The Astronomical Nights Program

Each Friday night, from 18:00 to 20:00 hours, children, youth, and adults have the opportunity to become familiar with the wonders of the Universe by visiting CASAO/NAUH. Most of the visitors, on average 40 persons, usually walk to the astronomical observatory; but others come by car or buses, especially family groups or schools groups coming from different parts of the capital city or even from inner towns. On these occasions, the number of visitors may go up to 100 and sometimes even to 200 people.

The featured talks offer visitors current information about the seasonal night sky, constellations, and the main features of planets, their moons, comets, asteroids, stars, nebulae and galaxies, and in general about the visible Universe. For each academic period, talks are organized as a tour through the Universe, starting at our varied neighborhood, the Earth, the

Solar System, and our Galaxy. Some of the talks presented on Friday nights include "Getting to know the Solar System," "Space Exploration," "Binary Stars," "Birth and Death of a Star," and "Galaxy Formations."

As the year progresses, visitors have the opportunity to observe the Moon, Mercury, Venus, Mars, Saturn, Jupiter, comets, asteroids, bright stars, open and globular clusters, nebulae, and bright galaxies. They also have the opportunity to observe with the naked eye accompanied by an astronomer who familiarizes them with the patterns of the seasonal constellations; teaches them the names of the most brilliant stars; helps them to distinguish between a star and a planet, and sometimes between stars or planets and artificial satellites or the International Space Station; and clarifies how weather may affect the observations. On those occasions when the sky is cloudy or rain is falling, an alternative plan of activities is ready in advance, so that – regardless of weather – visitors can gain knowledge conveying the excitement of astronomy and of science in general. In case of inclement weather, visitors may have their curiosity stimulated by talks about eclipses; meteor showers; or equinoxes, solstices, and zenith Sun passages.

As is the case with the "Visiting Program to CASAO/NAUH," the quality and level of the "Astronomical Nights Program," which draws significant audiences each Friday, have made it into a permanent outreach program of the National Autonomous University of Honduras.

37.2.3 Astronomers in Honduras and the impact of the above-mentioned programs

The community of astronomers in Honduras, although small, is growing. Currently all astronomers work at CASAO/NAUH. There are national and regional astronomers and also foreign astronomers who regularly collaborate in education, research, and outreach programs and projects. Astronomers who have completed the master's program in astronomy and astrophysics at CASAO/NAUH, supported by undergraduate students working as assistants, are responsible for the development of research projects, the introductory astronomy courses, and for the activities offered in the Visiting Program to CASAO/NAUH and the Astronomical Nights Program. Visiting professors mainly collaborate as thesis directors for the master's program in astronomy and astrophysics, and in research projects and graduate courses.

An evaluation of the impact of the CASAO/NAUH outreach programs reveals that most students come from Tegucigalpa, the capital city, which has transportation facilities for teachers, students, and parents. Fewer people, in groups of 50 to 100 persons, teachers, students, parents, travel from inner cities or towns to visit the astronomical observatory. When major astronomical events, such as eclipses or zenith Sun passages, occur, the media (radio, television, and press) stimulate such great interest that visitors wish to come from beyond the capital city.

Willing to contribute in a different way to impact the national educational system, and aware that national indicators reveal that in Honduras average schooling reaches a maximum of 9 years (UNAT, 2006), astronomers at CASAO/NAUH decided to reach out to populations beyond the capital city borders. We are now working on a new project with the ambitious goals of raising the level of science education and of reducing poverty in Honduras. A proposal of financial support to increase and improve astronomical equipment at CASAO/NAUH has already been presented to the International Cooperation (CASAO/NAUH, 2006). We hope to work in four municipalities of Honduras with low human development indices: Cabanas in the Department of Copan (average number of years in school: 3.5), San Francisco in the Lempira Department (average number of years in school: 4.2), San Antonio in the

Cortes Department (average number of years in school: 4.7), and La Libertad in the Comayagua Department (average number of years in school: 5.1). The social, economic, cultural, and environmental conditions of these municipalities are well known because some of our CASAO/NAUH postgraduate students come from these areas.

It is expected that, in each of the four municipalities where the project will be undertaken, children and youth in general will (a) be motivated to continue their studies and, by themselves, look for ways to acquire new knowledge and abilities to maintain themselves updated in the fields of astronomy and space sciences in general; (b) acquire knowledge and abilities in the use of equipment which will be of great help during their years in school. It is also expected that local schools will have at hand an astronomy room, where updated equipment and materials will be available to students and to teachers. In the chosen towns, the target of our activities will be the children and youth, who are the majority of the population of Honduras. We want them to learn to carry out observations, as children in Tegucigalpa do at our astronomical observatory. We want them to learn how to use (a) computers and software to prepare sky maps for use during their observations; (b) telescopes, filters, and Web cameras to obtain images of the Sun, the Moon, and planets; and (c) the Internet, to display their results in the Web pages of the astronomical observatory, in a special gallery designed by themselves for this purpose. We also want them to use a portable planetarium, as an alternative means for studying the sky, especially when the sky is cloudy or during the tropical rainy season. We want our children and youth, whether or not poverty keeps them from attending school, to appreciate the marvels of the Moon, the planets, the stars, and the galaxies; to learn what these objects are, how they were formed and how they will evolve.

Here is where assistance and support to elementary and secondary school teachers becomes useful. For natural and social science teachers, help comes mainly from the Visiting Program to CASAO/NAUH. They are also assisted by the Central America Astronomy and Astrophysical Courses (CAAAC), which are held each year in a different country in the region, and by the workshops developed at CASAO/NAUH for groups of 30 to 50 teachers. Lectures, practical activities, astronomical observations, solution of numerical problems, use of didactical material and data, and computer software use comprise these courses and workshops.

Even recognizing the great value of these courses and workshops, which strongly motivate teachers especially to include practical activities in their courses and lectures, there are some limitations. One is that only a few teachers can attend. Another limitation is that these workshops are not frequently offered. Therefore, in order to be more effective in training and in supporting elementary and secondary school teachers, other complementary forms should be considered.

37.2.4 The Introduction to Astronomy @ Internet Program

Introduction to Astronomy (AN-111) is a general course that the National Autonomous University of Honduras offers to all matriculated students as part of their general education. It may be chosen from among a list of natural science courses offered to university students so that they may fufill the science requirement. At CASAO/NAUH we have offered the course for about ten years to an average of 400 students per semester. This is a 4-credit course that meets four hours per week and includes theoretical lectures, practical activities, solution of numerical problems, and astronomical observations.

Intended to present the broad panorama of Astronomy, it is divided into four units: (I) Observations and Models, (II) The Solar System, (III) Stars and Interstellar Medium, and (IV) Galaxies and Cosmology. Each academic period, students must take four exams, one per unit, and are assigned a grade based on exam results together with the reports on their practical activities.

As a technological platform to support lectures and practical activities, AN-111 students can download from the Internet, at the CASAO/NAUH Web page (www.oacs-unah.edu.hn): the course program, general information for each section (classrooms, schedules, dates of the exams and practical activities), the guides for each practical activity, data sheets and report formats, instructions for teachers, and a set of slides explaining the practical activities. Two practical activities are included for each unit of the course, to make up eight in total for the academic term. The titles of these practical activities are: Unit 1: (i) Apparent Movements of Stars, (ii) The Virtual Telescope; Unit 2: (iii) Planetary Geology, (iv) Measuring Jupiter's Mass; Unit 3: (v) Following the Sunspots, (vi) Classification of Stellar Spectra; Unit 4: (vii) Distance to the Center of the Galaxy, (viii) The Hubble Law and the Expansion of the Universe.

Students enjoy AN-111. They enjoy the lectures, the practical activities, the astronomical observations; and, when they finish the course, they are well trained and motivated to understand Earth's place in the Universe, the role of science and space technologies, and the value of astronomy education. These are among the reasons we strongly urge elementary and secondary schoolteachers to take this course.

Because we are aware that not everybody has the opportunity to attend university and take AN-111, however, we are working on an online version of this course, the Introduction to Astronomy @ Internet Program. On the Internet, it can be accessible any time, not only to Honduran but also to Central American university students and, of course, to elementary and secondary school teachers. This on-line course, first presented in Panama, at the X-CAAAC held in February 2006 (Universidad de Panamá, 2006), includes four units: (1) Foundations, (2) The Neighborhood, (3) Stars and More, (4) The Universe. The contents for each unit and the themes and sub-themes are as follows.

> Unit 1: Foundations. Our place in the Universe: our city, our planet, our neighborhood, our Galaxy and the known Universe. The sky and the constellations: the night sky; the constellations; zodiacal constellations; apparent magnitudes of stars. The celestial sphere: a definition; the sky of one day and Earth's rotation; the sky of a year and the Earth's revolution; the sky of the centuries and the precession of Earth's axis. Apparent movements of the Sun: diurnal motion; annual motion; the seasons of the year. Apparent movement of the Moon: daily motion; Moon phases; sidereal and synodic months; eclipses.
>
> Unit 2: The Neighborhood. Planetary motion: the planets Mercury and Venus; the planets Mars, Jupiter, and Saturn; direct and retrograde motion; the geocentric and the heliocentric models. Galileo, Kepler, and Newton: the telescope; Galileo's observations: Kepler's three laws; Newton's laws of motion and universal gravitation. The Solar System planets (*): the Earth, Mercury, Venus and Mars; Jupiter and Saturn; Uranus and Neptune. The dwarf planets and small bodies (*): Ceres, Pluto, Eris, and trans-Neptunian objects; asteroids; comets; meteors and meteorites; the Oort cloud. Solar systems origins: the origins of the Solar System; other solar systems and their origins. (*: Modified after the XXVI IAU GA Resolutions).

Unit 3: Stars and more. Atoms and light of stars: brightness, colors and temperatures of stars; laws of radiation; interaction of light and matter; spectra of stars; spectral classifications. Observation of stars: distances to stars; stellar magnitudes; size of stars; the H-R diagram; stellar systems. The Sun: our star. The solar atmosphere; solar activity and its influence on the Earth; the solar interior. Life and death of stars: birth of a star; star models and evidence for them; stars of the main sequence; evidence of star evolution; death of stars like the Sun; death of massive stars. The interstellar medium: nebulae; observations in different wavelengths; components of the interstellar medium.

Unit 4: The Universe. The Milky Way: our Galaxy: structure; origins of the Milky Way. Normal and active galaxies: the discovery of galaxies; classifications and properties of galaxies; active galaxies and super–massive black holes; quasars. The big structure of the Universe: space telescopes; clusters of galaxies; collisions of galaxies; evolution of galaxies. A history of the Universe: Einstein and Hubble; the expansion of the Universe; model of the Universe. Are we alone?: the origins of life; life in the Solar System; are we alone in the Universe?

On the Internet, pages are arranged for each of the topics to include links to other Web pages where different universities, observatories, organizations, and space agencies present well-tested, freely available educational online resources and materials.

We expect that this Introduction to Astronomy @ Internet Program will be a useful and frequently visited site on the Internet, especially because, written in Spanish, it will be useful for Honduran, Central American, and also Latin American students and teachers.

37.3 Discussion and summary

At this point it is necessary to summarize the CASAO/NAUH Educational Programs in terms of the levels of the national educational system, and also in terms of the numbers of people for whom they are intended. Towards this end, let us consider first those specialized studies on astronomy and astrophysics that now comprise a regional program in astronomy and astrophysics, at the undergraduate and graduate levels, for the benefit of Central America. Clearly this career does not yet attract a high number of students because it is a new academic field. But it represents a great effort to train autochthonous, or indigenous, astronomers within the region.

The introductory astronomy course (AN-111) is a general and optional course that is part of the National Autonomous University of Honduras curriculum. It thus represents an opportunity for university students to become part of a scientifically literate generation, trained by professional astronomers.

The Visiting Program to CASAO/NAUH is intended for pre-school, elementary, and secondary school students and teachers. It represents an important contribution for the national educational system in which the majority of students and the Honduran population are registered. Topics presented during the visits, because they are adapted to the specific elementary and secondary levels, are useful material for teachers and also to be included as part of the national curriculum. This program deserves additional support to enable it to reach more cities within the country.

The Astronomical Nights Program is intended for pre-school, elementary, and secondary levels, and for university students. But also it is intended for professional and even for those not attending school or university at all. Potentially, it represents a real opportunity to impact the national population, if well supported. It is a program that helps convey the excitement of astronomy and of science in general.

The Introduction to Astronomy @ Internet Program represents a new opportunity and an innovation in teaching and learning astronomy within the region, and also in assisting elementary and secondary school teachers to obtain better access to existing and future training resources and in astronomy.

The Honduran National Representative to the IAU is Chair of the Central America Astronomical Observatory at the National Autonomous University of Honduras. Together with national, regional, and foreign astronomers working at the CASAO/NAUH, we are all calling the attention of the Honduran national educational system to the resources provided in astronomy.

Another competent body is the Central America Assembly of Astronomers (CAAA) , whose current president is also the chair of the CASAO/NAUH, and whose current secretary is the Nicaragua National Representative of the National Autonomous University of that country. The CAAA is responsible for organizing the Central America Astronomy and Astrophysics Courses (CAAAC). Between 1995 and 2006, ten such courses have been developed, each time in a different country: I-CAAAC (1995) and VII- CAAAC (2002) in Honduras; II-CAAAC (1996) and VIII-CAAAC (2003) in El Salvador; III-CAAAC (1997) and IX-CAAAC (2004) in Guatemala; IV-CAAAC (1998) and X-CAAAC (2006) in Panama; V-CAAAC (1999) in Nicaragua; VI-CAAAC (2001) in Costa Rica; and XI-CAAAC (2007) in Nicaragua. From CASAO/NAUH, seat of the CAAA, we are working to join efforts with the Local Organizing Committee of the XI-CAAAC, other Central America universities, the IAU, and the international community of astronomers to make these events real opportunities to strongly support astronomy and astronomers in each country of Central America.

37.4 Conclusions

From the point of view of CASAO/NAUH and related to the implementation of the Resolution on the Value of Astronomy Education passed by the IAU General Assembly 2003, we conclude that:

- with the Visiting Program to CASAO/NAUH and the Astronomical Nights Program we are implementing the inclusion of astronomy at both primary and secondary levels of education;
- with the Introduction to Astronomy @ Internet Program and also with the workshops for teachers, we are assisting elementary and secondary schools teachers to obtain better resources and to enhance effective teaching and learning methods in science;
- as national representative to the IAU, we are calling the attention to resources provided nationally and worldwide in astronomy;
- astronomers graduated at CASAO/NAUH, together with those participating in our Programs and Projects, are collaborating towards the training of a new scientifically literate generation by assisting and training educators at all levels.

Acknowledgments

I want to thank the International Astronomical Union and the Members of the Commission responsible for organizing this Special Session (SpS2), who allowed me to share with everybody what we have achieved at the Central America Suyapa Astronomical Observatory of the National Autonomous University of Honduras, especially this year when, after a rigorous evaluation process, our master's program in astronomy and astrophysics has been accredited as a regional one.

References

CASAO/NAUH, 2006, *Project: Space and Technology at the Service of Reducing Poverty in Honduras*, Tegucigalpa, Honduras.

IAU Commission 46, 2003, *IAU Resolution on the Value of Astronomy Education passed by the General Assembly.*

Instituto Nacional de Estadística (INE), 2001, *Censo 2001*. Tegucigalpa, Honduras.

Pineda de Carías, M. C., 2001, Astronomy for developing countries. In Alan H. Batten (ed.), *The Central American Master's Program in Astronomy and Astrophysics*, IAU Special Session at the 24th General Assembly.

Unidad de Apoyo Técnico (UNAT), 2006, *Mapa Temático del Indicador de Anos de Estudio Promedio para 2001*, Secretaría de Estado del Despacho Presidencial. Tegucigalpa, Honduras.

Universidad de Panamá, 2006, *X Curso Centroamericano de Astronomía y Astrofísica (X-CURCAA)*, Ciudad de Panamá, Panamá.

Comments

A. H. Batten: I think that it is seven or eight years since I was with you in Honduras, and your report shows encouraging signs of progress.

38

Towards a new program in astronomy education in secondary schools in Turkey

Z. Aslan and Z. Tunca

Physics Department, Akdeniz University, Antalya, Turkey; TUBITAK National Observatory, Antalya, Turkey and Department of Astronomy and Space Sciences, Ege University, Izmir, Turkey

Abstract: It has been of great concern for Turkish astronomers that the teaching of astronomy, which is a part of the physics course in secondary schools, is not very effective, mainly because a majority of the physics teachers have had no formal education in astronomy. TUBITAK National Observatory (TUG) proposed to the Ministry of Education in 2005 that a national meeting for physics and astronomy teachers be held during the opportune time of the total solar eclipse of March 29, 2006 with the subject matter "teaching of astronomy and using astronomy to teach physics." The meeting, with participants from all over Turkey, was very successful. The speakers included astronomers and physicists from Turkish universities, educators from the Ministry, and three educators from abroad. The minutes of the meeting have been submitted to the Ministry of Education. The details and their relevance to the 2003 IAU Resolution and the role TUG has undertaken are presented below.

38.1 Introduction

In the Turkish school curriculum, astronomy was taught on its own at the secondary level (high school or lycée, age 15–18 years) before 1974. Since then, however, it has been taught at secondary schools as part of the physics course and at primary schools as part of the elementary science course. In recent years, the Turkish Ministry of Education (ME), in collaboration with universities, has held a series of meetings entitled "Education in Physical Sciences and Mathematics," discussing better ways of teaching and improving the school science education as a whole.

TUBITAK National Observatory (TUG) decided first to study the elementary and secondary school curricula and find out the extent of the implementation, if any, of the results of these meetings and of the 2003 IAU Resolution, and take an action towards implementation, at least of the latter. We report our work here.

38.2 Astronomy in Turkish schools

Primary school in Turkey starts at age 7 (Grade 1) and ends at age 14 (Grade 8). The secondary schools are of three types: technical lycée, general lycée, and science lycée, all starting at age 15 and lasting for three years. We will not be concerned with the technical lycée here. In primary school, astronomy is taught as topics in a geography and elementary science course:

Grade 4 (age 10): our planet – shape and size, day and night, seasons;
Grade 7 (age 13): discovering space – Solar System bodies, stars, galaxies.

In the secondary school program, no astronomy topics are present explicitly but basics are included in the physics course:

Innovation in Astronomy Education, eds. Jay M. Pasachoff, Rosa M. Ros, and Naomi Pasachoff. Published by Cambridge University Press.

Grade 9 (age 15): topics in physics, including mass and weight, gases;
Grade 10 (age 16): topics in physics, including Newton's laws, Kepler's laws, Earth's magnetism;
Grade 11 (age 17): topics in physics, including light, optics.

In the science lycée program, in addition to the general lycée program just mentioned, some advanced topics in mathematics and physics are taught, including elementary calculus, the Doppler effect, black body radiation, constituents of matter, and special relativity. We note in passing that these schools take the top few percent of the primary school graduates, none of whom chose astronomy at university! They normally chose medicine, engineering, or industrial managment. Many of them, after graduating from a Turkish university, go to western countries, mainly to the USA, and quite a sizeable fraction never come back home!

38.3 Effectiveness of astronomy teaching

The following items reflect the situation in 2006.

- The results of the meetings mentioned in the Introduction have not yet been effectively implemented in schools far as teaching physics and astronomy is concerned.
- Only a small percentage of the teachers are astronomy graduates. The teachers who teach astronomical topics are elementary science teachers in primary schools and physics teachers in secondary schools, the majority of whom have had no formal education or training in astronomy.
- No observations are made as part of the astronomy education. There is limited freely available educational resource material in astronomy at any level.
- Astronomy topics are generally scheduled at the end of a particular course; the school year, or the semester, often ends before there is any time left for them!

38.4 Meeting for the teachers

The situation being what it is, TUG decided to make use of the opportune time of the total solar eclipse of March 29, 2006, to bring the question of astronomy education to the attention of the authorities and of the educators themselves, the teachers. The Ministry of Education immediately approved our application for a meeting aimed at schoolteachers during the total solar eclipse, with the subject "teaching astronomy and using astronomy to teach physics." The aims of the meeting, in brief, were:

- introduction to the facts about astronomy education and training as currently practiced in the country,
- description of the physics and geometry of the TSE,
- observation of the TSE,
- making several scientific observations and experiments by pre-selected teachers during the eclipse with participation of pre-selected school pupils in the experiments,
- delivering lectures on how to use astronomical events such as the TSE to teach astronomy,
- delivering lectures on how to use astronomy to teach physics,
- preparing a list of suggestions for participants on how to share the results of the meeting with other schoolteachers upon their return home,
- preparing a list of questions and proposals from the lecturers and teachers for the Ministry of Education on how to teach astronomy topics as part of a science course.

The meeting was a success. First, we had a beautiful sky to see the eclipse and to carry out the experiments! About 120 pre-selected schoolteachers from all over Turkey who were involved in the teaching of astronomy, and 10 pre-selected pupils, participated in the meeting. The speakers were astronomers and physicists from Turkish universities and educators from the Ministry plus three educators from abroad. The presentations (albeit mostly in Turkish) can be accessed at www.tug.tubitak.gov.tr/OGRSEM2006/sunumlar/.

The titles of the presentations by the educators from abroad were:

- "Training Science Teachers in the 21st Century," by Nahide Craig, Space Sciences Lab., University of California – Berkeley.
- "Astronomy Education in Schools and Mass Media," by Magda Stavinschi, Astronomical Institute of the Romanian Academy, Bucharest, Romania.
- "Introducing University Research Topics in Secondary School," by Alex van den Berg, University of Groningen, The Netherlands.

The lectures are on the website www.tug.tubitak.gov.tr/ogrsem2006/ogrsem2006_son.htm. The minutes of the meeting included the following:

- Astronomy must be used in teaching physics.
- ME should collaborate with TUG in providing educational material to schools on, e.g., dimension, distance, time, gravity, electromagnetic radiation, energy, magnetism, etc.
- TUG can provide educational material either from its own telescopes and/or from observatories abroad with larger telescopes.
- ME should form a team to adapt this material into material to be used for school projects and experiments and should provide a budget for the work.
- ME should collaborate with TUG and with university departments of astronomy to hold summer schools and in-service training for primary and secondary school teachers, where the teachers should be trained in astronomical topics to be used in teaching physics, learn how to use the educational material provided by TUG, or by others, be trained in laboratory experiments or observation planning, learn to obtain better access to existing and future training resources in astronomy.

38.5 A new program in astronomy education

Recognizing that the Turkish school system educates only a handful of students well but leaves the majority not well educated, and aiming at a scientifically and technologically literate generation, ME is working on a new program of teaching science at schools. The program was outlined by M. Yildiz during the teachers' meeting. A draft program was published in 2005 in a book entitled *Science and Technology Course*. In this book, four broad areas of science teaching are described, one of which is "The Earth and the Universe." All the astronomy topics previously taught in the primary years as parts of other subjects are collected under "The Earth and the Universe" in the new program. The book gives the curriculum and explains how the teaching should be conducted. As an example, here we give a copy of the so-called "concept map" of the broad topic "Solar system and beyond" (Figure 38.1), where all the boxes have been translated into English. In the book, this map is followed by a table explaining what the pupils are supposed to learn, and giving examples of observations or activities to be carried out, materials to be used, and brief explanations related to each activity. Fourteen hours are allocated for the subject at grade 7 (age 13), which is 9.7%

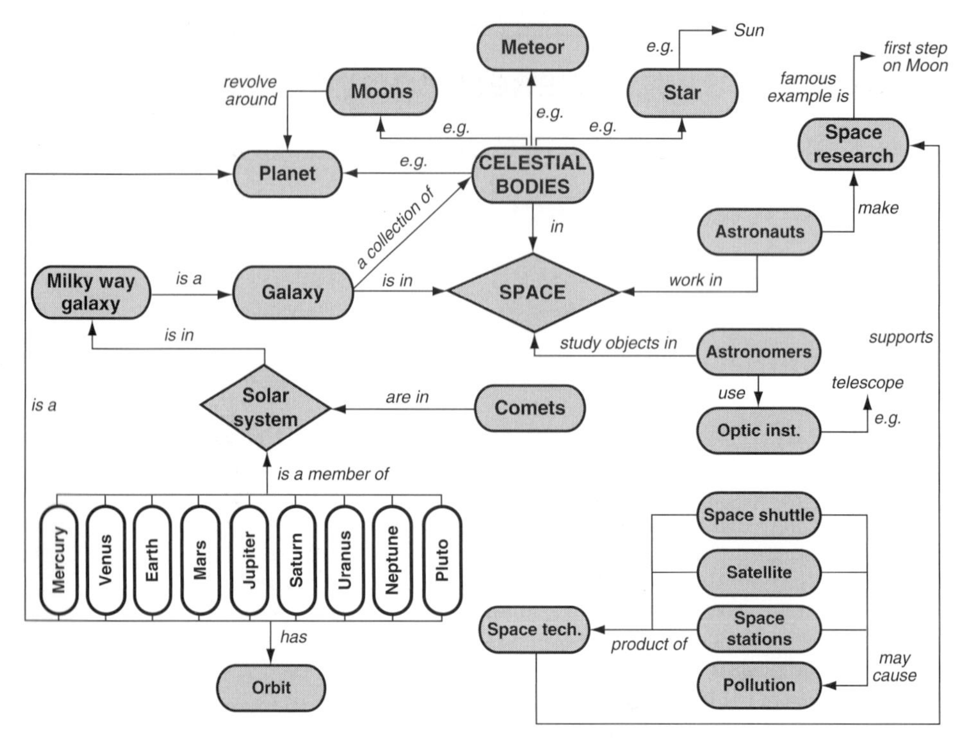

Figure 38.1 Concept map of "Solar System and Beyond"

of the whole Science and Technology Course. A student's book and a teacher's guide book are being prepared. The ME is working on a similar program for secondary schools, ages 15–18, with an updated map according to the 2006 IAU resolutions.

38.6 Conclusion

A national meeting for physics and astronomy teachers, organized by TUBITAK National Observatory during the total solar eclipse of 29 March 2006 has shown collective enthusiasm of teachers towards in-service training in astronomy. A text containing minutes of the meeting was submitted to the Ministry of Education, where it has been welcomed. We expect that the minutes will be used in implementing the new Science and Technology Course in primary and secondary education, which, in turn, means implementing the 2003 IAU Resolution.

Acknowledgments

The meeting for teachers was made possible by the support of the Turkish Ministry of Education and TUBITAK (Scientific and Technological Research Council of Turkey). We are grateful to the lecturers for accepting our invitation to take part in the meeting.

39

Universe awareness for young children: some educational aspects and a pilot project

Cecilia Scorza, George Miley, Carolina Ödman, and Claus Madsen

ZAH (Zentrum für Astronomie Heidelberg der Universität Heidelberg) Astronomisches Rechen-Institut, Mönchhofstrasse 12–14, 69120 Heidelberg, Germany; Leiden Observatory, PO Box 9513, NL-2300 RA Leiden, The Netherlands; ESO, Karl-Schwarzschild-Strasse 2, D-85748 Garching bei München, Germany

Abstract: Universe Awareness (UNAWE) is an international initiative for economically disadvantaged young children aged four to ten. UNAWE will expose children in developing countries and in underprivileged communities in Europe to the inspirational aspects of astronomy. UNAWE is also a network of Astronomy outreach/education professionals and volunteers worldwide. Here we discuss some aspects that will be taken into account during the development phase of the program and describe the Venezuelan pilot project.

39.1 Introduction

Universe Awareness (UNAWE) is an international program that will expose economically disadvantaged young children between ages 4 and 10 to the inspirational aspects of modern astronomy (Miley *et al.*, 2005). Astronomy is a unique discipline for motivating and informing young children and imbuing them with an appreciation of both science and culture. Astronomy is also one of the oldest human activities, with cultural roots in many ancient civilizations.

Universe Awareness is motivated by the premises that access to simple knowledge about the Universe is a birthright and that the formative ages of 4 to 10 years play an important role in the development of a human value system. This is also the age range in which children can readily appreciate and enjoy the beauty of astronomical objects and can learn to develop a feeling for the vastness of the Universe. Exposing young children to the wonders of the Universe can help to broaden their minds and stimulate their world-view. Universe Awareness is aimed at very young children because cognitive disparities among children that depend on background increase with age. UNAWE will focus on economically disadvantaged children because they are less likely to gain knowledge of the Universe by conventional means and are therefore the most needy.

UNAWE is being developed as a "bottom-up" program that will carry out or participate in projects in several countries starting in 2009, designated as the International Year of Astronomy. Goals include (i) the production of entertaining material in several languages and cultures, (ii) organizing training courses for teachers and others involved in the delivery of the program, and (iii) providing a network where teachers and other professionals can exchange ideas and experiences. During 2006 pilot projects were carried out to investigate the feasibility of UNAWE in Venezuela and Tunisia.

39.2 Diverse environments

Young disadvantaged children, the target group for Universe Awareness, live in diverse environments, including isolated rural villages and the centres of large cities. Three types

Innovation in Astronomy Education, eds. Jay M. Pasachoff, Rosa M. Ros, and Naomi Pasachoff. Published by Cambridge University Press.

of environment have been identified, for which different materials and methods need to be developed. In a very basic environment, children would be exposed to very little schooling, with minimal or no infrastructure. Children in an advanced environment would typically all go to school with well-trained teachers and have regular access to the Internet. Some infrastructure would be present in an intermediate environment. Television would be widespread, the Internet would be available sporadically. The development of UNAWE will proceed according to the demands of active coordinators in the participating countries.

UNAWE will assist professionals in each country in developing a programme suited to their special circumstances (environments, languages and cultures). In appropriate cases a modular programme will be developed, tailored to the ages and capabilities of the children.

39.3 Relevant educational aspects

The emphasis of UNAWE is on *inspiration and entertainment* rather than on imparting dry facts. Songs, games, toys, and animated films will play a key role in the UNAWE programs. These will be developed by professionals with experience of children's needs.

UNAWE will also encourage twinning activities between countries sharing the same language. There are Spanish-speaking populations in many cultures, spread throughout countries located at various latitudes. Astronomical themes like the seasons, the orientation of the Moon, and the constellations can be discussed; myths, stories, and songs about the sky can be exchanged. In some cases the observations of the Moon and its relation to local calendars will be exploited. It will often be essential and desirable to involve parents and community leaders in the process.

39.3.1 Astronomical framework

Astronomical aspects of Universe Awareness that will be considered during the development of the program include the following.

Awareness of the sky

Observing the sky is one obvious and inexpensive manner to convey the wonder of the Universe to young children. Looking at the Sun and the Moon, at the planets and stars visible with the naked eye, can be an excellent inspirational tool. In many environments simple observations require little effort. The Moon was a popular topic of conversation with children in Venezuela. Stories about the Sun, the Moon, and the constellations will play a significant role. The wealth of mythologies and legends about the sky provides a source of entertaining stories for UNAWE. The inclusion of stories from different parts of the world will illustrate cultural diversity.

Earth awareness

Awareness of the Earth is an important stepping stone to developing an awareness of the Universe. Convincing young children that the Earth has a spherical shape is important, but not trivial, because children believe what they see (Vosniadou and Brewer, 1992; Nobes *et al.*, 2003). The effect of gravity and the interaction between the Earth and the Moon could be mentioned in this context, when dealing with older children.

Solar System awareness

The Solar System is the next step in promoting a sense of vastness of the Universe. The program could exploit the diversity in shapes and colors of objects in the Solar System. The relatively modest size of the Earth compared to its neighbours often comes as a surprise to young children and awakens their environmental conscience.

The Milky Way and other galaxies
The zoo of diverse and exotic objects in our Galaxy provides a treasure of beauty and fascination for stimulating children's wonder, a possible source for exciting adventure stories, and a tool for the development of cognitive skills. The message that the Sun is just a typical star in our Galaxy of more than a billion stars and that galaxies come in various beautiful shapes and sizes is obviously inspirational for older children. The effect of gravity on interactions between celestial bodies is one of several phenomena for which the elegance of rational thought could be demonstrated inspirationally to older children.

39.3.2 Didactical approaches and complementary activities
Each of the topics described above could be approached in different didactical ways and using different approaches. These include the use of (i) direct observations, (ii) myths, stories, games, and songs about the Sun, Moon, and constellations, (iii) hands-on activities, (iv) interactive educational software, and (v) joint projects involving class twinning. Several participants in the Universe Awareness network are studying these possibilities in more detail.

39.4 The Venezuelan pilot project

After contact with the Venezuelan UNESCO National Commission, one of us (CS) went to Venezuela in January and March 2006 to conduct a limited pilot project that incorporates some of the methods described above.

Venezuela is multicultural owing to its heritage and colonial history. Because of its geography, some communities are very isolated; thus a range of urban and rural environments is present. There is considerable support in Venezuela for new educational initiatives and the importance of preserving the country's cultural legacy is widely recognized. This atmosphere contributed to favourable conditions for the pilot project.

Venezuela also has a large community of professional and amateur astronomers. The pilot activities were based at the Centro de Investigaciones de Astronomía (CIDA) in Mérida. The activities of the pilot project were advertised efficiently through the Venezuelan UNESCO ASPnet network of schools, which has coordinators throughout the country.

One of the key ingredients of UNAWE is the training of coordinators and people involved with its implementation. As part of the Venezuelan pilot project, 87 teachers from all parts of Venezuela attended two teacher training courses in Mérida. The training included several of the topics described in Section 39.3.1. The teachers built a kit to take back to their schools with the materials brought to the workshops and collected playful activities linked to the cultural roots of the country. Children were involved in the teacher training and their participation contributed to enriching the experience. See Figure 39.1.

Two representatives of an Amazonian Ye'kuana indigenous village traveled from the southernmost province of the country to attend the first teacher course. They explained their lunar calendar and astronomical traditions. The participants learnt about the importance of certain constellations in the Ye'kuana culture. The ceiling of Ye'kuana houses depicts the Milky Way, and their hunting season is determined by the annual appearance of the constellation of the "Danta" (tapir) in the night sky. They celebrate solar and lunar rituals. The morning and evening stars (both Venus) play a part in the running of their daily lives.

Figure 39.1 Teacher training in Mérida

Figure 39.2 Children at the UNESCO associated school "Flor de Maldonado" in Mérida

In addition to the teacher training courses, several schools were visited. Children of various ages reacted enthusiastically after exposure to several of the topics mentioned in Section 39.3. Finally, ad hoc contacts were made with children in a remote coastal village. They were excited by pictures of the Moon, planets and galaxies and entertained by a game that demonstrated the "dance of the planets." See Figure 39.2.

39.5 Conclusions and further work

It is clear that every region in which the Universe Awareness philosophy is adopted will require its own approach to implementing the program. Aspects that need individual attention include the relation of UNAWE to belief systems and the balance between reality and fantasy. The further development of the Universe Awareness program will be challenging, but the pilot projects have demonstrated the importance of a program such as Universe Awareness in inspiring and stimulating young children and their teachers throughout the world.

More information about the Universe Awareness program can be found at www.unawe.org/.

Acknowledgments

The Venezuelan pilot project was kindly supported by the European Organisation for Astronomical Research in the Southern Hemisphere (ESO) and the Centro de Investigaciones de Astronomía" (CIDA).

References

Miley, G., Madsen, C. and Scorza de Appl, C., 2005 Universe Awareness for Young Children, *The Messenger*, **121**.

Nobes, G., Moore, D., Martin, A., Butterworth, G., Panagiotaki, G., and Siegel, M., 2003, Mental models or fragmented knowledge? Children's understanding of the Earth in a multicultural community. *Developmental Science*, **49** (1), 74–87.

Vosniadou, S., and Brewer, W. F., 1992, Mental models of the earth: a study of conceptual change in childhood, *Cognitive Psychology*, **24**, 535–585.

Comments

Stewart Eyres: Much of the pedagogic research is warped by the exposure of young children to formal schooling. I would be interested to see if children who have not benefited from formal schooling understand concepts differently.

Toshihiro Handa: Do you use Web cameras located at many places to understand "the Earth is global"? Watching images through the Web cameras helps to understand it. There are at least three Web cameras, called "I-CAN," with which you can observe constellations. I recommend using them when the Internet is available.

John Mattox: Software twinning sounds like what in the pre-Internet era in the USA we called penpals. Can you elaborate on your plans for this?

Cecilia Scorza: Indeed. The plan is to use e-mail between children in areas with computer infrastructure, largely to discuss the appearance of the sky.

John Mattox: Sounds like a great idea!

Kala Perkins: Children can, it seems, understand much more than the pedagogic literature credits them with. A 5th–6th grade group with whom I worked, on request for "quasars and black holes" from the teacher after the presentation, said the multifrequency Universe was their favorite part of the presentation.

40

Education in Egypt and Egyptian response to eclipses

Ahmed A. Hady

Department of Astronomy and Meteorology, Faculty of Science, Cairo University, Giza, Egypt

Abstract: Since 1939 astronomy and space science courses have been offered at the university level in Egypt at the Department of Astronomy and Meteorology, Cairo University. This paper will discuss astronomy education in Egypt at the undergraduate and graduate levels. The astrophysics research groups at Cairo University and Helwan Observatory are interested in the fields of solar physics, binary stars, celestial mechanics, interstellar matter, and galaxies. The paper will also discuss the Egyptian response to two total solar eclipses (February 25, 1952 and March 29, 2006). The results of observations and photos will be discussed.

40.1 History

In 1840 an astronomical observatory was constructed at Boulac, West Cairo, Egypt. This observatory was closed in 1860. In 1868 another observatory was built at Abbasya, east of Cairo. It continued to function until the end of the century, when the site became unsuitable for astronomical observations, as light pollution increased around Abbasya following the introduction of the electric tram way in Cairo. In 1903 astronomical observations at the still-operational Helwan Observatory (www.nriag.sci.eg) began, with a 30-inch reflecting telescope.

Modern astronomy and space science education in Egypt began at the university level in 1936, at the Department of Astronomy, Cairo University. Students earn a B.Sc. over a four-year period, with astronomy introduced in the third year, after a firm foundation is laid in math and physics during the first two years. The university offers the B.Sc. in astronomy and physics, as well as in astronomy (special). It also offers the M.Sc. in space science (special). The master's program is designed to be completed in 11 months, followed by at least one additional year for thesis preparations in one of the following topics: astrophysics, theoretical astronomy, and space science. Doctoral studies in Egyptian universities are offered at the international level. About five students yearly obtain their Ph.D. degrees in astronomy from Egyptian universities.

The astrophysics research groups at Cairo University and Helwan Observatory are interested in the fields of solar physics, binary stars, celestial mechanics, interstellar matter, galaxies, relativistic astronomy, cosmology, and space sciences. Most of their research is published in national scientific journals, and some in international journals (Hady, 2002).

From February 1914 to December 1954, Helwan Observatory, in collaboration with the astronomy department of Cairo University, recorded the solar radiation at normal incidence using equipment installed in the Sant Catrine mountains in the Sinai. Other Helwan

Innovation in Astronomy Education, eds. Jay M. Pasachoff, Rosa M. Ros, and Naomi Pasachoff. Published by Cambridge University Press.

Figure 40.1 Zeiss–Coudé refractor (6-inch) in Helwan, with solar and lunar cameras, and its dome.

Observatory measurements of solar radiation at normal incidence were done from February 1914 to December 1927. In addition, using other equipment in the Sant Catrine mountains, Professor Charles Greeley Abbot, director of the astrophysical laboratory of the Smithsonian Institute, also measured solar radiation at normal incidence at Helwan Observatory from 1935 to 1954 (Abbot, 1958). From 1989 until 1998, solar radiation equipment (pyrheliometers pyranometers, ultraviolet, radiometers, etc.) was installed at selected locations in Egypt, including Helwan, Abu Simbel, Hurghada, and Marsa Matrouh.

The first solar station in Egypt was erected at Helwan Observatory in 1957. It consisted of a 25-cm horizontal-image coelostat, a solar camera, and a spectrograph. The photographic and spectroscopic observations of International Quiet Sun Year (IQSY, 1964–1965) were done by using this station. The 16-cm-diameter image of the solar disk was taken using this system, with an exposure time of 1/50 of a second.

In 1960 Albert George Wilson, the former director of the Lowell Observatory in Flagstaff, Arizona (1954–1956), visited this system and suggested changing the camera and the base of the coelostat to remove the effects of base reflections (Galal, 2001).

In 1964 a 15/225-cm Zeiss–Coudé refractor equipped with solar and lunar cameras was installed. Daily maps and photographs of the full disk and the transient solar phenomena have been regularly taken by the system since then. DayStar™ interference filters of 0.6 Å for H-alpha and 1.0 Å at the calcium K-line have been adapted to this system. Figure 40.1 shows the telescope and its dome.

We continue to use this telescope to record daily solar disk morphology and solar sunspot maps. The daily data are transferred to the Royal Observatory of Belgium (Galal, 2001). Some photographic observations of the solar disc and sunspots are shown in Figure 40.2.

In 1962 Kottamia Observatory was established. Kottamia lies in the northeastern desert 80 kilometers from Cairo. The observatory houses a 74-inch telescope equipped with both Cassegrain and Coudé spectrographs, plus a camera at the Newtonian focus. The observatory's coordinates are latitude 29° 55′ 48″ N and longitude 31° 49′ 30″ E. Its altitude is 476 meters.

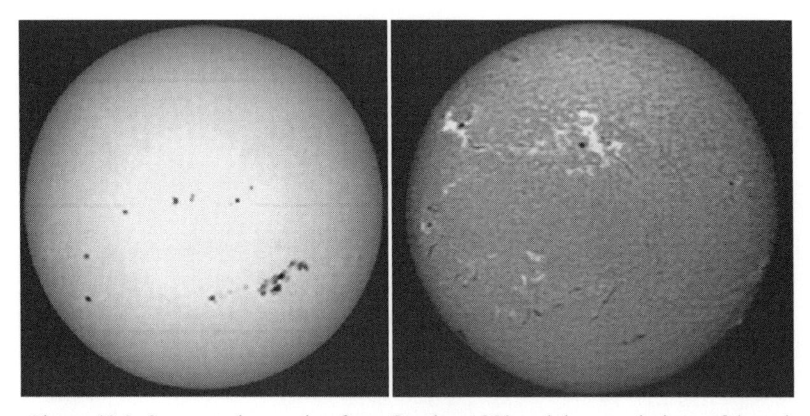

Figure 40.2 Sunspot observation from October 1989 and the morphology of the solar disk by using the hydrogen filter.

40.2 Total solar eclipse in Khartoum, February 25, 1952

The total solar eclipse on February 25, 1952, in Khartoum (the capital of Sudan), was observed by an Egyptian–French group. This group was divided into four subgroups. The first subgroup, under the supervision of Professor Reda Madwar, observed the solar corona during totality (duration about 3 minutes) with a Worthington camera and a coeleostat. The second subgroup, under the supervision of Professor Oort, worked on photoelectric observations of the solar corona. The Egyptian astronomer Professor Helmy Abdel Rahman was a member of this group. The third subgroup, under the supervision of the famous French astronomer Professor Bernard Lyot, and the Egyptian Professor M. K. Aly, observed the solar spectral lines emitted during totality in the visible and ultraviolet regions. Dr. Etkineson and Mr. Cordwel were members of the fourth subgroup.

One of the images of the total solar eclipse on February 25, 1952 appears in Figure 40.3.

Professor Bernard Lyot observed the emitted solar spectral lines, as shown in Figure 40.4.

40.3 Total solar eclipse of March 29, 2006 in El-Saloum City, Egypt

A coordinated effort involving French–Egyptian scientific cooperation permitted joined simultaneous eclipse observations of the solar corona (Koutchmy *et al.*, 2006). Several ground-based instruments and space-borne EIT and Lasco experiments of SOHO were used.

White-light images were processed to show the magnetic structure of the corona. Polarization analysis was performed to analyze the F-corona in the outer corona. Several filtergrams were obtained to show the distribution of the emission measures of the inner and middle corona. Spectra were obtained over several emission lines. Figure 40.5 shows the white-light corona; the chromosphere is visible at "second contact," when the Moon first entirely covers the Sun.

The large-scale coronal structures observed are produced by the interaction of the solar wind with the coronal magnetic field. The coronal modeling technique now allows realistic simulation of this interaction, based on MHD equations. This comparison is very useful to show us the level of understanding reached in this magnetic field and how it appears in the form of a forecast, as shown in Figure 40.6.

Figure 40.3 The solar corona during the Khartoum total solar eclipse of 1952, observed by an Egyptian–French group.

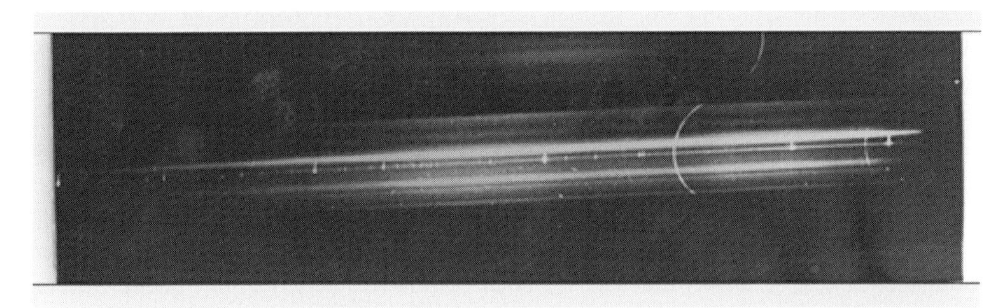

Figure 40.4 Emission lines in a spectrum taken during the solar total eclipse in Khartoum, Sudan, February 25, 1952.

There is a correlation between the SOHO/EIT 304 structure and white-light structure observed at the E limb, at Sidi Barany, where the EIT image was made at the same time, as shown in Figure 40.7.

Selected spectra (extract of just one frame with 1/50-exposure) of the spectral sequence obtained during totality is shown in Figure 40.8.

40.4 Summary

The history of astronomy and space science education and research in Egypt was reviewed in this study.

Figure 40.5 Solar corona, from El-Saloum observations of the March 29, 2006 total eclipse.

Figure 40.6 Computed magnetic structures, using the measured magnetic field at the surface of the Sun during the month before the eclipse. The second figure shows the generated brightness.

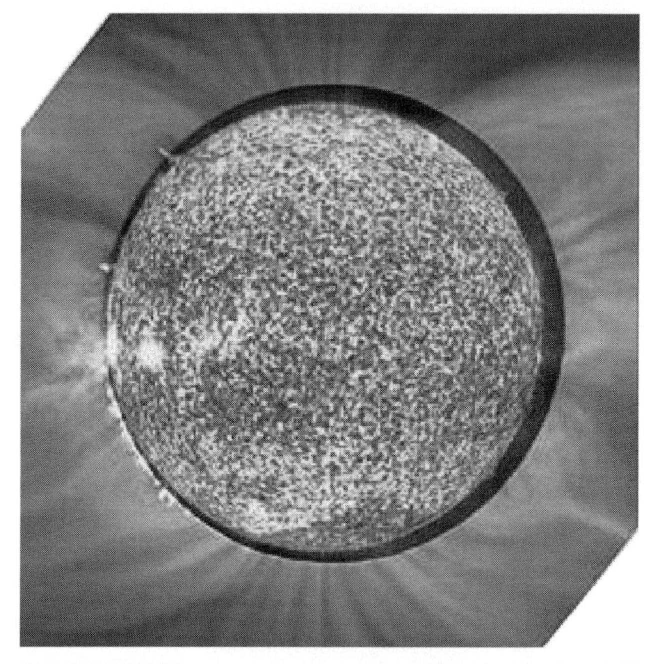

Figure 40.7 SOHO/EIT image at an ultraviolet wavelength of ionized-helium emission at 30.4 nm superimposed on a coronal photograph taken during the eclipse.

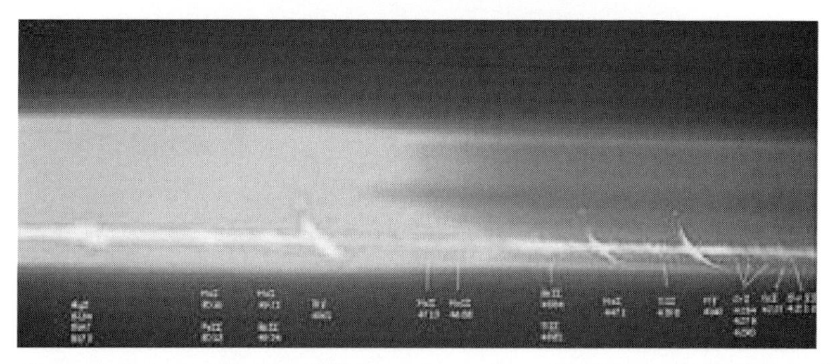

Figure 40.8 A spectrum from El-Saloum observations of the March 29, 2006 total eclipse.

An overview of astronomy education on the undergraduate and graduate levels in Egypt was presented, along with a description of the the research facilities and groups at Cairo University and Helwan Observatory.

Solar corona and photoelectric observations by an Egyptian–French group at the total solar eclipse of 25 February 1952 in Khartoum were summarized, as were observations of the emitted solar spectral lines in the visible and ultraviolet regions during totality.

The total eclipse observations in Egypt by an Egyptian–French group on March 29, 2006 were summarized.

References

Abbot, C. G., 1958, The constancy of the solar constant, *Smithsonian Contrib. Astrophysics.*, **3**, 13.
Galal, A., 2001, 100 years of astronomical observations in Helwan Observatory (in Arabic), Helwan Observatory.

Hady, A. A., 2002, Analytical studies of solar cycle 23 and its periodicities, *Planetary and Space Science Journal*, **50**, 89–92.
Koutchmy, S., Daniel, J. Y., Mouette, J., Vilinga, J., Noëns, J.-C., Damé, L., Faurobert, M., Dara, H., Hady, A., Semeida, M., Sabry, M., Domenech, A., Munier, J. -M., Jimenez, R., Legault, Th., Viladrich, Ch., Kuzin, S., and Pertsov, A. (The O. A. Team), 2006, Preliminary results from the March 29, 2006 total eclipse observations in Egypt, in Barret, D., Casoli, F., Lagache, G., Lecavelier, A., and Paganini, L. (eds.), *SF2A-2006: Proceedings of the Annual Meeting of the French Society of Astronomy and Astrophysics*, p. 547.

Comments

Jan Vesely: You told us there are five people earning their Ph.D. in astronomy per year. What are their career opportunities after the Ph.D.?
Ahmed Abdel Hady: They find work in the Egyptian universities. A new Egyptian astronomy Ph.D. typically begins as a lecturer and undertakes some research to find a more senior position as assistant professor at one of the universities.

41

Astronomy in the cultural heritage of African societies

Paul Baki

Department of Physics, University of Nairobi, PO Box 30197, 00100 Nairobi, Kenya

Abstract: African perspectives on astronomy are explored. Some traditional methods used to study astronomical phenomena as a way of meeting environmental challenges, including food insecurity, droughts, and floods, are considered. Local ethnic groups forecast weather and climate for the coming seasons by observing stellar positions as well as changes in plant and animal behavior. In academic terms we might refer to these interactions of astronomy and the local environment as a form of interdisciplinary discourse, namely, astrobiology. This interdisciplinary discourse has the potential to benefit the community not only through environmental protection measures but also by boosting the tourism industry in some African countries.

41.1 Introduction

Most African societies practice astronomy largely for understanding and predicting the weather and climatic changes for seasons. Because these communities depend mostly on rain-fed agriculture for subsistence farming, they use their knowledge of the sky to forecast the rainfall and predict major natural disasters such as floods and droughts. The UNESCO initiative "Astronomy and World Heritage" – the 1972 convention concerning the protection of cultural and natural world heritage – provides a unique opportunity to preserve exceptional properties world-wide and to raise awareness about scientific concepts linked to these properties. The aim of this initiative is to establish a link between science and culture on the basis of research aiming at the recognition of cultural and scientific values of properties connected with astronomy.

In this paper we discuss some traditional tools that are used by some ethnic communities in East Africa to interpret astronomical phenomena for solving their local problems. Since these traditional methods rely on the interaction of plants and animals with the terrestrial environment, their scientific value needs to established and recognized.

41.2 Interaction of life systems with the solar–terrestrial environment

A fundamental principle that governs the transfer of matter and energy in both natural and artificial systems is the second law of thermodynamics. It says that the capacity for a system to do useful work (move something) decreases with time, unless usable energy is pumped into it. Looked at this way, systems tend to become more disordered with time.

A measure of the system's ability to do work on its environment is called entropy. The greater the entropy, the more disordered the system and the lower the system's capacity to do work. The entropy of a system increases with time unless an external device is applied to

Innovation in Astronomy Education, eds. Jay M. Pasachoff, Rosa M. Ros, and Naomi Pasachoff. Published by Cambridge University Press.

pump more usable energy into it, and then the external device gains entropy. In fact a fundamental principle of thermodynamics is that the total entropy of the Universe is increasing with time.

One situation in which entropy is not increasing but is instead constant is that of a system in equilibrium with its surroundings; this is a state in which there is no tendency to exchange matter or energy with the surroundings. Such systems in which entropy is not increasing can do no work.

The requirement of the second law of thermodynamics that there be an increase of entropy applies to isolated systems, which when the maximum entropy is reached are in an equilibrium state. If the system is not isolated, we may, for instance, extract heat from it and cool it and observe phase transitions to equilibrium states, which minimize the free energy by trading a decrease in internal energy. The system below some critical temperature may display order at equilibrium.

A living system, whether plant or animal, is not isolated and is not in equilibrium (Cerdonio and Noble, 1986). It is rather an open system, which can exchange matter and energy with the environment. It is in a stationary state, which is not in equilibrium, but which must display stability for times that are short with respect to a lifetime and long with respect to the characteristic times of the internal processes of the system. It can evolve continuously to other stationary states of slightly different structure and function in times comparable to its lifetime. The Sun is the source of life and almost all physical and chemical phenomena on Earth. Heat and light from the Sun are not only the basis of the origin and existence of life but also the reason for all changes in Earth's atmosphere and hydrosphere. Thus life systems interact with the terrestrial environment and also with the Sun directly. Because of the variability of solar–terrestrial relationships, there are epochs with high and low levels of solar activity. This variability also affects the behavioral patterns of plants and animals, and, because nature has a habit of repeating itself, these behavior changes have been observed by man since time immemorial and used to predict weather and climatic changes and to predict disasters such as droughts and floods. These behavioral changes are called **traditional indicators**. In a nutshell, the various African ethnic communities use the ecosystem to monitor astronomical effects.

41.3 The traditional astronomical indicators

The local African communities combine their knowledge of plant and animal behavioral changes together with their knowledge of the sky to predict the weather for the coming season. The communities recognize that some plants and animals, including human beings, are more sensitive to changes in the atmospheric conditions than others.

Traditional forecasting complements meteorological forecasting and is still the major source of weather and climate information for farm management in the rural areas. This discussion focuses on the traditional forecasting methods used by the Luo community, who live around Lake Victoria in Kenya and Tanzania. In this part of Kenya and Tanzania, there are two distinct wet seasons. Short rains occur from October through December and long rains from March through May.

41.3.1 Plant indicators

Certain types of plants are known to shed their leaves to signal the onset of dry conditions and to flower before a wet season begins. The shedding of leaves is an indication of water stress conditions associated with dry conditions. The trees shed their leaves to reduce evapotranspiration and flower when the rains approach. These behavior changes have been used to predict the weather for the coming season. Among the plants with these observed properties are the following.

(a) Those plants that shed leaves to indicate an impending dry season are (by scientific names): ***Terminii browni, Ficus sur, Kigelia africana*** – trees that shed leaves twice a year to mark distinct dry conditions around the Lake Victoria region. The plants flower at the onset of a wet season.
(b) Those that flower to indicate an impending change of season are:
    (i) ***Zephranthus*** – a field flower that appears a week or two before the onset of rains. The flower appears white during a rainy season and pinkish during rainfall deficiency.
    (ii) **Blue lotus (or water lily)** – this plant grows in water but will never blossom during a dry season. Its flowering is normally an indication that a wet season is approaching and that the rains will be adequate. If the coming rains will be poorly distributed, that plant does not flower at all.

41.3.2 Animal indicators

Certain seasonal bird cries are believed to be communicating changes in weather, e.g.

(a) The bird **Robin Chat**, which disappears for several months and only reappears when a season begins. Also **Hirundo Abyssinia** and **Hirundo Smithic** are common swallows which make circular movements in the sky when the weather changes.
(b) Absence of **frogs** and **toads** indicate a dry season. When frogs stop croaking during the rainy season even if it is still raining, it is an indication that they have disappeared.
(c) Movements of **ants** indicate that a wet or rainy season is approaching.
(d) The appearance of **snakes**, **other reptiles**, and other wild animals around homes is an indication of the prevalence and continuity of a dry spell.

41.3.3 The stars and the Moon

The movement of stars has also been related to the weather and change of seasons. The Luos classify the constellation **Orion** as the "male" constellation and the Sisters as the "female" constellation. Their appearance in the sky is well understood to be linked with an impending change of season. The constituents of the female constellation are many, but usually only seven can be seen. They are observed to move from east to west followed by the male ones.

(a) The appearance of the female constellation indicates the cultivation season, while the appearance of the male constellation signals a decline in rains, showing the start of dry season or harvesting.
(b) It has been noted that the appearance and positioning of the **Milky Way** (called *Rip*- in Luo), especially in April, is normally an indication of an impending onset of a dry spell or dry season. The Milky Way normally appears to be crossing the sky from north to south and changes its position every three months.

These traditional indicators are still the most widely used methods for farm management and food production. An understanding of the link between these traditional indicators of weather changes and astronomy has the potential to change people's view of astronomy.

Rather than perceiving it an esoteric science, as is currently common, they might come to understand it as a practical discipline that can help put food on the table.

41.4 The cultural value value of astronomy and its economic benefit to society

In the African perspective it seems that the best way to spread knowledge in astronomy is to begin by appreciating its cultural value. A possible path to follow is to:

(a) establish the scientific value of these traditional indicators (in an interdisciplinary project involving both astronomers and biologists – i.e., through astrobiology), and then incorporate them into the standard astronomy curriculum;
(b) support the African perspective in astronomy to continue addressing local problems of food insecurity and natural disasters like floods and droughts.

These kinds of initiatives will:

(a) encourage the protection of the *flora* and *fauna* used as traditional indicators; this will be a good strategy for sustainable environmental protection and thus astronomy will find its application in environmental protection measures;
(b) translate to a boost in revenue collection, especially for some African countries, like Kenya, that rely on eco-tourism for their foreign exchange earnings; thus most African governments might find it beneficial to support programs in astronomy if only to bring that much needed capital for development.

41.5 Conclusion

African astronomical traditions need to explored and recognized. Most, if not all, African ethnic groups have their own usage of sky knowledge, and it would be interesting to find out the scientific basis of the traditional indicators that the Luos of Kenya and Tanzania use to predict the weather, climate, and natural disasters such as droughts and floods.

For purposes of spreading astronomy education in Africa, it would be necessary to tap these traditional values of astronomy and incorporate them into the standard astronomy curriculum, so that the role of astronomy in tackling local challenges continues to be recognized. This process might in turn lead to the conservation of the environment, increased crop yields, and a boost in the tourism sector, and hence get recognition and possible funding from the various African governments.

Acknowledgments

I would like to thank the organizers of Special Session 2, particularly Jay Pasachoff and Rosa M. Ros, for inviting me to give a presentation, and the IAU in general for sponsoring me at the 26th General Assembly.

Reference

Cerdonio, M. and Noble, R. W., 1986, *Introductory Biophysics*, Singapore: World Scientific Publishing.

Comment

Derek McNally: Are steps being taken to record and publish the use of astronomy in determining seasons (are there records of the Kenyan constellation system?) and the use of animal and plant behavior in weather prediction?

Paul Baki: There are no elaborate steps being taken to use astronomy in determining seasons, but the meteorological department in Nairobi is taking steps to use animal and plant behavior for weather prediction. In general these things are already being practised by the various Kenyan ethnic groups anyway. Nonetheless, it is a project that can be undertaken if funding is available.

42

Education at the Pierre Auger Observatory: movies as a tool in science education

Beatriz García and Cristina Raschia

Universidad Tecnológica Nacional, Regional Mendoza–CONICET, Sarmiento 440, 6 to. piso, (1347) Buenos Aires, Argentina, and member of the International Collaboration at Pierre Auger Observatory; independent film-maker

Abstract: The broad mission of the International Collaboration at Pierre Auger Observatory (see Figure 42.1) in education and public relationships is devoted to encouraging and supporting a wide range of outreach efforts to link schools and audiences with the Auger scientists and the science of cosmic rays, particle physics, astrophysics, and associated technologies. This presentation focuses on a very recent professional production of three educational videos (for children, teenagers and general audiences) and the use of new resources, such as 2-D and 3-D animation techniques, to teach and learn in sciences.

42.1 Introduction

We believe in cinema as a tool for the dissemination of science. The work we introduce in this paper had as its objectives the production of three videos for scientific outreach for the Pierre Auger Observatory. In Figure 42.1 we can see the two types of detector at the observatory. Figure 42.2 shows the director and cameraman. Each video was directed at a different audience: initial level (children from 6 to 11); medium level (teenagers from 12 on and general audiences), and technical level (educational and professional).

In order to produce an audiovisual product with scientific contents, we identified three disciplines we wished to play and interact with each other:

- education: desired goal,
- science: contents that we wanted to transmit,
- technology: tool we chose to use.

Our challenge was to unite these disciplines in a creative way in order to obtain an attractive audiovisual product for the student and simultaneously to achieve its objectives. If the audiovisual product embodies conceptual contradictions among these three disciplines, it is very likely that the outcome will not meet the desired goal. In other words, the educational experience will not take place.

42.2 The videos

We identified five unifying axes of concepts through the three disciplines that take part in this project (see Table 42.1). In order for the educational experience to take place, we considered the following aspects to be fundamental:

- contents
- creativity
- solidarity

Innovation in Astronomy Education, eds. Jay M. Pasachoff, Rosa M. Ros, and Naomi Pasachoff. Published by Cambridge University Press.

Figure 42.1 Surface detector (front) and fluorescent detector (back) at the Pierre Auger Observatory.

Figure 42.2 Cristina Raschia (Directress) and Mauricio Arias (cameraman).

- democracy
- humanism.

It is possible to describe all these interactions and give evidence of a connection among the subjects; if we can identify and use them, it will be possible to improve the audiovisual product and establish a real educational connection with the students.

Table 42.1 *Interactions between education, science, and technology*

Education	Science	Technology (movies)
Transmission of knowledge	Tools for life	Contained (documentary + fiction)
Teach to think	Solve problems	Originality in the language
Think together	Interdisciplinary collaboration	Work with collaborator
Ethical and civilian formation	The Universe is a living organism in an interrelated network	Movies as a parable of human behavior
My place in the world	Scientific vocation	The human being is the actor

42.2.1 About the contents

Transmission of contents is the excuse that puts us in contact with the student–public and starts the operation for the educational experience. The main value of the scientific content is its contribution as tools necessary for life. We search for knowledge in order to improve our quality of life; the simple pleasure of knowing leads to a better quality of life. The contents of our videos must always have this as the underlying theme.

Cinema is used in two parallel narrative ways to transmit contents: documentary and fictional. The limits of each become more and more diffuse and the audiovisual language becomes richer day after day with the contribution of new technologies that force movies to reformulate their narrative paradigms. We are always articulating a language that defines a certain way to approach reality, a way that is usually more attractive than reality itself. This leads us to the vast territory of subjectivity, where human fascination about storytelling forms a nexus with the objectivity of perception. In educational terms, the spectator's fascination for the audiovisual story is a priceless tool, but one that may become counter-productive if we do not manage to create a narrative structure suitable for the content we want to transmit in each instance. For this reason we used simulations (Figure 42.3 and 42.4) and animations (like the proton, one of the main characters in the video, see Figure 42.5, or several cartoon versions of real-life scientists, as in Figure 42.6) as important tools in order to connect the audience with the main concepts in the video.

42.2.2 About creativity

Education is not only about transmitting contents but about teaching how to think as well. Teaching audiences to think is the real goal of contemporary educational challenge. Science provides us with tools that must help us to solve problems. This fact must become evident in the approach we take to each subject to be taken into consideration.

Regarding the five concepts identified above, if we want to help the student–public to think and be able to solve problems, the audiovisual language must be characterized by its originality and its search for innovative expressiveness to stimulate imagination.

42.2.3 About individual experience and group work

Although individual experience is important in the knowledge process, experience shows that group work enriches the attainment of knowledge and nourishes thinking capacities. The scientific research approach is similar: interdisciplinary work can help researchers to reach optimal results and take into consideration new challenges.

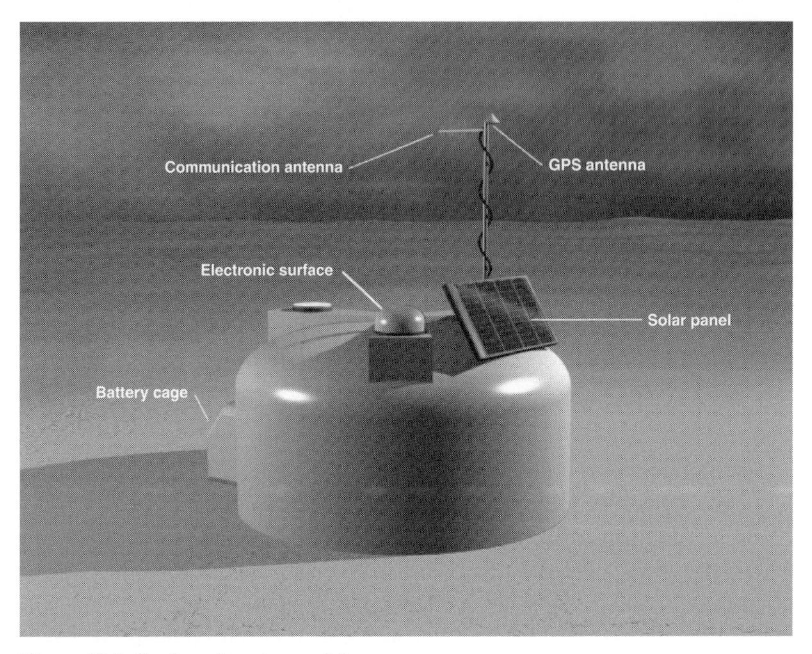

Figure 42.3 Surface detector model.

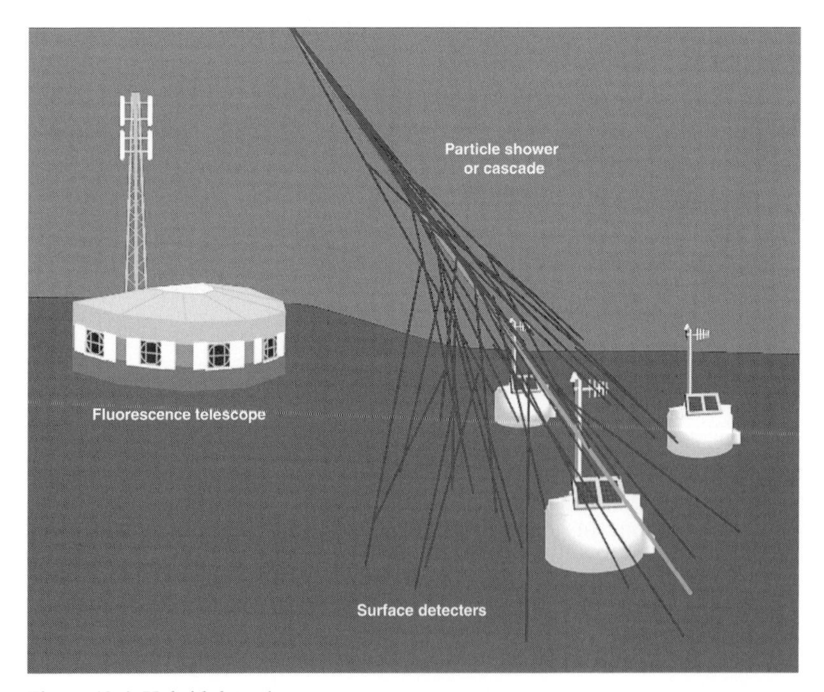

Figure 42.4 Hybrid detection.

Figure 42.5 The Proton, a creation of Jaime Suárez.

Figure 42.6 Cartoon of Victor Hess, Pierre Auger, and James Cronin by Jaime Suárez.

Audiovisual language puts into play many technical and artistic disciplines. Many people working together converge in the elaboration of an audio visual product.

42.2.4 About democracy

A complete educative experience implies that ethical and civilian formation crosses the teaching–learning process of all contents.

The science developed in the last decades confronts us with a Universe that mimics a gigantic network of interdependencies, a living organism in which each action exerts an inevitable influence on the operation of the entire network. More and more, science drives us towards an ethical responsibility for the life of the Universe.

The audiovisual tale is always a story that occurs in time and that needs a beginning, a conflict, and a resolution. Although we may not be specifically searching for it, the internal structure of the story always leads us to a lesson about human behavior.

42.2.5 About humanism

When we confront the educational experience, we are helping our student–public to find its place in the world: "What do I want to do and why?"; "How can I help to integrate the complex clockwork of human society?"

Scientific outreach is an important way to encourage people to pursue scientific vocations, especially in countries where scientific professions are not perceived in accordance with high social status.

In the audiovisual story, the human being is always the actor (see Figure 42.7). In this lies his or her enormous human value. The very structure of this narration identifies life as a challenge, a permanent fight between people and their environment, underlining humankind's capacity to transform reality.

Figure 42.7 Children and scientists during a video presentation.

Figure 42.8 The authors and James Cronin.

Acknowledgments

The authors (in Figure 42.8 with James Cronin) specially thank the Information and Communication Center's (CICUNC) technical team of the National University of Cuyo and the Pierre Auger Observatory's personnel for their help during the production of the videos. Contributions from Dinoj Surendran and Randy Landsberg of the Cosmus Project at the University of Chicago (proton shower simulations), Jaime Suárez (proton's design, cartooning and 2-D animations), Sebastián Bertoldo (3-D animation), Gustavo Parra (original music score), Sergio Gras (proton's off voice and commentaries), Maria del Mar Gatti (English translation), Rosanna Ortiz (French translation), and Alexia Salguero (directress's assistant) are deeply appreciated.

This work was supported by the Pierre Auger Foundation of Argentina and by the National Technological University Foundation of Mendoza. Website: www.auger.org.

43

Freshman seminars: interdisciplinary engagements in astronomy

Mary Kay Hemenway

Department of Astronomy, University of Texas at Austin, Austin, TX 78712, USA

Abstract: The Freshman Seminar program at the University of Texas is designed to allow groups of fifteen students an engaging introduction to the university. The seminars introduce students to the resources of the university and allow them to identify interesting subjects for further research or future careers. An emphasis on oral and written communication by the students provides these first-year students a transition to college-level writing and thinking. Seminar activities include field trips to an art museum, a research library, and the humanities research center rare book collection. This paper will report on two seminars, each fifteen weeks in length. In *The Galileo Scandal* students examine Galileo's struggle with the church (including a mock trial). They perform activities that connect his use of the telescope and observations to astronomical concepts. In *Astronomy and the Humanities* students analyze various forms of human expression that have astronomical connections (art, drama, literature, music, poetry, and science fiction); they perform hands-on activities to reinforce the related astronomy concepts. Evaluation of the seminars indicates student engagement and improvement in communication skills. Many of the activities could be used independently to engage students enrolled in standard introductory astronomy classes.

43.1 The freshman seminar program at the University of Texas

Students entering a university may come from a wide range of backgrounds and educational preparation. The University of Texas at Austin attracts a diverse population of students, some of whom come from cities whose population is considerably smaller than the university enrollment of almost 50 000. To aid students in both their social and their academic transition to college, the Freshman Seminar series was organized ten years ago. Only students in their first semester of college may enroll. The academic purpose of the course is to provide a small-class (15 student) experience that will assist the students in their transition from high school to college-level writing and thinking.

Instructors are selected by invitation only. These instructors may choose any topic for their seminar, but must follow regulations concerning the number and types of writing assignments that students must complete. Activities that introduce students to the academic richness and resources of the university, such as events or tours at the Blanton Museum of Art, Harry Ransom Center (humanities research center), and the Texas Natural Science Center are recommended. Also required of all students is attendance at sessions on time-management and using the library. Students who attend only two hours of formal seminar each week are also required to attend a "third hour" event each week. The third hour events range from lectures (presented by other Freshman Seminar instructors) to special tours of facilities; also included under the third hour category are department-sponsored events such as star parties or music recitals.

Innovation in Astronomy Education, eds. Jay M. Pasachoff, Rosa M. Ros, and Naomi Pasachoff. Published by Cambridge University Press.

43.2 *The Galileo Scandal*

One example of a seminar with astronomical origins is *The Galileo Scandal*. Offered from 1999 to 2004, the seminar included a critical examination of the life of Galileo, including an overview of the history of astronomy prior to his use of the telescope in 1609. Among the classroom activities were:

(a) "kinematical astronomy" (i.e., students model rotation and revolution of Solar System objects with their own bodies) (StarDate, 2001);
(b) experiments with lenses to learn focal length, magnification, field of view, and principles of refracting telescopes. The lenses have identical diameters but different focal lengths (Sneider and Gould, 1988);
(c) comparing positions of Jupiter's moons as drawn from *Sidereus Nuncius* to positions calculated by the *Starry Night* computer planetarium program for the dates and location of Galileo's original observations;
(d) a dramatic reading of Bertolt Brecht's play *Galileo* (1966).

Outside the classroom, a visit to the art museum to see Italian art produced during Galileo's lifetime (1564–1642) allows students to see evidence of patronage on a different scale. They also see how art was influenced by the religious struggles of his era. A field trip to the Harry Ransom Center provides an opportunity for an overview of the history of astronomy through examination of original volumes, such as the first printed Latin version of Ptolemy's *Almagestum* (1515), Copernicus's *De Revolutionibus Orbium Coelestium* (1543), the 1602 printing of Tycho Brahe's *Astronomiae Instauratae Mechanica*, several works by Kepler – *De Stella Nova* (1606), *Astronomia Nova* (1609), *Harmonices Mundi* (1619), *Tabulae Rudolphinae* (1627) with its famous frontispiece, and, of course, works by Galileo, including *Istoria e Dimostrazioni Intorno Alle Macchie Solari* (1613), *Il Saggiatore . . . di Lotario Sarsi* (1623), *Dialogo* (1632), *De Sacrae Scripturae Testimoniis* (1635), and *Discorsi a due nuoue scienze* (1638). Although the students' knowledge of Latin is minimal, the illustrations and even the binding of these rare books play a role in the description of their role in the history of astronomy. The tour also makes them aware of the availability of the center's resources on a wide range of topics for their own eventual use.

For many students, the highlight of the semester is a mock trial in which Galileo has the benefit of something he lacked in reality – a defense team. Students are assigned particular areas to research, with the topics for the prosecution and defense teams matched (the students' resources are not revealed to each other). A role is even assigned to a ghostly Cardinal Bellarmine character who, although he passed away before the trial of 1633, can testify to the events of 1616. All students, both those on the teams and those not, take part in the judging phase. Although students usually begin in a straightforward manner listing "facts," many eventually exhibit an emotional response to the situation. The trial begins to take on the aura of a television drama.

43.3 *Astronomy and the Humanities*

The second example of a freshman seminar has the title *Astronomy and the Humanities*. Although the aim of the class is similar to that of the Galileo course, the content is different. For centuries, artists, authors, and composers have often used astronomical concepts in their works. Inspired by a session with a similar title at an American Astronomical Society meeting (Fraknoi and Greenstein, 2004), I began to collect references (including Fraknoi, 2002 and 2003) that could be used in introductory astronomy classes and realized that enough existed for an entire course.

43.3.1 Astronomy and science fiction

Although some may not consider it to be literature in the common sense, science fiction offers a good entry into literary analysis. Using guidelines established by Lebofsky and Lebofsky (2002), students examine brief passages with the following questions.

(a) Is this passage science fiction or science fact?
(b) Is there science fact in the passage? If so, is it probable, possible, or impossible?
(c) Does it make mistakes in science facts?
(d) If the story is on a future Earth, is it believable? Would you want to live there?
(e) Guess when and where the story might have been written.
(f) Does the author's point of view reflect misconceptions based on when it was written?

Similar questions can be asked as the students read short stories selected for their science content. Entire short stories were selected from a list of science fiction with appropriate content (Fraknoi, 2006a). For these stories, the science content in each story was elucidated. Among the stories that students selected were: Isaac Asimov's *Nightfall* (multiple star systems, day and night), Arthur C. Clarke's *Summertime on Icarus* (asteroid, orbits, rotation), Larry Niven's *At the Core* (structure of the Milky Way), and Tom Godwin's *The Cold Equation* (Newton's laws of motion).

43.3.2 Astronomy and literature

A vast array of astronomically linked poetry exists, and student interest can take this phase of the course in different directions. The usual mode is to pick several poems that exhibit a theme and examine their astronomical roots. Thus, attitudes about astronomers and their work are seen in poems such as Walt Whitman's 1867 poem "When I heard the Learn'd Astronomer" and Anne Perlman's 1982 poem "The Specialist." Examples of related poems are those about astronomers from history (Cedering, 1985 or Dillard, 1983) or astronomical objects (Ackerman, 1991; Dickinson, 1985; Frost, 1928; or Williams, 1917).

Under the heading of literature, students have found astronomical links to works by famous authors such as Dante Alighieri (1265–1321), Geoffrey Chaucer (1343–1400), William Shakespeare (1564–1616), John Milton (1608–1674) and William Blake (1757–1827). For the earlier authors, the clash between Ptolemaic and Copernican astronomy is most apparent. Olson *et al.* (e.g., 1989 and 1998) have documented some specific astronomical events related to literature.

Galileo provides the most common example for an astronomically linked drama, with plays by Brecht (1966), Bentley (1977), and Goodwin (1998). But, the most readily available drama is the one used in *The Galileo Scandal* class. Therefore, this class turns to a dramatic reading of Brecht's *Galileo*, but without the extensive examination of his life. It is placed in the semester following other topics that have introduced the conflict between different world-views and how society in general was affected.

43.3.3 Astronomy and music

Exposure to music offers another avenue to introduce astronomy. Music arrives in many varieties – from pop culture to opera. Fraknoi (2006b) produced his suggestions for music too late to be included in the class reported here; he has more stringent conditions for inclusion on his lists than students generally consider. For example, he omits *The Planets* by Gustav Holst because of its astrological viewpoint. Students, however, find this composition very approachable and can use it as a springboard to considering the components of the solar system.

Vangelis's *Mythodea, Music for the NASA mission: 2001 Mars Odyssey* links myth and science. Although the CD contains only the music, the DVD adds visuals of the concert from Athens with several astronomical images as well as an introduction by a scientist, Scott Bolton.

Seasonal variations for the northern and southern hemispheres can be examined when students hear the familiar work of Antonio Vivaldi and compare it to that of Astor Piazzolla in *Eight Seasons*.

When asked to supply astronomical music, students are more likely to submit common songs based on their titles (e.g., "Night and Day") rather than choose a song with real astronomical content (e.g. Monty Python's "Galaxy Song" or "Why does the sun shine?" by They Might be Giants). Less commercial songs with real astronomy content are usually outside the students' realm of knowledge (e.g., music by The Chromatics, who include subjects such as Doppler shift, planets, Sun, and radio astronomy in their albums).

Among the related activities is using solar motion demonstrators; these simple devices allow students to manipulate the model to study the appearance of the Sun for different dates and latitudes, and thus to explore seasons. (Astronomical Society of the Pacific)

43.3.4 Astronomy and art

Art offers its own challenge since the variety is so large. In one way, the challenge is similar to that presented by music. Students are likely to pick anything with an astronomical object appearing, or with an astronomical "name." (This brings up the possibility of their locating pieces based totally on mythology with no astronomical connection at all.) Bringing together several works of art with the same theme or similar titles means that one can compare, for example, works with the title "Starry Night" by both Vincent van Gogh (1853–1890) and Edvard Munch (1863–1944). The title "Evening Star" has been used by Joseph Mallord William Turner (1775–1851), Frederic Edwin Church (1826–1900), and Georgia O'Keefe (1887–1984) and by other artists as part of their titles.

Just as schoolchildren can be asked to draw an image of an astronomer, and often show an older white male with facial hair, one can survey artists' perceptions of astronomers. Examples include "The Astronomer" by Albrecht Dürer (1471–1528) and by Johannes Vermeer (1632–1675). Portraits of astronomers (e.g., Tycho, Kepler, Galileo, Newton, Struve) can also be considered as indicative of society's view of astronomers and their view of themselves. Some art is quite accurate concerning the astronomical details, such as the depiction of the Milky Way in "Flight into Egypt" by Adam Elsheimer (1578–1610) or "Moonlight View over Table Bay Showing the Great Comet of 1843" by astronomer Charles Piazzi Smyth (1819–1900) (Olson and Pasachoff, 1998).

Internet resources allow easy access to many forms of art including paintings, drawings, woodcuts, and sculpture. Suggested resources include: ArtCyclopedia (www.artcyclopedia.com/), Bridgeman Art Library (www.bridgeman.co.uk/), Web Gallery of Art (www.wga.hu/), World Images (worldimages.sjsu.edu/), and proprietary resources such as ArtStor (www.artstor.org/info/).

Field trips to the university's Blanton Museum of Art to view original works of art relating to astronomy and to the Harry Ransom Center also occur during this course. A smaller number of astronomical first editions are viewed so that some time can be allocated to viewing first editions of materials relating to the humanities (such as poetry or science fiction) and original works such as a drawing by Blake or the hand-written diaries of Caroline Herschel.

43.4 Relevance of the freshman seminars to general introductory astronomy courses

Although these courses were developed with different goals from standard introductory astronomy courses, the materials and activities they use can be incorporated into such courses. The combination of presentations, activities, student reports, and discussion provide students with a rich experience in connecting astronomical concepts to many areas of human culture. The *AAS Report of 2003* (Partridge and Greenstein, 2003) included in its goals for introductory astronomy classes the concept of science as a cultural process (i.e., an acquaintance with the history of astronomy and the evolution of scientific ideas). Several of the themes in these courses provide this acquaintance as well as a cosmic perspective about the contents and scale of the Universe. Students extend their science knowledge base while they make linkages to cultural and historical topics. The course evaluations indicated that students felt that the course communicated information effectively and was educationally valuable.

References

Ackerman, D., 1991, Halley's Comet, in *Jaguar of Sweet Laughter* (New York: Random House).

Asimov, I., 1969, Nightfall, in *Nightfall and Other Stories* (Garden City, NY: Doubleday).

Bentley, E., 1977, The recantation of Galileo Galilei, in *Rallying Cries: Three Plays* (Washington: New Republic Book Company).

Brecht, B., 1966 (English version by Laughton, C. Edited and with an introduction by Eric Bentley) *Galileo* (New York: Grove Press).

Cedering, S., 1985, Letters from Astronomers, in Gordon, B. B. (ed.), *Songs from Unsung Worlds: Science in Poetry* (Boston, MA: Birkhäuser) 10–14.

Clarke, A., 1967, Summertime on Icarus, in *The Nine Billion Names of God* (New York: Harcourt, Brace and World).

Dickinson, E., 1985, Arcturus is his other name, in Gordon, B. B. (ed.), *Songs from Unsung Worlds: Science in Poetry* (Boston, MA: Birkhäuser) 165.

Dillard, R. H. W., 1983, How Copernicus Stopped the Sun, in *The First Man on the Sun* (Baton Rouge, LA: Louisiana State University Press).

Fraknoi, A., 2002, Astronomy and poetry: a resource guide, *Astronomy Education Review*, **1**, 114–116.

Fraknoi, A., 2003, Teaching astronomy with science fiction: a resource guide, *Astronomy Education Review*, **1**, 112 119.

Fraknoi, A., 2006a, *Science Fiction Stories with Good Astronomy and Physics: a Topical Index*, www.astrosociety.org/education/resources/scifi.html.

Fraknoi, A., 2006b, The music of the spheres in education: using astronomically inspired music, *Astronomy Education Review*, **5**, 139–153.

Fraknoi, A. and Greenstein, G., 2004, Astronomy teaching through the humanities: literature, the visual arts and more, *BAAS*, **36**, 1552.

Frost, R., 1928, Canis Major, in *West-Running Brook* (New York: Holt).

Godwin, T., 1954, The cold equation, in *Astounding Science Fiction*, August, p. 62–84.

Goodwin, R. N., 1998, *The Hinge of the World* (New York: Farrar, Strauss, and Giroux).

Holst, G., 1914–16, *The Planets* (various recordings).

Lebofsky, L. A. and Lebofsky, N. R., 2002, Using science fiction in the classroom, *BAAS*, **34**, 894.

Niven, L., 1962, At the core, in *Neutron Star* (New York: Ballantine Books).

Olson, D. W. and Jasinski, L. E., 1989, Chaucer and the Moon's speed, *Sky and Telescope*, **77**, 376–377.

Olson, D. W., Olson, M. S., and Doerscher, R., 1998, The stars of Hamlet, *Sky and Telescope*, **96**(5), 68–73.

Olson, R. J. M. and Pasachoff, J. M., 1998, *Fire in the Sky: Comets and Meteors, the Decisive Centuries, in British Art and Science* (New York: Cambridge University Press).
Partridge, B. and Greenstein, G. (American Astronomical Society), 2003, *Goals for "Astro 101": Report on a Workshop for Astronomy Department Leaders*, www.aas.org/education/publications/workshop101.html.
Perlman, Anne S., 1985, The Specialist, in Gordon, B. B. (ed.), *Songs from Unsung Worlds: Science in Poetry* (Boston, MA: Birkhäuser) 168.
Python, M., 1991, The Galaxy Song, from *Monty Python Sings*.
Snyder, J. L., 1992, *Solar Motion Demonstrator Kit*, Astronomical Society of the Pacific.
Sneider, C. I. and Gould, A., 1988, *More than Magnifiers* (Berkeley: Lawrence Hall of Science).
StarDate, 2001, Modeling the Night Sky, in *StarDate/Universo Teacher Guide* (Austin, TX: University of Texas McDonald Observatory).
The Chromatics, 1998, *Astrocappella*. (www.astrocappella.com)
The Chromatics, 2002, *Astrocappella 2.0*.
They Might Be Giants, 1998, Why does the Sun shine? (The Sun is a mass of incandescent gas), from *Severe Tire Damage*.
Vangelis, 2001, *Mythodea: Music for the NASA Mission – 2001 Mars Odyssey* (CD and DVD).
Vivaldi, A. and Piazzolla, A. 2001, *Eight Seasons* (Nonesuch).
Whitman, W., 1867, When I heard the learn'd astronomer, in *Leaves of Grass* (New York: Wm. E. Chapin & Co.).
Williams, W. C., 1917, Peace on Earth, in Monroe, H. and Henderson, A. C. (eds.) *The New Poetry: an Anthology*.

Comments

Richard Gelderman: Please discuss the logistics – how many hours per meeting, how many meetings per semester, how much credit is awarded? How do students find this course? Do you promote the course?
Mary Kay Hemenway: 15 meetings of two hours per semester plus one hour per week of outside activities organized by the Freshman Seminar program office. Three hours of college credit is awarded. Students are anxious to take a freshman seminar course because they are small discussion courses taught by instructors who are invited to be presenters; also most of their other courses are rather large, so this is an attractive elective course. The course is listed in the course schedules and no promotion is needed to fill all the spaces available.

Mary Kadooka: Have you used this material for secondary teacher workshops?
Mary Kay Hemenway: I have done public lectures on Galileo. The history of astronomy is required for Texas science standards.

Jay Pasachoff: From the illustrations in my junior-high-school book *Scott, Foresman Earth Science*, students can readily see that there are non-white male astronomers and female astronomers.
Mary Kay Hemenway: In spite of these examples in your book, and in textbooks from other publishers, students persist in their preconception about astronomers. At McDonald Observatory, in spite of examples in the audiovisual representations in the Visitor's Center and pre-visit materials, over 1000 students per year are asked to describe an astronomer. The vast majority hold to the misconception of astronomers as white, male, sometimes with a lab coat, often looking nerdy.
Jay Pasachoff (ed.) I have never liked the anti-science tone of Walt Whitman's "When I Heard the Learn'd Astronomer," so I was happy to find a rejoinder by the MIT professor Scott Aaronson called "When I Heard the Learn'd Poet," www.scottaaronson.com/writings/whitman.html.

44

Astronomy for teachers

Julieta Fierro
Instituto de Astronomía, UNAM, Apdo. Postal 70-264, C. P. 04510, Mexico, D. F.

Abstract: Professional astronomers in developing nations can aid basic education by implementing a few strategies that have been put into effect in Mexico. Recently the number of years of compulsory education in Mexico was doubled, from six to twelve. Astronomy was included in the high-school curriculum as an optional subject, so there are several ways in which one can contribute to the national education system. We must find ways to train people in a more effective way. Magazines and books dedicated to teachers can play an important role, as can brief courses for educators and organized conferences for science teachers.

44.1 Learning from the primates how to teach

Educational research is being carried out in several nations, with many interesting results, but unfortunately it is difficult to disseminate practical information for teachers. I encourage researchers to post their educational materials on the World Wide Web and to write for an audience of educators.

In Mexico we have several magazines for teachers. *El Correo del Maestro*, for instance, includes astronomy articles and publishes 17 000 copies, which is a large number for a developing nation but a small one compared to the number of elementary school teachers we have: over a million.

I believe primates have much to teach us about the basics of good education. If we analyze how an ape mother aids her child to use a stick to break nuts, we can observe how she uses the following steps:

(1) she exposes him to interesting surroundings (she takes him to a site where other large apes are using tools to break nuts);
(2) she lets him experiment on his own (she watches while he tries to break nuts with an improper tool, does not succeed, and even gets slightly injured);
(3) she intervenes only at the precise moment when the infant is about to give up, and she then teaches him how to grab the correct tool;
(4) she allows her child to practice till he masters nut-breaking (the mother follows closely and only helps in case of necessity).

In other words, the student must learn on his own, and the teacher should go way beyond lecturing, set an example, and become a facilitator.

These same steps can be used for good teaching: expose, experiment, facilitate, practice, and succeed. We should, of course, add another step: creation. We want our students to go beyond current knowledge. Creation occurs when items that seemed disconnected are joined.

Innovation in Astronomy Education, eds. Jay M. Pasachoff, Rosa M. Ros, and Naomi Pasachoff. Published by Cambridge University Press.

For instance, painters or composers put together colors, shapes, sounds, and rhythms in ways nobody has done before. So we must expose our students to an interesting and varied set of ideas and allow them to ponder them and give them challenging assignments.

44.2 Pre-school in Mexico

Pre-school in Mexico has recently been increased from one to three years. This means many new teachers must be trained. In Mexico, education is mainly state-run. Luckily, a group of elementary school educators contacted me with a list of 600 science questions that students had asked and they did not how to answer. So a group of educators answered them, and five books were published with the answers and related topics. To make sure the responses got to the teachers who were interested in them, the group of original teachers sold the books personally.

Pre-school teachers have several problems in answering science topics. Twenty-one percent of households have a woman as the sole provider. Mothers, therefore, have less time to teach their children how to talk, and teachers are assuming this responsibility. Instructors have to understand science questions asked in poor language, translate the questions so that they are comprehensible to scientists, and translate the scientists' answers into language the children can comprehend.

In order to contribute to the national effort I wrote an alphabetically arranged wordbook of scientific words. I tried to make them difficult to pronounce, choosing words of three syllables or those with sounds (like "rui") that are hard for small children to pronounce. Each page has the word, its definition, and a short sentence, written by a poet (Alberto Vital), which uses the word. The book is to aid both mothers and teachers: "a word a day keeps poor learning skills away." As we all know, words are necessary to build ideas.

I also wrote an astronomy book in small format. Then I realized teachers wanted large-format books so that children can see the pictures. So large-format books were printed for teachers and small-format books for students. When experience showed that students also wanted large-format books, I began a series of science books for pre-school students in large format. The topics include light, sound, and time, all with an astronomical component. I am preparing one on water and will continue with the sky.

44.3 Middle school

Three years of middle school, for children ages 12 to 15, have been added to compulsory education in Mexico. I believe only those subjects that are meaningful for children should be taught. Part of the present curriculum seems to me extremely boring. It includes little astronomy in spite of the attraction astronomy holds for both teachers and students. To fill the gap, I have written several books for that age level. These books have been included in the classroom library, which is furnished with books chosen by the teachers. They can select about ten books every year.

To make a point about reading – namely, that students read if they are interested – I wrote a book on scatology. Students read it in groups, laugh, and talk about the book. This is the kind of response that all books should elicit. My book on scatology has held the top place in the list of most bought books in Mexico's most popular bookstore.

I am presently writing more books for this age level. An advantage to the omission of astronomy from the curriculum is that it won't be taught in a boring way. Because of its multidisciplinary nature it is ideal for informal reading.

44.4 High-school teacher training

For many years there was an optional course on positional astronomy in the geography department of Mexican high schools. It was changed to a general astronomy course in the physics department. A 180-hour teacher training course was designed. At first professional astronomers lectured, and little by little the high-school teachers have learned to train each other. Some of them have prepared written materials and placed them online together with sets of images for lectures. I believe one learns when one teaches, and having high school instructors train themselves has doubtless made them feel more confident about what they teach.

44.5 Symposia

The Mexican Academy for Science Teachers is an association whose members include teachers from all levels, from preschool to graduate. Its main activity is a biannual conference attended by about 1500 teachers. There are lectures, workshops, and demonstrations, and books on astronomy are always available. (www.ampcn.org.mx)

In developing nations, where there is no tradition of fund-raising, such meetings are hard to organize. Teachers are not used to presenting posters, and it is not always easy to convince authorities to authorize the meetings and sponsor teacher participation. Nevertheless, teachers benefit immensely by going to meetings. Multilevel bilingual teachers in remote areas can share experiences with their peers. Teachers feel proud of what they are achieving when they compare their work with efforts made elsewhere. They can purchase books, magazines, and teaching materials that they might not otherwise know about.

44.6 Books

I believe it is impossible to overstate the importance of books. Mexico has taken three important steps to promote reading by setting up small libraries in school rooms, book fairs, and a new major library that includes scholars on the board.

Classroom libraries hold books selected by teachers from a group of 700. As mentioned above (Section 44.3), teachers can choose about ten books per year for their classroom library. The 700 titles come from a previous selection among the 13 000 that are printed in Spanish in different countries and are presented to the selection committee by the editors.

A new large library was built in Mexico City in a section that has few cultural activities. There are several scholars on the board who have contributed not only by suggesting titles but also by participating in other activities such as workshops, lectures, use of the web, etc.

44.7 Conclusions

(a) High-quality science education is needed by a large segment of the population of developing nations. Curricula reforms are only a very small portion of what is needed. Children must be well-nourished and healthy in order to learn.

(b) In Mexico astronomy is not taught as such at any level, but if good written materials are made available for teachers and students, they will attract readers.

(c) If one wants to improve the quality of education one must take time to work with teachers in order to understand their need and difficulties.

Further reading

Delgado, H. and Fierro, J., 2004, *Volcanes y Temblores en México*. (Mexico City: Editorial SITESA).
Domínguez, H. and Fierro, J., 2005, *Albert Einstein: Un Científico de Nuestro Tiempo*. (Mexico City: Editorial Lectorum).
Domínguez, H. and Fierro, J., 2006, La luz de las estrellas, *Correo del Maestro*, (Mexico City: Ediciones La Vasija).
Fierro, J., 2004, *El Sol, la Luna y las Estrellas*, (Mexico City: DGDC, Colección Ciencia para Maestros).
Fierro, J., 2001, *La Astronomía de México*, (Mexico City: Lectorum).
Fierro, J., 2005a, *Lo Grandioso del Sonido, Gran Paseo por la Ciencia*, (Mexico City: Editorial Nuevo México).
Fierro, J., 2005b, *Lo Grandioso de la Luz, Gran Paseo por la Ciencia*, (Mexico City: Editorial Nuevo México).
Fierro, J., 2005c, *Lo Grandioso del Tiempo, Gran Paseo por la Ciencia*, (Mexico City: Editorial Nuevo México).
Fierro, J. and Domínguez, H., 2007, *Galileo y el Telescopio, 400 Años de Ciencia*, (México: Uribe y Ferrari Editores).
Fierro, J. V., 2006, *Palabras para Conocer el Mundo*, (Mexico City: Editorial Santillana).
Fierro, J. and Sánchez Valenzuela, A. C. A., 2006, *Un Romance Científico del Tercer Tipo* (Mexico City: Editorial Alfaguara).
Tonda, J. and Fierro, J., 2005, *El Libro de las Cochinadas*, Ilustraciones de José Luis Perujo (Mexico City: ADN Editores).

45

Daytime utilization of a university observatory for laboratory instruction

John R. Mattox

Department of Natural Sciences, Fayetteville State University, Fayetteville, NC 28304, USA

Abstract: Scheduling convenience provides a strong incentive to fully explore effective utilization of educational observatories during daylight hours. I present two compelling educational activities: daylight observation of Venus, and the use of a CCD camera to determine the surface temperature of a sunspot.

45.1 Alignment of the telescope mount for daytime observation

This work utilized the Observatory at Fayetteville State University (FSU – see astro.uncfsu.edu/observatory for more information). With our digital pointing system, it is possible to retain the alignment from a previous night-observing session for subsequent use during the day. Alternatively, the Sun may be used for daytime alignment of the telescope mount. Note that sunlight intensified by telescope optics can cause permanent damage to vision. Appropriate precautions must be observed, and have been published by others (e.g., Chou, 1981a, 1981b, 1998, 2004; Marsh, 1982; Pasachoff and Covington, 1993).

At FSU, sunlight is projected through a co-mounted refracting telescope, and pointing adjusted to center the Sun's image in the shadow of the telescope, as shown in Figure 45.1. Once the telescope is pointed at the Sun, the setting circle for right ascension is moved to match the current right ascension of the Sun, which is determined with digital planetarium software (*Starry Night* is used at FSU). Thus, the system is ready to be accurately pointed during the day.

45.2 Observing Venus

Venus is readily apparent in the daytime sky once the telescope is pointed at it when its elongation exceeds approximately 10°, however, it is very difficult to find unless the telescope can be pointed at it a priori. When observing Venus close to the Sun, it is essential that unfiltered sunlight be never viewed – see discussion above. The required minimum elongation angle depends upon how effectively off-axis light is baffled by the telescope – it has been observed at FSU at an elongation of 8°. Once Venus has been acquired, a digital tracking system can be set using Venus, and is then available for observing other objects, see below.

After the telescope pointing alignment, the declination and right ascension are obtained from digital planetarium software and the telescope is moved to the position of Venus. After moving the dome slit to permit the viewing of Venus and to block as much sunlight as possible from falling on the telescope, the dust caps are removed from our 2-inch sighting telescope and our 16-inch Cassegrain telescope. If the telescope mount has been properly aligned, this brings the telescope to point within approximately 1° of Venus – often too far away to be seen

Innovation in Astronomy Education, eds. Jay M. Pasachoff, Rosa M. Ros, and Naomi Pasachoff. Published by Cambridge University Press.

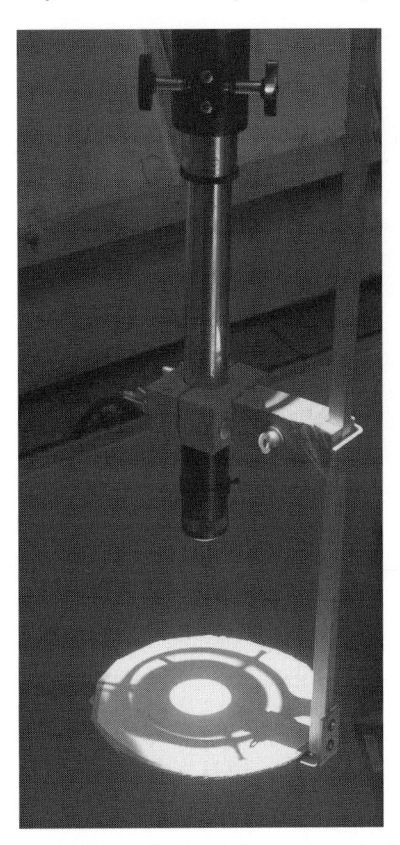

Figure 45.1 The Sun being projected with a 4-inch refracting telescope onto a white screen for student examination.

with the 16-inch telescope (using a 60 mm eyepiece with a 0.5° diameter field of view), but always within the approximately 3° field of the 2-inch sighting telescope where it is readily apparent as a bright white spot. After moving the telescope mount to center Venus in the finding telescope, it can be seen through the large telescope and the phase of Venus is then apparent. It is appropriate to pre-focus the large telescope for a celestial object (during a night session, or with the Sun through an appropriate filter) but not essential – with Venus located in the field, it is possible to search for focus with the large telescope. A yellow or orange filter will enhance contrast relative to the sky, but is not essential.

Students at FSU are directed to sketch the appearance of Venus through the large telescope. Unless Venus is near superior conjunction (where its phase is full), it is suggested that an image erecting prism be mounted ahead of the eyepiece, and that the student be asked to sketch how the sunlit crescent of Venus is oriented in relationship to the Sun and to write an explanation of this orientation, i.e., why is the portion of Venus opposite the direction of the Sun not apparent. If a crescent Moon is visible, students should be asked to similarly analyze the alignment of the crescent Moon, and compare and contrast the phases of the Moon and Venus. The objective is for the student to realize that for both Venus and the Moon, the portion of the surface facing the Sun is lit, and thus obtain a deeper appreciation of the phenomena. If appropriate mathematical training is a prerequisite for the class, students may be asked to print

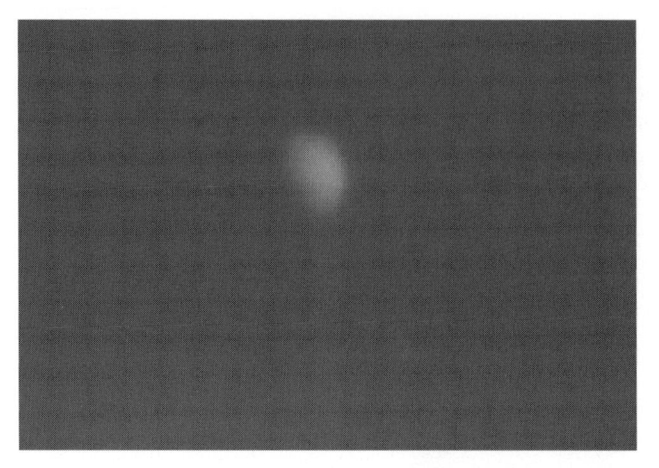

Figure 45.2 An image of Venus obtained during daylight on September 28, 2005.

and analyze a CCD image of Venus – see Figure 45.2. They are directed to determine the angular size of Venus (given the pixel size and focal length of the telescope), and compare this to the expected angular size calculated with the small-angle formula, the diameter of Venus (from their text), and the distance of Venus at the time the image was acquired.

Once the telescope is pointed, if the sky is very clear, the telescope can be used to guide the eye to make a naked eye sighting of Venus. To succeed, one must be able to focus the eyes for a distant object. The Moon or a distant horizon may be used for focus if relatively close to Venus. Usually this will not be the case, and substantial effort may be required to have the eyes correctly focused when looking in the right direction. Many students succeed – and it appears to be a remarkable experience for them. It is recommended that a naked eye sighting of Venus not be required of students; some students are observed to be unable to see it despite reasonable effort with good conditions.

45.3 Observing other objects during daytime

It is also possible to see the Moon, Jupiter, Saturn, Mercury, Mars, and bright stars during the day. The surface brightness of Mars, Jupiter and Saturn is less than Venus, so these objects will not have a compelling appearance unless the sky is very clear (with high atmospheric transparency). Mercury and bright stars (magnitude less than 0) provide a good contrast with the sky using a 16-inch telescope but are usually not apparent with a 2-inch telescope because they are not nearly as bright as Venus.

Thus, with at least a 12-inch aperture and a high-quality mount, it will be possible to point with sufficient accuracy to easily sight any bright object that is at least 15° above the horizon in a clear daylight sky. The large telescope should be pre-focused for a celestial object. A short focal-length eyepiece is recommended since it provides a larger field of view making pointing less critical. At FSU, we can normally only view an object in daylight if it is located close to Venus (so that the digital pointing system can be set using Venus, and will be accurate to 0.2° in the vicinity).

45.4 Observing the Moon

Until about three days before or after New Moon, the Moon may be easily observed in the sunlit sky, either with or without a telescope. Under marginal conditions, one must know where it is. For a thin crescent Moon, the technique described for Venus pertains. Under very good conditions, it can be seen by an experienced observer about one day before or after New Moon. Under the worst conditions (e.g., low atmospheric transparency or haze), it will not be visible for two to three days before and after New Moon. Schaefer, Ahmad, and Doggett (1993) report that the earliest known sighting after New Moon occurred 15.4 hrs afterward unaided, and 13.4 hrs afterward with optical aid.

45.5 Observing sunspots

Pointing the telescope at the Sun is easily accomplished. Extreme caution should be exercised to avoid risk of injury to the vision of the telescope operator or observatory guests – see discussion above. Sunspots are frequently (but not always) visible even at solar minimum. The Internet can be consulted for a contemporary solar image (e.g., umbra.nascom.nasa.gov/images/latest.html).

In addition to allowing students and guests to view the limb of the Sun or sunspots with an appropriate filter for solar viewing through the FSU 16-inch telescope, or 4-inch telescope, it has proven useful to display a projection of an unfiltered solar image with a co-mounted 4-inch refracting telescope. A clamp has been fabricated that attaches to the back of the telescope and holds a screen consisting of clean white paper cemented to cardboard (see Figure 45.1). It is also worthwhile to have students view the Sun using only "eclipse shades" during the same session.

Once a sunspot has been selected, and centered in the eyepiece with the 16-inch telescope, a CCD camera is substituted for the eyepiece. The filter remains in place. The appropriate exposure has been found to be about 1 millisecond. A rapid series of exposures is used to focus, and frame the desired exposure. An example is provided in Figure 45.3.

After FSU students view the Sun with "eclipse shades" and in projection using a 4-inch refractor, they analyze a CCD image of a sunspot, preferably obtained during their visit and with their assistance. If no sunspots are apparent, or if observing conditions aren't good, an archival image is used. The students use the SBIG CCD software to determine the bias and background (for an identical exposure with the telescope aperture covered), and the average CCD counts/pixel in the center of the darkest sunspot, and in a region with a normal solar surface. After subtracting the estimate of bias and background, the ratio of the emission intensity at the center

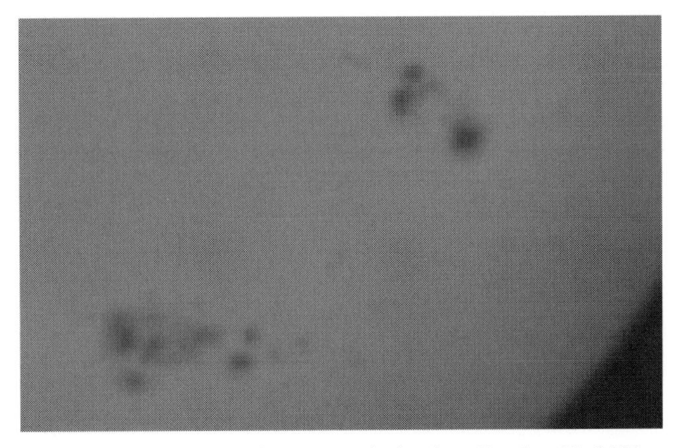

Figure 45.3 An image of sunspots obtained on October 31, 2003.

of the sunspot to that of the normal surface is formed. Students are then asked to use the Stefan–Boltzmann law and the nominal temperature of the surface of the Sun (5780 K) to determine the temperature of the center of the sunspot (assuming it has the same albedo as the normal solar surface, and that the CCD detection efficiency is not significantly affected by the temperature difference). If the Stefan–Boltzmann law is not covered in their text, it is explained in a "pre-lab exercise" completed by students before the beginning of the lab.

Consider the following example. The average of bias and background is 31 counts per pixel. At the center of the sunspot, 322 counts per pixel are observed. For the normal solar surface near this spot, an average of 677 counts per pixel is detected. The corresponding intensity ratio is 291/646 which is 0.45. Using the Stefan–Boltzmann law, the temperature of the center of the sunspot is determined to be 5780 times $0.45^{1/4}$, which is 4735 K.

45.6 Mitigation of dome seeing

Because the FSU Observatory is integrated into its building, it can't be completely isolated thermally and this exacerbates dome seeing. It can be partially isolated from the rest of the building by closing the entrance door. Also, we have installed a fan rated at 2830 cubic feet per minute to blow outside air into the observatory. This helps to bring the temperature inside the dome to the outside temperature during both day and night observing. Often the temperatures still differ substantially, and about 5 arc-seconds of seeing is apparent. Sometimes the inside and outside temperatures will match and our seeing is then observed to be about 2 arc-seconds.

References

Chou, B. R., 1981a, *Sky and Telescope*, **62**(2), 119.

Chou, B. R., 1981b, Retinal protection from solar photic injury. *Am. J. Optom. Physiol. Opt.*, **58** (7), 570–580.

Chou, B. R., 1998, Solar filter safety. *Sky and Telescope*, **95** (2), 36–40.

Chou, B. R., 2004, Eye safety and solar eclipses. In Espenak, F. and Anderson, J. (eds.), *Total Solar Eclipse of 2006 March 29*, NASA TP-212762 (Greenbelt, MD: National Aeronautics and Space Administration), 14–16, 69.

Marsh, J. C. D., 1982, *J. Brit. Ast. Assoc.*, **92**, 6.

Pasachoff, J. M. and Covington, M. A., 1993, *The Cambridge Eclipse Photography Guide* (Cambridge: Cambridge University Press).

Schaefer, B. E., Ahmad, I. A., and Doggett, L. E., 1993, *Quarterly Journal of the Royal Astronomical Society*, **34**, 53.

Poster highlights

Astronomy education in the Republic of Macedonia

O. Galbova and G. Apostolovska
Institute of Physics, Faculty of Science, PO Box 162, 1000 Skopje, Republic of Macedonia

Abstract

We discuss the general structure of astronomy education and activities for the improvement of astronomy education in the Republic of Macedonia. In 2005 for the first time astronomy was introduced as a new area of specialization at the Institute of Physics, the faculty of Natural Sciences at the University Kiril i Metodij. Selected topics in astrophysics are included in the last year of the newly revised secondary school curriculum. Bearing in mind that most physics teachers lack background in astrophysics, the staff of the Institute of Physics has been organizing courses and preparing educational materials for the continuing education of physics teachers. In terms of public education, most of the activities are undertaken by the Macedonian Astronomical Society (MAS), established in 1996. By presenting the cultural and scientific values of astronomy; supervising the establishment of astronomical clubs in schools; and overseeing the clubs' initial activities, the society is helping raise the educational and professional level. Young astronomers from two regional astronomical societies (from Skopje and Bitola) are very active during interesting astronomical phenomena. Given that there is no astronomical observatory in the country and that formal astronomy instruction began only recently, it is not surprising that astronomical literacy in the country is very low. On the other hand, the interest in astronomy is surprisingly high, as evidenced by the huge number of children who participate in the winter astronomy schools organized by the MAS at the Institute of Physics in Skopje. The help of astronomers from neighboring countries is crucial for the success of our efforts to raise astronomy literacy. Currently, our contacts with the Bulgarian astronomers from the National Astronomical Observatory Rozhen are extremely beneficial.

L'Aula del Cel: communicating astronomy at school level

A. Teresa Gallego, Amelia Ortiz-Gil, and Miquel Gómez Collado
Observatorio Astronomico, Universidad de Valencia, E-46980, Spain

Abstract

The *Aula del Cel* (Valencian for *Sky Classroom*) is a project carried out by the Astronomical Observatory of the University of Valencia, Spain. Its aim is teaching and spreading astronomy to students aged from 10 to 17. In some cases, we also prepare sessions for audiences with

Innovation in Astronomy Education, eds. Jay M. Pasachoff, Rosa M. Ros, and Naomi Pasachoff. Published by Cambridge University Press.

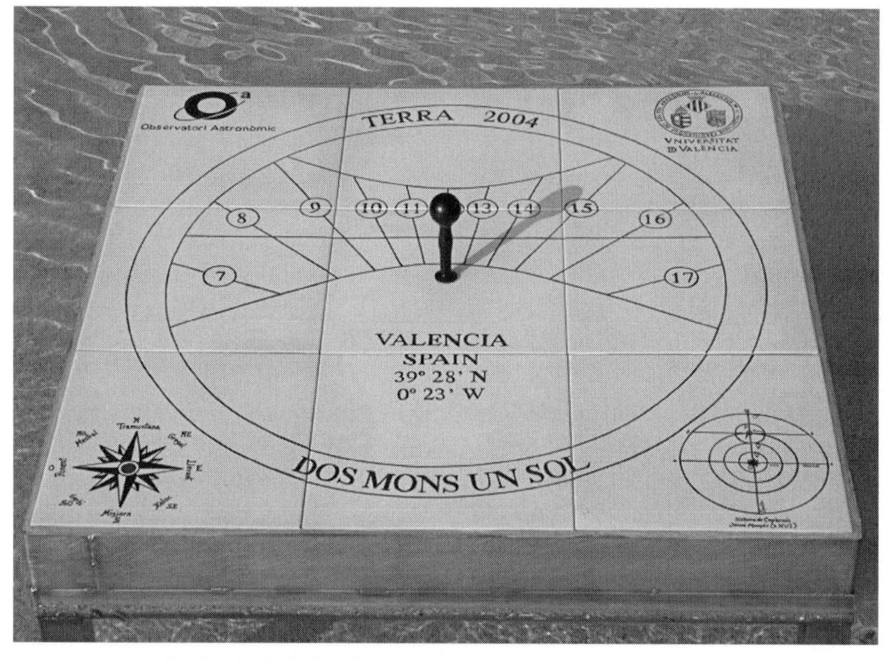

A sundial for the latitude of Valencia, used in Aula del Cel (Sky Classroom).

special needs, such as five-year-olds, over 55-year-olds, or disabled children. We describe different experiments we have carried out with standard and special groups, not only in our Sky Classroom but also in the students' regular schools, as well as the resources we used, and positive (or negative) results we have obtained.

Gemini Observatory's innovative education and outreach for 2006 and beyond

Janice Harvey
Gemini Observatory, 670 N. A'ohoku Place, Hilo, Hawaii, HI 96720, USA

Abstract
In preparation for its bold informal science plan for the next five years, Gemini Observatory is making an exciting effort to bring education outside of our typical borders and into our partner countries and beyond. We will introduce some of the innovative outreach programs that have been highly successful in Hawaii in particular and the US in general. We describe how Gemini's PIO efforts will further develop new concepts as we move forward in our next phase of education and outreach. The poster highlights a few of Gemini's highly successful initiatives, including the *Stars Over Mauna Kea* newspaper tabloid, Hawaii's "Journey through the Universe" program, StarTeachers International teacher's exchange program, the newly developed partnership between the Imiloa astronomy Education Center's planetarium and Gemini's StarLab programs, extension of our Family Astro programs internationally, Gemini's student "Time on the Telescope" mentor program, "Gemini Live" (a videoconference between our astronomers at the Gemini control room and classrooms around the world). We also suggest how other astronomical facilities

might emulate the partnership between Gemini's PIO department and the Department of Education in Hawaii.

A history of astronomy teaching in Serbian schools

S. Vidojevic and S. Segan

Department of Astronomy, Faculty of Mathematics, Studentski trg 16, 11001 Belgrade, Republic of Serbia

Abstract

The history of astronomy teaching in Serbia, which dates back to the mid-nineteenth century, is discussed in detail. Among the important dates are 1896, when astronomy appeared as a separate subject at the Department of Mathematics and Physics in the Grand School, which had been founded 16 years earlier; 1905, when the university was established, with astronomy from the outset as an elective subject; 1925–26, when astronomy for the first time was considered an educational discipline in its own right; 1961, when the astrophysical laboratory at the department of astronomy was officially inauguarated, not only for student training but also for research in radio astronomy; 1971, when the department of astronomy became the Institute of Astronomy; and 1995, when the Institute was renamed the department of astronomy. Although in 1964 astronomy was made an obligatory subject in the fourth year of high school, in 1990 the Education Council of the Republic of Serbia decided that astronomy would no longer be a subject taught in high schools. A review of this tumultuous history underscores how important it is for astronomers to be included in broad curriculum planning. Currently astronomers are working on using virtual teaching to improve coverage of astronomy in elementary and high schools. Public astronomy education is supported by two planetaria, public observatories, and 14 astronomical socieities, as well as through books and journals.

News from the Cosmos: daily astronomical news web page in Spanish

Amelia Ortiz-Gil

Observatorio Astronómico–Universidad de Valencia, Edifici Instituts d'Investigació, Pol. La Coma, s/n E-46980 Paterna, Valencia, Spain

Abstract

A good way to introduce topics in astronomy classes is to make use of recent discoveries. In Spain and other Spanish-speaking countries, teachers might find it difficult to implement this approach, however. Most press releases and published news are written in English, making it difficult for our students and teachers to read and understand them. *Noticias del Cosmos* (News from the Cosmos) (www.uv.es/obsast/es/divul/noticias/) is a web page where we daily publish brief abstracts about the most recent discoveries and breakthroughs in astronomy and astronautics. To make this news more readily accessible to Spanish-speaking audiences – not only teachers and students but also the general public with little specific knowledge in the field – the page is written in Spanish. Each abstract is linked to the original press release from the corresponding space agency, university, or research institution sponsoring the research. The interested reader can then obtain all the details from the original releases. Weekly, monthly, and complete archives of the news are kept, allowing for their use even long after their initial publication. Although about three-quarters of our readers are based in Spain, since June 2003, *Noticias del Cosmos* has received more than 19 000 visits from readers around the world. Our readership includes the general public, teachers, and professionals who are interesting in knowing what is going on in fields different from their own.

Reproduction of William Herschel's metallic mirror telescope

N. Okamura, S. Hirabayashi, A. Ishida, A. Komori, and M. Nishitani
Mito Second High School, Ibaraki, Japan

Abstract

Following the reproduction of Cassini's open-air telescope, which took us almost three years to complete, our club decided to reproduce the metallic mirror telescope invented by William Herschel, which is a telescope of the subsequent generation. We based our design on the 7-foot telescope that he used to discover Uranus in 1781, using the same ratio of seven to three of copper and tin. The surface of the cast mirror had many imperfections, such as hollow portions and bubbles. These were removed by using the rock grinder at our school; the mirror was later polished at the Hidaka Optical Institute. The tube enclosing the mirror was also made up of eight polygons, just like the original's. When we observed stars with the metallic mirror telescope, they were a little bit dark, but we could observe them well and observe the Cassini division in Saturn's rings. We also succeeded in observing Uranus with this telescope. We are reproducing the telescope mount with nearly the same design as the original. We have learned through the reproduction that the unique design of the mount allows us to make observations with precise tracking accuracy while in a comfortable observing position.

Students from Mito High School, Japan, with the reproduction of Herschel's telescope that was used to discover Uranus in 1781. A metal mirror is installed.

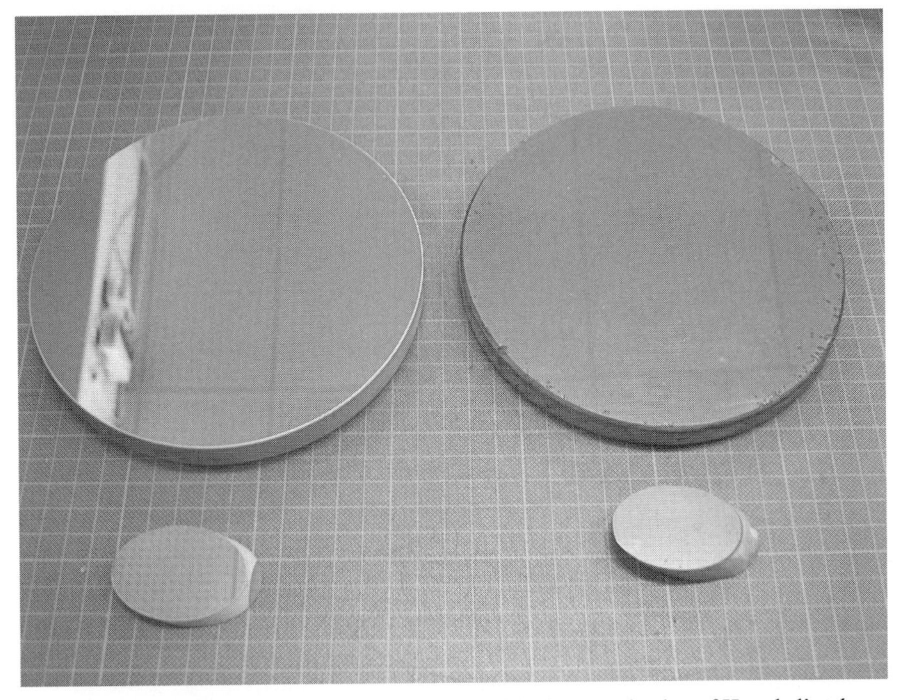

The mirrors, one glass and one metal, used in the reproduction of Herschel's telescope.

Uranus, as observed with the 7-foot metallic-mirror telescope.

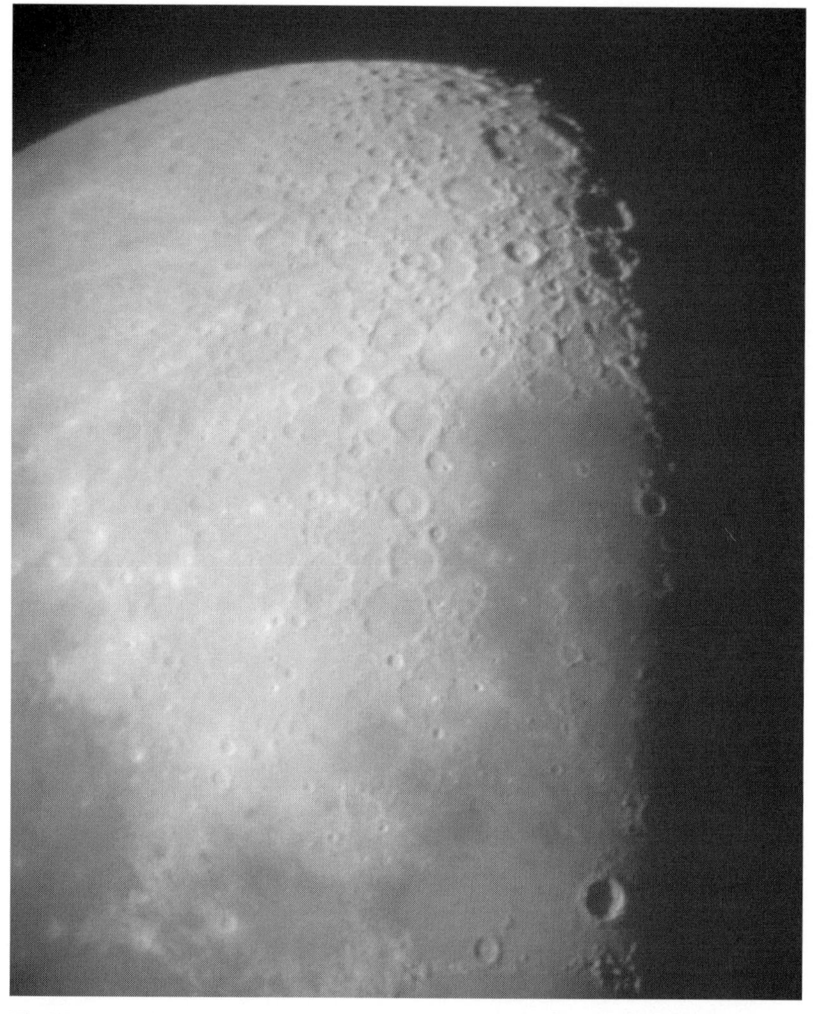

The Moon, as observed with the 7-foot metallic-mirror telescope.

Mito High School, students with the reproduction of Her school's 7-foot telescope.

History of Ukrainian culture and science in astronomical toponymy

Iryna B. Vavilova

Dobrov Center of Science – Technical Potential and History of Science Research, NAS of Ukraine, 60 Shevchenko Blvd., Kyiv 01032, Ukraine, and Institute of Space Research, NAS of Ukraine, 40 Acad. Hlushkov Av., Kyiv 187 03680, Ukraine

Abstract

In 2003 a beautifully illustrated encyclopedia called *The Names of Ukraine in Space* was issued. It contains about 300 articles devoted to names immortalized in astronomical toponymy, including historical events, geographical objects, and well-known persons from around the world who, regardless of nationality, continued to consider Ukraine their Motherland. The unusual approach developed for this book has not only helped advertise astronomy in Ukraine but also allowed scholars and amateurs to get a wide knowledge about the world and its customs. In 2006 another book, *Ukraine in the Constellation of Space Nations of the World* (Paton *et al.*, 2001), was selected by the Ministory of Education in Ukraine as required reading for secondary school students in order to fulfil the Ukrainian literature requirement. As a result, about 200 000 graduates extended their knowledge in space sciences and astronomy.

Reference

Paton, B. E., Vavilova, I. B., Negoda, O. O., and Yatskiv, Ya. S., 2001, *Ukraine in the Constellation of Space Nations of the World* (Kyiv: VAITE).

The Universe: helping to promote astronomy

Rosa M. Ros and Javier Moldón Vara

Applied Mathematics 4, Technical University of Catalonia, Jordi Girona 1-3, Modul C3, 08034 Barcelona, Spain; Departamenta d'Astronomia i Meteorologia, Universitat de Barcelona, E-08028 Barcelona, Spain

Abstract

Very few sciences are able to affect the everyday lives of people as much as astronomy. Some natural events, such as solar and lunar eclipses, can stimulate the curiosity of the public without any additional input from professionals; others, such as the transit of Venus, need to be promoted a bit in order to attract the interest of the general public. In recent years, we have been able to see that, with relatively little effort, people have been "turned on" to astronomy by such events. We present several examples of efforts proposed by different institutions, astronomical societies, and groups of people to promote science in general and astronomy in particular, using natural astronomical events as a springboard. We describe the impact on society of such events during recent years marked by a succession of such special events. In particular we consider the use of contests to stimulate interest among primary and secondary school students and the efforts of astronomical associations and universities to keep the public informed about upcoming events.

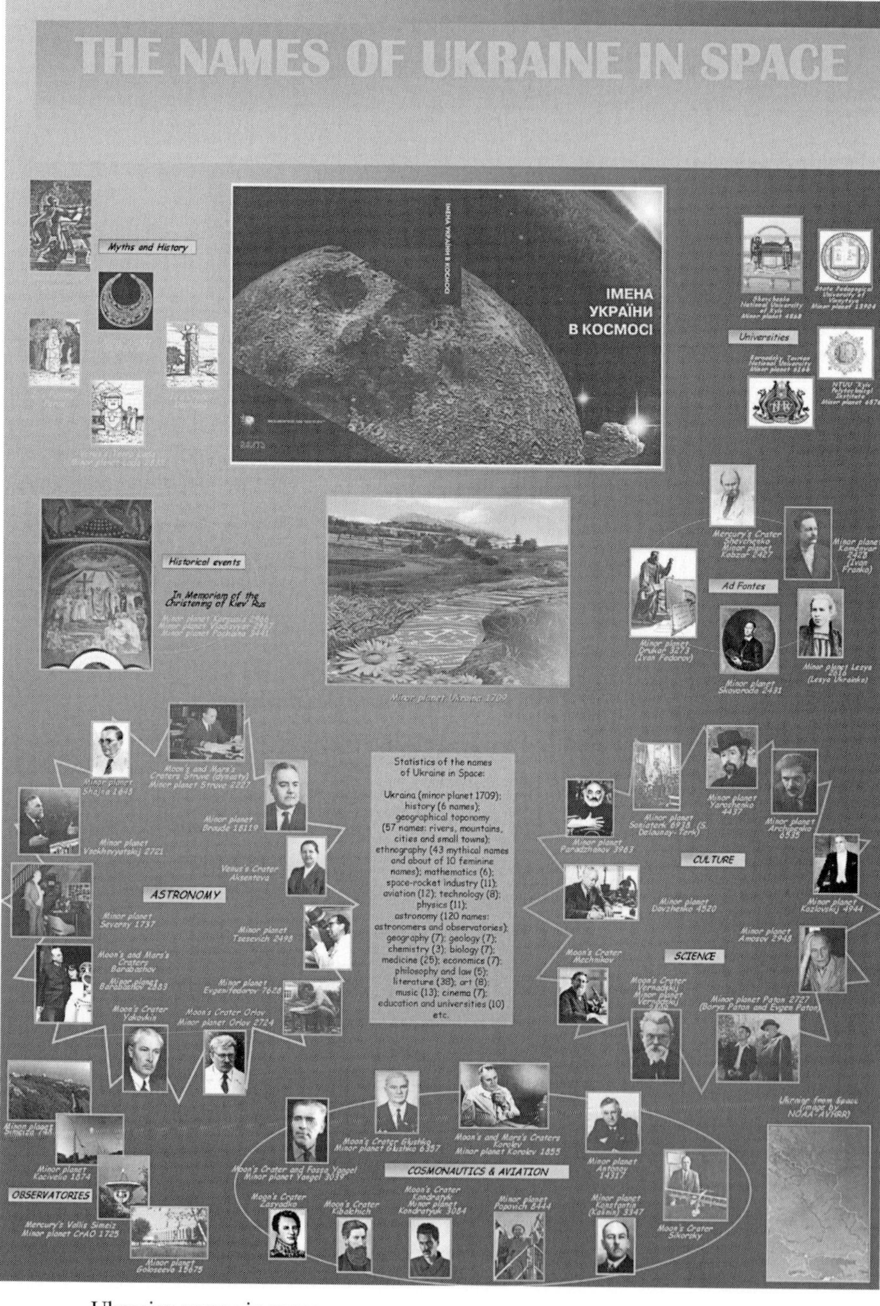

Ukranian names in space.

Astronomy education in Ukraine, the school curriculum, and a lecture course at Kyiv Planetarium

N. S. Kovalenko, K. I. Churyumov, and E. V. Dirdovskaya
Kyir Planetarium, V. Vasylkivska str. 57/3, Kyiv 03150, Ukraine

Abstract

After having been banished for some years from the school curriculum, astronomy was re-introduced into the Ukrainian school system in the late 1990s as a result of an initiative of the Ukrainian Astronomical Association. Since the academic year 2001–2002 astronomy has been taught in the final year, using a specially prepared manual, which builds on the background the students have already had in physics, mathematics, chemistry, and geography. The goal of the manual, which is divided into eight thematic parts, is to shape the students' scientific perception of the world. Since the official school course is limited to a mere 17 hours, the Kyiv Planetarium offers a set of educational programs to support the school curriculum. In addition to a set of seven lectures for those in the last year of high school, the planetarium also offers programs for pupils of all ages.

Conclusions

It is inspiring to meet with colleagues from around the world who share our interests in astronomy education. The hundred or so contributors to our special session in Prague in 2006 came from North America, South America, Europe, Africa, and Asia – so perhaps next time we should collaborate with the International Astronomical Union's sessions on Astronomy in Antarctica to complete the sweep.

The papers in this book flesh out ideas we discussed at our meeting, and they provide many concepts and concrete examples that we hope will be of widespread interest to educators of all kinds, not only those already working in astronomy. We think that the very favorable view that people all around the world have of astronomy makes our science a good basis for examples that can be used in other fields. And we hope that some of the concepts and techniques discussed in this volume help and inspire teachers everywhere.

As we ready ourselves for the International Year of Astronomy, 2009, approved by the International Astronomical Union, UNESCO, and the United Nations' General Assembly, we hope that our current book succeeds in advancing science literacy and helps current and prospective teachers of astronomy in their interesting tasks.

Jay M. Pasachoff
Rosa M. Ros
Naomi Pasachoff

Author index

Index